# THE BONOBOS
## BEHAVIOR, ECOLOGY, AND CONSERVATION

# DEVELOPMENTS IN PRIMATOLOGY: PROGRESS AND PROSPECTS

Series Editor: Russell H. Tuttle, University of Chicago, Chicago, Illinois

This peer-reviewed book series melds the facts of organic diversity with the continuity of the evolutionary process. The volumes in this series exemplify the diversity of theoretical perspectives and methodological approaches currently employed by primatologists and physical anthropologists. Specific coverage includes: primate behavior in natural habitats and captive settings; primate ecology and conservation; functional morphology and developmental biology of primates; primate systematics; genetic and phenotypic differences among living primates; and paleoprimatology.

## PRIMATE BIOGEOGRAPHY
Edited by Shawn M. Lehman and John Fleagle

## REPRODUCTION AND FITNESS IN BABOONS: BEHAVIORAL, ECOLOGICAL, AND LIFE HISTORY PERSPECTIVES
Edited By Larissa Swedell and Steven R. Leigh

## RINGAILED LEMUR BIOLOGY: *LEMUR CATTA* IN MADAGASCAR
Edited by Alison Jolly, Robert W. Sussman, Naoki Koyama
and Hantanirina Rasamimanana

## PRIMATES OF WESTERN UGANDA
Edited by Nicholas E. Newton-Fisher, Hugh Notman, James D. Paterson
and Vernon Reynolds

## PRIMATE ORIGINS: ADAPTATIONS AND EVOLUTION
Edited by Matthew J. Ravosa and Marian Dagosto

## LEMURS: ECOLOGY AND ADAPTATION
Edited by Lisa Gould and Michelle L. Sauther

## PRIMATE ANTI-PREDATOR STRATEGIES
Edited by Sharon L. Gursky and K.A.I. Nekaris

## CONSERVATION IN THE 21ST CENTURY: GORILLAS AS A CASE STUDY
Edited by T.S. Stoinski, H.D. Steklis and P.T. Mehlman

## ELWYN SIMONS: A SEARCH FOR ORIGINS
Edited by John G. Fleagle and Christopher C. Gilbert

## THE BONOBOS: BEHAVIOR, ECOLOGY, AND CONSERVATION
Edited by Takeshi Furuichi and Jo Thompson

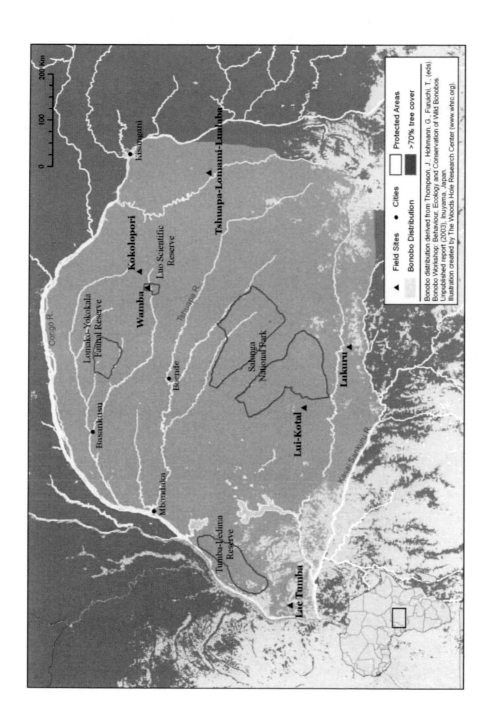

Bonobo distribution derived from Thompson, J., Hohmann, G., Furuichi, T., (eds).
Bonobo Workshop: Behaviour, Ecology and Conservation of Wild Bonobos
Unpublished report (2003), Inuyama, Japan.
Illustration created by The Woods Hole Research Center (www.whrc.org).

# THE BONOBOS
## BEHAVIOR, ECOLOGY, AND CONSERVATION

Edited by

## Takeshi Furuichi
*Kyoto University*
*Japan*

and

## Jo Thompson
*Lukuru Project*
*Democratic Republic of Congo*

 Springer

Takeshi Furuichi
Primate Research Institute
Kyoto University
Inuyama, Aichi, 484-8506
Japan
furuichi@pri.kyoto-u.ac.jp

Jo Thompson
Lukuru Project
Democratic Republic of Congo
c/o Lukuru Wildlife Research Foundation
129 Pinckney Street, PO Box 875
Circleville, Ohio 43113
USA
lukuru@gmail.com, jat434@aol.com

ISBN: 978-0-387-74785-9          e-ISBN: 978-0-387-74787-3
DOI: 10.1007/978-0-387-74787-3

Library of Congress Control Number: 2007936621

Printed on acid-free paper

9 8 7 6 5 4 3 2 1

springer.com

# Contents

# Contributors

Mbenzo Abokome
World Wide Fund for Nature, WWF-DRC Program, 06 Lodja Av., Gombe,
Kinshasa, Democratic Republic of Congo

Claudine André
Lola Ya Bonobo, "Petites Chutes de la Lukaya", Kimwenza,
Mont Ngafula, Kinshasa, Democratic Republic of Congo

Matungila Bewa
World Wide Fund for Nature, WWF-DRC Program, 06 Lodja Av., Gombe,
Kinshasa, Democratic Republic of Congo

Ndouzo Bokomo
Zoological Society of Milwaukee, Av. Lilas, No.4, Gombe, Kinshasa,
Democratic Republic of Congo

Barbara Fruth
Max Planck Institute for Evolutionary Anthropology, Department of Primatology,
Deutscher Platz 6, 04103 Leipzig, Germany

Takeshi Furuichi
Primate Research Institute, Kyoto University, Inuyama, Aichi, 484-8506 Japan

Falk Grossmann
Wildlife Conservation Society, Africa Program, c/o K.C.M.C.,
Private Bag, Moshi, Tanzania

Patrick Guislain
Zoological Society of Milwaukee, 10005 W. Blue Mound Road,
Milwaukee, Wisconsin, 53226 United States

Brian Hare
Duke University Department of Biological Anthropology and Anatomy,
Box 90383, Biological Sciences Bldg. Durham, NC 27708-0383 Unites Sates

John A. Hart
Lukuru Project, Tshuapa-Lomami-Lualaba Landscape, 1235 Ave. Poids Lourds,
Kinshasa, Democratic Republic of Congo

Chie Hashimoto
Primate Research Institute, Kyoto University, Inuyama, Aichi, 484-8506, Japan

Emi Hibino
Primate Research Institute, Kyoto University, Inuyama, Aichi, 484-8506, Japan

Gen'ichi Idani
Great Ape Research Institute, Hayashibara Biochemical Laboratories,
Inc. 952-2 Nu, Tamano-shi, Okayama, 706-0316 Japan

Hiroshi Ihobe
School of Human Sciences, Sugiyama Jogakuen University, 37-234,
Tekenoyama, Iwasaki, Nisshin, Aichi, 470-0131 Japan

Omari Ilambu
World Wide Fund for Nature, WWF-DRC Program, 06 Lodja Av., Gombe,
Kinshasa, Democratic Republic of Congo

Jose Ilanga
Ministry of the Environment, Nature Conservation, Water and Forests, Ave des
Cliniques, Kinshasa, Gombe, Democratic Republic of Congo

Bila-Isia Inogwabini
World Wide Fund for Nature, WWF-DRC Program, 06 Lodja Av., Gombe,
Kinshasa, Democratic Republic of Congo

Edmond Isomana
Zoological Society of Milwaukee, Av. Lilas, No.4, Gombe, Kinshasa,
Democratic Republic of Congo

Annette Jeneson
Department of Psychology, University of California, San Diego,
9500 Gilman Drive, La Jolla, California, 92093-0109 United States

Richard Bovundja Kabanda
Congolese Institute for Nature Conservation, Anga Headquarters Post, Salonga
National Park, Democratic Republic of Congo

Crispin Kamate
Lola Ya Bonobo, "Petites Chutes de la Lukaya", Kimwenza, Mont Ngafula,
Kinshasa, Democratic Republic of Congo

Mbende Longwango
World Wide Fund for Nature, WWF-DRC Program, 06 Lodja Av., Gombe,
Kinshasa, Democratic Republic of Congo

Lubuta Mbokoso Nestor
Congolese Institute for Nature Conservation, Lukuru Project Focal Point,
Commune de la Gombe, Avenue des Cliniques 13, BP 868,
Kinshasa 1, Kinshasa, Democratic Republic of Congo

Pierrot Mbonzo
Lola Ya Bonobo, "Petites Chutes de la Lukaya", Kimwenza, Mont Ngafula,
Kinshasa, Democratic Republic of Congo

T.D. Mboyo Bolinga
Zoological Society of Milwaukee, Av. Lilas, No.4, Gombe, Kinshasa,
Democratic Republic of Congo

Meike Mohneke
Rheinische Friedrich Wilhelms University, Department of Zoology, 53115
Bonn, Germany

Dominique Morel
Lola Ya Bonobo, "Petites Chutes de la Lukaya", Kimwenza, Mont Ngafula,
Kinshasa, Democratic Republic of Congo

Balemba Motema-Salo
Research Center for Ecology and Forestry, Ministry of Scientific
Research and Technology, Mabali, Equateur, Democratic Republic of Congo

Mbangi Mulavwa
Research Center for Ecology and Forestry, Ministry of Scientific
Research and Technology, Mabali, Equateur, Democratic Republic of Congo

Ndunda Mwanza
Research Center for Ecology and Forestry, Ministry of Scientific
Research and Technology, Mabali, Equateur, Democratic Republic of Congo

Mafuta Ngamankosi
Congolese Institute for Nature Conservation, Commune de la Gombe,
Avenue des Cliniques 13, BP 868, Kinshasa 1, Kinshasa,
Democratic Republic of Congo

Elisabetta Palagi
Natural History Museum, University of Pisa, Via Roma 79, 56011 Calci, Pisa,
Italy

Tommaso Paoli
Department of Ethology, Ecology and Evolution, Anthropology Unit,
University of Pisa, Via Roma 79, 56011 Calci, Pisa, Italy

Amy S. Pollick
Association for Psychological Science, 1010 Vermont Ave. NW, Suite 1100,
Washington, DC, 20007 United States

Gay E. Reinartz
Zoological Society of Milwaukee, 10005 W. Blue Mound Road,
Milwaukee, Wisconsin, 53226 United States

Jeroen Stevens
Centre for Research and Conservation, Royal Zoological Society of Antwerp,
Koningin Astridplein 26, 2018 Antwerp, Belgium
University of Antwerp, Department of Biology, Universiteitsplein I,
2610 Antwerp, Belgium

Osamu Takenaka
Primate Research Institute, Kyoto University, Inuyama, Aichi, 484-8506, Japan

Yasuko Tashiro
Great Ape Research Institute, Hayashibara Biochemical Laboratories,
Inc., 952-2 Nu, Tamano, Okayama, 706-0316 Japan

Jo Myers Thompson
Lukuru Project, Democratic Republic of Congo,
c/o Lukuru Wildlife Research Foundation, 129 Pinckney Street, PO Box 875,
Circleville, Ohio, 43113 United States

Linda Van Elsacker
Centre for Research and Conservation, Royal Zoological Society of
Antwerp, K. Astridplein 26, 2018 Antwerp, Belgium

Hilde Vervaecke
Centre for Ethology and Animal Welfare, KaHoSint-Lieven, Hospitaalstraat 23, 9100
Sint-Niklaas, Belgium

Ashley Vosper
Lukuru Project, Tshuapa-Lomami-Lualaba Landscape, 1235 Ave. Poids Lourds,
Kinshasa, Democratic Republic of Congo

Miezi Vuvu
World Wide Fund for Nature, WWF-DRC Program, 06 Lodja Av., Gombe,
Kinshasa, Democratic Republic of Congo

Frans B. M. de Waal
Living Links, Yerkes National Primate Research Center, Emory University,
954 N. Gatewood Road, Atlanta, Georgia, 30322 United States

Lisalama Wema Wema
Congolese Institute for Nature Conservation, Commune de la Gombe,
Avenue des Cliniques 13, BP 868, Kinshasa 1, Kinshasa,
Democratic Republic of Congo

Cosma Wilungula Balongelwa
Congolese Institute for Nature Conservation, Commune de la Gombe,
Avenue des Cliniques 13, BP 868, Kinshasa 1, Kinshasa,
Democratic Republic of Congo

Richard Wrangham
Department of Anthropology, Harvard University, Peabody Museum 50B,
11 Divinity Avenue, Cambridge, Massachusetts, 02138 United States

Mikwaya Yamba-Yamba
Research center for Ecology and Forestry, Ministry of Scientific
Research and Technology, Mabali, Equateur, Democratic Republic of Congo

Kumugo Yangozene
Research Center for Ecology and Forestry, Ministry of Scientific
Research and Technology, Mabali, Equateur, Democratic Republic of Congo

# Introduction

**Takeshi Furuichi[1] and Jo Thompson[2]**

Nearly eighty years have passed since the bonobo was officially designated as a unique species (Coolidge 1933). *Homo* and *Pan* share 98.8 % of some DNA sequences (Sibley and Ahlquist 1987, The Chimpanzee Sequencing and Analysis Consortium 2005), making the chimpanzees and bonobos our closest living relatives. Although we have known our cousin the bonobo in some capacity for more than 125 years (Thompson 2001), the bonobo is still considered the least known of the great apes. Although field work beginning in 1938 focused on bonobos specifically for collecting museum specimens, the first systematic field studies of living bonobos in their natural environment began in 1972, pioneered by Professor Toshisada Nishida. Though brief, Nishida conducted a survey of the region along the west bank of Lake Tumba. In July that same year, a team from Yale University, United States began a two-year study at Lake Tumba; the first field study site. Broader scientific research in nature commenced in 1973 when Professor Takayoshi Kano, the first scientist to study bonobos extensively in the wild, conducted a wide-reaching survey throughout the core forest block region south of the Congo River, resulting in the establishment of the first long-term field site for the study of bonobos since 1974.

Following preliminary studies in captivity, two teams of researchers, one led by Kano at Wamba and another led by Susman at Lomako, undertook long-term behavioral and ecological field studies of wild bonobos in the heart of the Congo Basin rain forest. Those studies brought new insights to our understanding of the ecological adaptation of great apes and the understanding of ourselves. In the early stages of study, research focused on the morphological similarity of bonobos with the oldest known human ancestor, and the similarity in sexual behaviors between bonobos and humans. As research progressed, scientists realized that bonobos exhibit other important features such as high social status of females and a society built around a nonviolent nature. Because chimpanzees and bonobos display marked differences in some aspects of behavior and ecology, studies of bonobos

[1] *Primate Research Institute, Kyoto University, Inuyama, Aichi, 484-8506 Japan*

[2] *Lukuru Project, Democratic Republic of Congo, c/o Lukuru Wildlife Research Foundation, 129 Pinckney Street, PO Box 875, Circleville, Ohio, 43113 United States*

helped to expand a range of possible strategies in which we can consider the characteristics of the common ancestor of great apes and humans, and the course of human evolution. Many of these early studies were presented in two influential texts: *The Pygmy Chimpanzee: Evolutionary Biology and Behavior*, edited by R.L. Susman in 1984 and *The Last Ape: Pygmy Chimpanzee Behavior and Ecology*, by T. Kano, first published in 1986 in Japanese, then translated into English in 1992.

However, just as these long-term field studies began to bear fruit, the bonobos' range became the arena of brutal human warfare. The global range of the bonobo is limited to a single area within the territorial limits of one government. This territory became eclipsed by human conflict beginning in 1991 and differentially affected safe access to study sites. Additional long-term studies were initiated in the early 1990s, one led by Hohmann at Lomako and another led by Myers-Thompson at Lukuru, which added new and exciting findings to our understanding of wild bonobos. Progress in bonobo research in these and other study sites in the 1990s were revealed in two additional important volumes: *Behavioural Diversity in Chimpanzees and Bonobos*, edited by C. Boesch, G. Hohmann, and L. Marchant in 2002 and *All Apes Great and Small Volume One: African Apes*, edited by B.M.F. Galdikas, N. Briggs, and A. Sheeran in 2001.

During the long-lasting civil war of 1996–2002, field researchers felt helpless to do anything that would effectively address concern over the perceived decline in numbers of bonobos, and to avoid their possible extinction in the wild. During this period, however, there was remarkable progress in two disciplines. First, during the absence of studies on wild bonobos, researchers carried out a number of studies on populations in captivity. Those studies revealed various features of female dominance, socio-sexual behavior, and the prolonged estrus of bonobos in more detail than ever before. Yet, those findings still need to be examined in the wild to know whether the features are indicative of the real nature of bonobos. Second, progress also occurred in the ecological study of other great apes. Though the main focus in the early stage of ape studies was on behavior, researchers began to direct more interest on ecological adaptation of great apes to the variable environments across their tropical forest homes. Scientists introduced new paradigms and hypotheses, and developments occurred in the analysis of methodology for the comparison between different species and different geographic sites. Those studies posed a lot of questions to be examined in bonobos.

Responding to the cessation of hostilities across the bonobos' range in 2002, field research and conservation efforts were infused and expanded. To facilitate information exchange, researchers working in active field sites and captive colonies gathered in two symposia at the 2006 International Primatological Society Congress, the first such meeting to be held in a great ape range country (Uganda). Researchers used this opportunity to present information on what they had found and what they plan to reveal in the current era of bonobo study. This book is a result of those symposia.

In this book, the first section introduces recent progress in studies of behavioral research of captive bonobos. The first two papers examine the social status of males and females by comparing results from various research facilities and across time periods, and by analyzing the relationship of dominance with other factors. The following two chapters deal with new topics concerning play behavior and multiple uses of gestures for communication, both of which present new aspects of bonobo intelligence.

The second section deals with the ecological study of bonobos that has been carried out after the resumption of field studies since 2002. The first three chapters report new findings from the Wamba study site on the peaceful nature of bonobo groups and the relationship between fruit abundance, party size, and ranging patterns that were examined by using a method enabling comparisons with studies of other great apes. The following three chapters present results of population censusing in the Salonga National Park, the largest protected area in Africa. These chapters report new methodologies and different aspects of relationships between bonobo density and habitat vegetation.

The third section focuses on studies for the conservation of bonobos. The first two papers analyze the human-bonobo interrelationship and give insight into realistic ways for conservation of bonobos in contemporary circumstances. The following two chapters report on the status of bonobos and efforts for conservation in a re-established bonobo site as well as in the longest running study site. The final chapter describes the contribution of a bonobo sanctuary to the varied perspectives.

In 1986, the IUCN/Species Survival Commission, Primate Specialist Group (Oates 1986) produced an Action Plan which identified research and conservation priorities for African primates, including the bonobo. Subsequent to the publication of this pivotal statement, numbers of action plans, workshops, international conferences and meetings have repeated the same call for field research priorities and conservation efforts on bonobos. But an underlying problem was that census-based legislation governing protection of bonobos requires reliable quantitative data. This book presents the first comprehensive effort to address that basic requirement, and the most recent findings of the wild bonobo status representing the post-conflict period. Although some areas recognized in past publications have identified sites where bonobos no longer occur, within this book we provide encouraging news on a far-reaching view of the breadth of field occurrence and newly identified populations.

Thus, this book updates readers with the most recent advances in various aspects of research and the integration of bonobo conservation. Following the publication of the previous comprehensive books on bonobos, we anticipate that this book will be another milestone to encourage further studies of bonobos, our least-known relative. As we proceed into a new era in the homeland of the bonobo, we face new challenges for conservation. By illuminating the current status of the bonobo and perspectives for its future, a critical framework is now emerging.

This book would not have been possible without the efforts of many people. First, we would like to thank Debby Coxe who gave us a large amount of time during the IPS Congress in Uganda, enabling us to have the comprehensive symposia for bonobo research and conservation. We would also like to thank Dr. Russell Tuttle who provided the venue for publication of papers presented in those symposia. Dr. Tuttle also read all the chapters in this book and gave invaluable suggestions for improvement. We thank Dr. Annette Lanjouw and many other researchers who gave us helpful comments and advice for revision of chapters. Dr. Frans de Waal, Dr. Richard Wrangham, and Pastor Cosma Wilungula Balongelwa kindly agreed to read chapters of each section and write very informative forewords to the sections. We are also most grateful for Ms. Andrea Macaluso, Ms. Lisa Tenaglia, and Ms. Cynthia Manzano of Springer, whose great efforts and patience helped us realize the publication of this book. Though quite unusual for this kind of book, they agreed to include French translations of the introduction and forewords. We are most grateful to Mr. Michel Hasson and Ms. Vanessa Anastassiou for translating the English to French and editing the French contribution. It is our hope that the French text will be most useful for people living in the range country of bonobos, Democratic Republic of Congo, to understand the scientific and conservation value of bonobos. We are indebted to Dr. Nadine Laporte and Jared Stabach of Woods Hole Research Center for the production of the introductory map. We graciously acknowledge the Lukuru Wildlife Research Foundation for covering the cost to produce the color figures. To all involved, we extend our gratitude.

# References

Boesch C, Hohmann G, Marquardt L (eds) (2002) Behavioral diversity of chimpanzees and bonobos. Cambridge Univ Press, New York
Coolidge HJJr (1933) *Pan paniscus*, pigmy chimpanzee from south of the Congo River. Amer J Phys Anthropol 18:1–59
Galdikas BMF, Briggs NE, Sheeran LK, Shapiro GL, Goodall J (eds) (2001) All Apes Great and Small, Vol 1: African Apes. Kluwer Academic/Plenum
Kano T (1992) The Last Ape: Pygmy Chimpanzee Behavior and Ecology. Stanford Univ Press, Stanford
Sibley CG and Ahlquist JE (1987) DNA hybridization evidence of hominoid: results from an expanded data set. Journal of Molecular Evolution 26:99–121
Susman RL (ed) (1984) The pygmy chimpanzee: evolutionary biology and behavior. Plenum, New York
The Chimpanzee Sequencing and Analysis Consortium (2005) Initial sequence of the chimpanzee genome and comparison with the human genome. Nature 437:69–87
Thompson JA Myers (2001) On the nomenclature of *Pan paniscus*. Primates 42:101–111

# Introduction

Takeshi Furuichi[1] et Jo Thompson[2]

Presque quatre-vingt ans ont passé depuis que le bonobo a été reconnu officielle-
ment comme une espèce à part entière (Coolidge, 1933). *Homo* et *Pan* partagent
98,8 % de certaines séquences d'ADN (Sibley et Ahlquist 1987, The Chimpanzee
Sequencing and Analysis Consortium 2005), faisant des chimpanzés et des bonobos nos plus proches parents en vie. Bien que, d'une certaine façon, nous connais-
sions notre cousin le bonobo depuis plus de 125 ans (Myers Thompson, 2001), le
bonobo est toujours considéré comme le moins connu des grands singes. Malgré le
fait que le travail de terrain commencé en 1938 se concentrait spécialement sur les
bonobos, dans le but de collecter des spécimens pour les musées, les premières
études de terrain systématiques de bonobos vivants dans leur habitat naturel ne
commencèrent qu'en 1972 avec le Professeur Toshisada Nishida. Bien que limitée,
Nishida conduisit une étude de la région qui borde la rive occidentale du lac
Tumba. En juillet de cette même année, une équipe de l'université de Yale aux
Etats-Unis commença une étude de deux ans au lac Tumba: le premier site d'étude
sur le terrain. En 1973, la recherche scientifique dans la nature prit de l'ampleur
lorsque le Professeur Takayoshi Kano, le premier scientifique à étudier les bonobos
dans la nature, conduisit une étude à large spectre à travers la totalité du noyau du
bloc forestier au sud du fleuve Congo. Le résultat fut la création en 1974 du premier
site dédié à l'étude sur le long terme des bonobos.

Après des études préliminaires en captivité, deux équipes de chercheurs, l'une
conduite par Kano à Wamba et l'autre par Susman à Lomako, entreprirent des
études de terrain sur le long terme à propos du comportement et de l'écologie de
bonobos sauvages vivant au cœur de la forêt pluviale du bassin du Congo. Ces
études apportèrent un nouveau regard sur notre compréhension de l'adaptation
écologique des grands singes ainsi que sur la compréhension de nous-mêmes. Au
début de l'étude, la recherche se focalisa sur la ressemblance morphologique entre
les bonobos et le plus ancien ancêtre connu de l'homme et sur la similarité des

[1] *Primate Research Institute, Kyoto University, Inuyama, Aichi, 484-8506 Japan*

[2] *Lukuru Project, Democratic Republic of Congo, c/o Lukuru Wildlife Research Foundation,
129 Pinckney Street, PO Box 875, Circleville, Ohio, 43113 United States*

comportements sexuels des hommes et des bonobos. Au fur et à mesure que la recherche progressait, les scientifiques réalisèrent que les bonobos présentaient d'autres particularités importantes comme le statut social élevé des femelles et une société construite autour de la non-violence. Du fait que les chimpanzés et les bonobos affichent des différences marquées dans certains aspects de leur comportement et de leur écologie, l'étude des bonobos a permis d'élargir la gamme des stratégies possibles parmi lesquelles nous pouvons estimer les caractéristiques principales de l'ancêtre commun des grands singes et de l'homme et le cours de l'évolution humaine. Beaucoup de ces études anciennes ont été publiées dans deux textes influents: THE PYGMY CHIMPANZEE: EVOLUTIONARY BIOLOGY AND BEHAVIOR, édité par R.L. Susman en 1984 et THE LAST APE: PYGMY CHIMPANZEE BEHAVIOUR AND ECOLOGY, par T. Kano, d'abord publié en japonais en 1986 puis traduit en anglais en 1992.

Cependant, juste au moment où ces études à long terme commençaient à porter leurs fruits, l'habitat des bonobos devint l'arène de combats armés entre humains. La totalité du territoire des bonobos est limitée à une aire unique à l'intérieur des frontières d'un seul pays. Ce territoire fut éclipsé par les conflits humains qui débutèrent en 1991 et affectèrent de manière différente l'accès aux sites d'étude. Des études complémentaires sur le long terme furent initiées au début des années 1990, l'une conduite par Hohmann à Lomako et l'autre conduite par Myers-Thompson à Lukuru, qui ajoutèrent des découvertes nouvelles et excitantes à notre compréhension des bonobos sauvages. Les progrès de la recherche sur les bonobos dans ces sites d'études et d'autres durant les années 1990 furent publiés dans deux importants volumes additionnels: BEHAVIOURAL DIVERSITY IN CHIMPANZEES AND BONOBOS, édité par C. Boesch, G. Hohmann, et L. Marchant en 2002 et ALL APES GREAT AND SMALL VOLUME ONE: AFRICAN APES, édité par B.M.F. Galdikas, N. Briggs, et A. Sheeran en 2001.

Durant la longue guerre civile de 1996 à 2002, les chercheurs de terrain se sentirent impuissants à faire quelque chose d'utile pour éveiller l'attention sur leur constatation du déclin du nombre de bonobos et pour éviter leur possible extinction dans la nature. Néanmoins, durant cette période, des progrès furent réalisés dans deux disciplines. Premièrement, pendant l'absence d'études sur les bonobos sauvages, les chercheurs conduisirent un certain nombre d'expériences sur les populations en captivité. Ces études révélèrent différentes particularités de la domination des femelles, du comportement socio-sexuel et du cycle menstruel prolongé des bonobos avec plus de détails que jamais auparavant. Cependant ces découvertes doivent encore être observées dans la nature pour savoir si elles sont véritablement indicatives de la nature réelle des bonobos. Deuxièmement, des progrès furent également faits dans l'étude écologique d'autres grands singes. Bien que, dans les premiers moments de l'étude des grands singes, l'intérêt principal se portait sur leur comportement, les chercheurs commencèrent à porter plus d'attention sur les adaptations écologiques des grands singes aux environnements variables à travers les forêts tropicales qui les abritent. Les scientifiques émirent de nouvelles hypothèses et paradigmes et des développements se firent dans l'analyse de la méthodologie utilisée pour la comparaison entre différentes espèces et différents sites géographiques. Ces études amenèrent un tas de questions à examiner chez les bonobos.

Suite à la cessation des hostilités dans le territoire des bonobos en 2002, la recherche sur le terrain et les efforts de conservation reprirent et s'étendirent. Pour faciliter l'échange d'information, les chercheurs qui travaillaient sur des sites actifs ou sur des colonies captives se réunirent lors de deux symposiums en 2006 au Congrès de la Société Primatologique Internationale. C'était la première fois qu'une telle réunion se tenait dans un pays qui abrite des grands singes (Ouganda). A cette occasion, les chercheurs présentèrent le résultat de leur recherche et ce qu'ils avaient l'intention de révéler en cette ère d'études sur le bonobo. Ce livre est un des résultats de ces symposiums.

Dans ce livre, la première section présente les récents progrès dans l'étude comportementale sur des bonobos captifs. Les deux premiers articles examinent le statut social des mâles et des femelles en comparant les résultats de différents centres de recherche, en les considérant à travers le temps et en analysant la relation de domination avec d'autres facteurs. Les deux chapitres suivants traitent de nouveaux sujets concernant le comportement ludique et l'usage multiple des gestes pour la communication. Tous ces sujets représentent des aspects nouveaux de l'intelligence des bonobos.

La deuxième section traite de l'étude écologique des bonobos qui a été conduite après la reprise des études de terrain en 2002. Les trois premiers chapitres rapportent des découvertes de l'étude du site Wamba sur la nature pacifique des groupes de bonobos et la relation entre l'abondance des fruits, la taille du groupe et les modèles de répartition qui ont été examinés en utilisant une méthode permettant les comparaisons avec des études sur d'autres grands singes. Les trois chapitres suivants présentent les résultats des travaux de recensement des populations dans le parc national de la Salonga, la plus grande aire protégée d'Afrique. Ces chapitres présentent des nouvelles méthodologies et différents aspects des relations entre la densité des bonobos et l'habitat végétal.

La troisième section se focalise sur les études traitant de la conservation des bonobos. Les deux premiers articles analysent la relation entre les humains et les bonobos et donnent un aperçu sur les moyens réalistes de conservation des bonobos dans les circonstances actuelles. Les deux chapitres suivants traitent du statut des bonobos et des efforts de conservation dans un site rétablie abritant des bonobos et dans un site d'etude de travail le plus longues. Le dernier chapitre décrit la contribution d'un sanctuaire de bonobos aux différentes perspectives.

En 1986, le *Primate Specialist Group* de la *Species Survival Commission* de l'UICN (Oates, 1986) émit un plan d'action qui identifiait les priorités pour la recherche et la conservation des primates africains, y compris le bonobo. A la suite de la publication de cette déclaration charnière, de nombreux plans d'action, d'ateliers, de conférences internationales et de réunions ont répété le même appel urgent pour une recherche sur le terrain et des efforts de conservation en faveur des bonobos. Cependant un problème sous-jacent consistait dans le fait qu'une loi, basée sur des recensements et régissant la protection des bonobos, demandait un grand nombre de données fiables. Ce livre présente la première tentative pour apporter une telle information de base, et les plus récentes découvertes concernant le statut des bonobos dans la nature après la période de conflits. Bien que certaines aires, que d'anciennes publications décrivaient comme abritant des bonobos, n'en comptent plus aujourd'hui, nous amenons dans ce livre des nouvelles encourageantes sur la vision à long terme des observations sur le terrain et sur des populations nouvellement identifiées.

Ce livre met à jour les connaissances des lecteurs avec les plus récentes avancées de divers aspects de la recherche et de son intégration dans la conservation des bonobos. Venant à la suite de la publication d'autres livres sur les bonobos, nous espérons que cet ouvrage sera un jalon supplémentaire qui encouragera d'autres études sur les bonobos, nos parents les moins connus. Alors que nous avançons dans une nouvelle ère dans la planète bonobo, nous faisons face à de nouveaux défis pour leur protection. En mettant à jour le statut actuel des bonobos et les perspectives de leur futur, on voit émerger un cadre assez critique.

Ce livre n'aurait jamais paru sans les efforts de nombreuses personnes. Nous voudrions d'abord remercier Debby Coxe qui nous a consacré énormément de temps durant le congrès IPS en Ouganda ce qui nous a permis de tenir les symposiums sur la recherche et la conservation des bonobos. Nous voudrions également remercier le Dr Russell Tuttle qui nous a fourni l'adresse pour la publication des articles présentés à ces symposiums. Le Dr Tuttle a également lu tous les chapitres de ce livre et nous a donné d'inestimables conseils pour les améliorer. Nous remercions le Dr Annette Lanjouw ainsi que de nombreux autres chercheurs qui nous ont donné des commentaires utiles et des conseils pour la révision de certains chapitres. Dr Frans de Waal, Dr Richard Wrangham et le Pasteur Cosma Wilungula Balongelwa ont gentiment accepté de lire les chapitres de chaque section et ont écrit des préfaces très instructives à ces sections. Nous sommes également reconnaissants à Mlle Andrea Macaluso, Mlle Lisa Tenaglia, et Mlle Cynthia Manzano de Springer dont les efforts considérables et la patience nous ont aidé à réaliser la publication de ce livre. Bien qu'inhabituel dans ce genre de livre, ils ont accepté d'inclure des traductions françaises de l'introduction et des préfaces des sections. Nous sommes très reconnaissants à M. Michel Hasson et Mlle Vanessa Anastassiou pour leur traduction d'anglais en français et pour la rédaction de la contribution en français. Nous espérons que ce texte en français sera utile aux populations vivant dans l'aire de distribution des bonobos, la République Démocratique du Congo, afin de leur permettre de comprendre la valeur scientifique et l'importance de la conservation des bonobos. Nous sommes redevables au Dr Nadine Laporte et à Jared Stabach du Centre de Recherche de Woods Hole pour la production de la carte géographique de l'introduction. Nous reconnaissons gracieusement la Lukuru Wildlife Research Foundation pour couvrir le coût de produire des chiffres de la couleur. A tous ceux qui se sont impliqués, nous redisons toute notre gratitude.

# Part I
# Behavioral Study Section

# Foreword to Behavioral Study Section

Frans B.M. de Waal[1]

In the minds of some, bonobos compete with chimpanzees as the best nonhuman primate model of our species. Given that there can only be one best model, and that chimpanzees were known first, bonobos are at a disadvantage. Moreover, anthropologists seem heavily invested in the chimpanzee as a model for human social evolution. Chimpanzees show male bonding, intergroup "warfare," proficient tool-use, power "politics," hunting, and meat-eating. There is no shortage of similarities, albeit mostly on the male side, with our own species. Consequently, a coherent picture of human social evolution has arisen around the chimpanzee as close relative, one emphasizing meat, violence, and male superiority. This picture fit well with post World War II developments, led by Konrad Lorenz in Europe and Robert Ardrey in the United States (Lorenz 1963, Ardrey 1963). Understandably perhaps, emphasis was put on *Homo sapiens* as a "mentally unbalanced predator" endowed with vigorous aggressive instincts (Cartmill 1993, p. 14).

Then the bonobo came along. It is good to realize, though, that long before this happened, Robert Yerkes wrote an entire book about an ape named "Prince Chim." This ape was notably more sensitive, altruistic and intelligent than other apes familiar to Yerkes. The great psychologist did not know it at the time, but Prince Chim was a bonobo. Yerkes was so struck by this individual's behavior that he entitled his book *Almost Human* (Yerkes 1925). Anatomically as well, bonobos were found to be strikingly human-like. Harold Coolidge, the anatomist who gave *Pan paniscus* its eventual taxonomic status (and who performed the post-mortem on Prince Chim), concluded that this ape "may approach more closely to the common ancestor of chimpanzees and man than does any living chimpanzee" (Coolidge 1933, p. 56).

We had to wait until the 1980s, however, for detailed studies of bonobo social behavior. We know now that, compared with chimpanzee society, bonobo society is less violent (i.e. lethal aggression has thus far never been observed, neither in captivity nor in the field, quite in contrast to the chimpanzee), bonobos seem sex- rather than power-oriented, and adult males tend to defer to adult females (de Waal 1997). Obviously, the virtual absence of hunting and "warfare" in this ape, combined with its relative peacefulness and female dominance, should raise

[1]*Living Links, Yerkes National Primate Research Center, Emory University, 954 N. Gatewood Road, Atlanta, Georgia, 30322 United States*

T. Furuichi and J. Thompson (eds.), *The Bonobos: Behavior, Ecology, and Conservation*
© Springer 2008

questions about earlier scenarios of human evolution built around themes of violence and predation. However, what we have seen thus far instead is a concerted effort to marginalize bonobos and keep them out of evolutionary scenarios. This is despite the fact that the split between bonobo and chimpanzee occurred well after their lineage split off from ours, meaning that both species are phylogenetically equidistant to us.

Thus, Wrangham and Peterson (1996) treated bonobos as an afterthought, a nice respite of the blood-soaked image of our primate heritage drawn in *Demonic Males*. Whereas these authors assigned bonobos secondary importance, Stanford (1998) simply declared them less special than often assumed, downplaying differences between both *Pan* species even though study after study has shown them to be quite different (Parish and de Waal 2000, Doran et al. 2002). Stanford's (1998) argument that reliance on captive data explains why bonobo sexual behavior is said to be so highly developed ignores the fact that chimpanzees in captivity do not show anything close to such behavior. If the same conditions affect two species so differently, the logical conclusion is that the difference is due to the species, not the conditions (de Waal 1988, 1998). The high-point of the sidelining attempt with regards to bonobos came with Konner's (2002, p. 199) suggestion that science can safely ignore them, because "chimps have done far better than bonobos, which are very close to extinction."

The earlier citation from Coolidge was not offered to make the point that the bonobo is a better model of human ancestors than the chimpanzee, even though one could argue this point based on a) the bonobo's body proportions (Zihlman et al. 1978), b) the probability that this forest ape has encountered fewer evolutionary pressures to change than the ancestors of both humans and chimpanzees (Kano 1992), and c) the recent discovery that bonobos and humans share genetic code in relation to affiliative behavior that is absent in the chimpanzee (Hammock and Young 2005). Despite these arguments, there is no urgent need to choose between bonobo and chimpanzee with regards to our ancestry. We can be sure that more discoveries are on their way, and they may again change the picture. What we need most at this point is behavioral and ecological data so as to develop better models of how and why the bonobo evolved into what it is today. In the last couple of years, scientists have been actively collecting new data in both captivity and the field. This volume's first section features some of the best zoo studies. By themselves, such studies cannot provide the full comparison, but they do permit behavioral records of greater detail than possible in the field, and are capable of generating testable hypotheses about what makes bonobos different.

The first two chapters address the issue of agonistic dominance. Female dominance is rare in primates, which explains the initial skepticism towards claims that female bonobos dominate males. Stevens et al. find that all captive bonobo groups are dominated by an alpha female rather than alpha male, but also note that not all females necessarily dominate all males. Males and females rather seem to have overlapping ranks, with females being disproportionally represented near the top. Paoli and Palagi do not reach an equally clear conclusion on male vs. female dominance, but they do report that an individual's sex hardly predicts its rank, which is by itself already a huge contrast with chimpanzees. Perhaps the best way to look at the dominance relation between the sexes is not so much in terms of individual ranks, but on the basis of the outcome of multi-party encounters. As shown by Stevens et al., bonobos seem characterized by collective female dominance, as expressed in alliances among females against males, the effectiveness of which is facilitated by the virtual absence of bonding and alliances among adult males (see also de Waal 1997).

In discussing the adaptive potential of bonobos, Stevens et al. stress the preponderance of alliances directed down the hierarchy. Such alliances are bound to reinforce the hierarchy, hence create a more rigid structure than found in chimpanzees, which do show frequent coalitions from below (de Waal and Luttrell 1988). This may explain the relatively stable hierarchies of bonobos, which apparently change only when one of the older matriarchs falls ill or dies (Kano 1992). Here we recognize two major differences with chimpanzee society as observed in captivity. The first is that all healthy adult male chimpanzees dominate all females, which clearly does not apply to bonobos. The second is that flexible male alliances in chimpanzees create a rather unstable hierarchy, hence a volatile social environment compared to the more predictable social structure of bonobos.

The latter may explain the relative peacefulness of bonobos society, but in addition, bonobos employ effective conflict prevention and resolution strategies as described by Palagi and Paoli. Not only do they counter conflict by sexual means (cf. de Waal 1987), they also have playful ways of diffusing tensions. This is entirely consistent with the reputation of bonobos as neotenous animals, that is, animals that (like humans) have evolved by retaining juvenile characteristics into adulthood.

Pollick et al., finally, explore the gestural communication of bonobos, comparing and contrasting it with their facial and vocal communication. Bonobos use many different gestures, and it was confirmed that this mode of communication is used more flexibly than facial and vocal displays, most of which are closely tied to specific contexts. The meaning of gestures likely depends on how they combine with the social context and other forms of communication, whereas the meaning of facial and vocal displays is relatively invariant, not only within each species, but even between species. Comparing communication across both species and groups of each species, Pollick et al. found that by knowing the usage of a facial expression or vocalization in one species, one also knows its usage in the other species. With gestures, on the other hand, not only are there few similarities between the species, but even within each species gesture usage varies greatly from group to group. This makes gesture the better candidate for early language evolution. If our ancestors used gestures in the same flexible manner, this may have provided a stepping stone for the evolution of symbolic communication, which may have originated in the gestural rather than vocal domain (e.g. Corballis 2002).

Even though more detailed studies are urgently needed, especially studies that compare the social behavior and cognition of the bonobo with those of the other apes, we have come a long way since the first zoo research on their behavior (Tratz and Heck 1954). Given that the bonobo's behavior is in some respects rather unique, it has the potential of greatly enriching scenarios of human evolution.

# References

Ardrey R (1963) African Genesis. Dell, New York

Cartmill M (1993) A View to a Death in the Morning: Hunting and Nature through History. Cambridge, MA: Harvard Univ Press, Cambridge

Coolidge HJ (1933) *Pan paniscus*: pygmy chimpanzee from south of the Congo River. Amer J Phys Anthropol 18: 1–57

Corballis MC (2002) From Hand to Mouth: The Origins of Language. Princeton Univ Press, Princeton

Doran DM, Jungers WL, Sugiyama Y, Fleagle JG, Heesy CP (2002) Multivariate and phylogenetic approaches to understanding chimpanzee and bonobo behavioral diversity. In: Boesch C, Hohmann G, Marchant LF (eds) Behavioural Diversity in Chimpanzees and Bonobos. Cambridge Univ Press, Cambridge, pp. 14–34

Hammock EAD, Young, LJ (2005) Microsatellite instability generates diversity in brain and sociobehavioral traits. Science 308: 1630–1634

Kano T (1992) The Last Ape: Pygmy Chimpanzee Behavior and Ecology. Stanford Univ Press, Stanford

Konner M (2002) Some obstacles to altruism. In: Post SG, Underwood LG, Schloss JP, Hurlbut WB (eds) Altruistic Love: Science, Philosophy, and Religion in Dialogue. Oxford Uni Press, Oxford, pp 192–211

Lorenz KZ (1966) On Aggression. Methuen, London

Parish AR, de Waal FBM (2000) The other "closest living relative": How bonobos (*Pan paniscus*) challenge traditional assumptions about females, dominance, intra- and inter-sexual interactions, and hominid evolution. In: LeCroy D, Moller P (eds) Evolutionary Perspectives on Human Reproductive Behavior, Annals of the New York Academy of Sciences 907, pp. 97–113

Stanford CB (1998) The social behavior of chimpanzees and bonobos: Empirical evidence and shifting assumptions. Current Anthropol 39: 399–420

Tratz EP, Heck H (1954) Der afrikanische Anthropoide "Bonobo", eine neue Menschenaffengattung. Säugetierkundliche Mitteilungen 2: 97–101

de Waal FBM (1987) Tension regulation and nonreproductive functions of sex in captive bonobos (*Pan paniscus*). Nat Geogr Res 3: 318–335

de Waal FBM (1988) The communicative repertoire of captive bonobos (*Pan paniscus*), compared to that of chimpanzees. Behaviour 106: 183–251

de Waal FBM (1997) Bonobo: The Forgotten Ape. Univ of California Press, Berkeley

de Waal FBM (1998) Commentary on C. B. Stanford. Current Anthropol 39: 407–408

de Waal FBM, Luttrell LM (1988) Mechanisms of social reciprocity in three primate species: symmetrical relationship characteristics or cognition? Ethol Sociobiol 9: 101–118

Wrangham RW, Peterson D (1996) Demonic Males: Apes and the Evolution of Human Aggression.: Houghton Mifflin, Boston

Yerkes RM (1925) Almost Human. Century, New York

Zihlman AL, Cronin JE, Cramer DL, Sarich VM (1978) Pygmy chimpanzee as a possible prototype for the common ancestor of humans, chimpanzees, and gorillas. Nature 275: 744–746

# Avant-propos à la Section d'Etude du Comportement

Frans B. M. de Waal[1]

Dans l'esprit de certains, les bonobos sont en compétition avec les chimpanzés pour le titre de meilleur modèle de primate non humain de notre espèce. Etant donné qu'il ne peut y avoir qu'un seul meilleur modèle, et que les chimpanzés furent connus les premiers, les bonobos partent avec un handicap. De plus, les anthropologues semblent s'être beaucoup investis pour prendre le chimpanzé en tant que modèle de l'évolution sociale des humains. Les chimpanzés affichent des liens entre mâles, des « guerres » entre groupes, un usage compétent d'outils, une « politique » de pouvoir, la pratique de la chasse et la consommation de viande. Il ne manque pas de similitudes, bien qu'essentiellement du côté des mâles, avec notre propre espèce.

De ce fait, une image cohérente de l'évolution sociale humaine s'est construite autour du chimpanzé, considéré comme proche parent, qui met l'accent sur la viande, la violence et la supériorité des mâles. Cette image convenait bien aux théories en vogue après la deuxième guerre mondiale, conduites par Konrad Lorenz en Europe et par Robert Ardrey aux Etats-Unis (Lorenz 1963, Ardrey 1963). De façon peut-être compréhensible, l'accent fut mis pour présenter une image de l'*Homo sapiens* considéré comme un « prédateur mentalement déséquilibré » doté de vigoureux instincts agressifs (Cartmill 1993, p.14).

Puis arriva le bonobo. Il est bon de réaliser que bien avant cette découverte, Robert Yerkes écrivit un livre entier sur un chimpanzé appelé « Prince Chim ». Ce primate était considérablement plus sensible, plus altruiste et plus intelligent que les autres chimpanzés familiers à Yerkes. Le célèbre psychologue ne le savait pas à cette époque, mais « Prince Chim » était un bonobo. Yerkes fut tellement impressionné par le comportement de cet animal qu'il intitula son livre « Almost Human » (Presque Humain) (Yerkes 1925). Sur le plan anatomique également, il apparut que les bonobos étaient étonnamment proches des humains. Harold Coolidge, l'anatomiste qui donna à *Pan paniscus* son statut taxonomique définitif (et qui pratiqua l'autopsie sur Prince Chim), arriva à la conclusion que ce primate « *pourrait*

---

[1]*Living Links, Yerkes National Primate Research Center, Emory University, 954 N. Gatewood Road, Atlanta, Georgia, 30322 United States*

T. Furuichi and J. Thompson (eds.), *The Bonobos: Behavior, Ecology, and Conservation*       15
© Springer 2008

*se rapprocher plus de l'ancêtre commun des chimpanzés et de l'homme, que ne le fait aucun autre chimpanzé vivant* » (Coolidge 1933, p.56).

Il nous fallut néanmoins attendre les années 1980 pour des études approfondies sur le comportement social des bonobos. Nous savons maintenant que, comparée à la société des chimpanzés, la société des bonobos est moins violente (par exemple, des agressions mortelles n'ont jamais été observées à ce jour, ni en captivité ni en milieu naturel, ce qui contraste notablement avec les chimpanzés), les bonobos semblent plus orientés vers le sexe que vers le pouvoir et les mâles adultes tendent à se soumettre aux femelles adultes (de Waal 1997). Visiblement, l'absence virtuelle de chasse et de « guerre » chez ce primate, alliée avec son tempérament relativement pacifique et la domination des femelles, devrait susciter des questions à propos des anciens scénarios de l'évolution humaine bâtis sur des thèmes de violence et de prédation. Au lieu de cela, ce que nous avons pu constater jusqu'à ce jour est un effort concerté pour marginaliser les bonobos et les maintenir en dehors des scénarios de l'évolution. Ceci en dépit du fait que la séparation entre les bonobos et les chimpanzés a eu lieu longtemps après que leur lignage se soit séparé du nôtre, ce qui implique que les deux espèces sont à égale distance de nous sur le plan phylogénétique.

Ainsi, Wrangham and Peterson (1996) ont traité les bonobos comme une réflexion, une belle alternative à l'image sanglante de notre héritage simien esquissé dans *Demonic Males (Mâles Démoniaques)*. Alors que ces auteurs n'accordent qu'une importance secondaire aux bonobos, Stanford (1998) les a simplement déclaré moins spéciaux qu'on ne le dit souvent, minimisant les différences entre les deux espèces du genre *Pan*, malgré le fait qu'études après études il a été démontré qu'ils étaient assez différents (Parish and de Waal 2000, Doran *et al.* 2002). L'argument de Stanford (1998) selon lequel la dépendance envers des données obtenues en captivité explique pourquoi le comportement sexuel des bonobos est dit tellement développé, ne tient pas compte du fait que les chimpanzés en captivité ne présentent aucun comportement approchant. Si des conditions semblables affectent deux espèces si différemment, la conclusion logique est que cette différence est inhérente aux espèces et non aux conditions (de Waal 1988, 1998). Le point culminant de cette tentative d'écarter les bonobos a été atteint avec la suggestion de Konner (2002, p.199) qui préconisait que la science pouvait franchement les ignorer étant donné que « *les chimpanzés ont bien mieux réussi que les bonobos, lesquels sont proches de l'extinction* ».

La citation ci-dessus de Coolidge n'a pas été proposée pour prouver que le bonobo est un meilleur modèle d'ancêtre de l'humanité que le chimpanzé, même si on peut défendre cette assertion en se basant sur, premièrement les proportions corporelles du bonobo (Zihlman *et al.* 1978), deuxièmement sur la probabilité que ce primate forestier a rencontré moins de pressions évolutives le portant au changement que les ancêtres de l'homme et du chimpanzé (Kano 1992), et enfin sur la découverte récente que les bonobos et les hommes partagent un code génétique en relation avec le comportement affectif que ne possède pas le chimpanzé (Hammock and Young 2005). En dépit de ces arguments, il n'y a aucune urgence à choisir entre les bonobos et les chimpanzés en ce qui concerne notre lignage. Nous pouvons être sûrs que d'autres découvertes suivront, qui pourraient encore changer notre vision des choses. Ce dont nous avons le plus besoin à ce stade, ce sont des données écologiques et comportementales en vue de développer de meilleurs modèles sur la façon et les motifs qui ont amené le bonobo à évoluer en ce qu'il est aujourd'hui. Ces dernières années, les scientifiques se sont attelés à collecter de nouvelles données, aussi bien en captivité que dans la nature. La première section de ce volume présente certaines des meilleures études effectuées dans des zoos. Par elles-mêmes, de telles études ne peuvent pas fournir une totale comparaison, mais elles permettent des enregistrements de données comportementales d'une plus grande finesse que celles qu'on peut obtenir dans la nature, et sont capables d'engendrer des hypothèses vérifiables sur ce qui rend les bonobos différents.

Les deux premiers chapitres traitent du sujet de la dominance agonistique. La domination des femelles est rare chez les primates, ce qui explique le scepticisme initial envers les affirmations que les femelles de bonobos dominent les mâles. Stevens *et al.* ont trouvé que tous les groupes de bonobos en captivité sont dominés par une femelle alpha plutôt que par un mâle alpha, mais notèrent également que toutes les femelles ne dominent pas nécessairement tous les mâles. Les mâles et les femelles semblent plutôt avoir des rangs qui s'entrecroisent, avec les femelles représentées de façon disproportionnée vers le sommet. Paoli et Palagi n'arrivent pas à une conclusion aussi claire sur la domination des femelles sur les mâles, mais ils rapportent que le sexe d'un individu peut difficilement laisser prévoir son rang, ce qui est déjà en soi-même une immense différence avec les chimpanzés. Peut-être que la meilleure façon de juger la relation de domination entre les sexes ne réside pas tant dans l'observation des rangs individuels mais dans l'issue des rencontres entre plusieurs individus. Comme l'ont démontré Stevens *et al.*, les bonobos semblent caractérisés par une domination collective des femelles, exprimée dans des alliances de femelles contre des mâles, l'efficacité de ces dernières étant favorisée par la quasi absence de cohésion et d'alliances entre les mâles adultes (voir aussi de Waal 1997).

Dans leur discussion sur le potentiel d'adaptation des bonobos, Stevens *et al.* soulignent la prépondérance des alliances dirigées vers le bas de la hiérarchie. De telles alliances sont tissées pour renforcer la hiérarchie et de ce fait créent une structure plus rigide que celle trouvée chez les chimpanzés, chez qui on observe souvent des coalitions à partir du bas de l'échelle (de Waal and Luttrell 1988). Ceci pourrait expliquer la relative stabilité des hiérarchies chez les bonobos, qui ne changent apparemment que lorsqu'une des plus vieilles matriarches meurt ou tombe malade (Kano 1992). Nous constatons ici deux différences majeures avec la société des chimpanzés telle que l'on peut l'observer en captivité. La première, c'est que tout chimpanzé mâle adulte en bonne santé domine toutes les femelles ce qui n'est clairement pas le cas chez les bonobos. La deuxième, c'est que les alliances changeantes entre mâles chez les chimpanzés créent une hiérarchie plutôt instable et de ce fait un environnement social volatile comparé à la structure sociale plus prédictible des bonobos.

La dernière constatation pourrait expliquer la paix relative qui règne dans la société des bonobos, mais en plus de cela, les bonobos utilisent des stratégies efficaces de prévention et de résolution des conflits comme l'ont décrit Palagi et Paoli. Ils n'utilisent pas uniquement des moyens sexuels pour contrer les conflits (cf. de Waal 1987), mais ont également des moyens ludiques pour apaiser les tensions. Ceci cadre tout à fait avec la réputation qu'ont les bonobos d'être des animaux néoténiques, c'est-à-dire des animaux qui (comme les humains) ont évolué en gardant des caractéristiques juvéniles à l'âge adulte.

Pour terminer, Pollick *et al.* explorent la communication gestuelle des bonobos, la comparant et l'opposant à leur communication faciale et vocale. Les bonobos utilisent de nombreux gestes différents et il a été confirmé que ce mode de communication est utilisé de manière plus flexible que les manifestations faciales ou vocales qui sont pour la plupart intimement liées à des contextes particuliers. La signification des gestes dépend vraisemblablement de la façon dont ils se combinent avec le contexte social et d'autres formes de communication, tandis que la signification des manifestations faciales et vocales est relativement invariable, non seulement à l'intérieur de chaque espèce mais également entre espèces. En comparant la communication entre les deux espèces et entre groupes de chaque espèce, Pollick *et al.* ont découvert que le fait de connaître l'utilisation d'une expression faciale ou d'une vocalisation chez une espèce, permettait de connaître son utilisation chez l'autre espèce. Avec les gestes par contre, non seulement il y a peu de similarités entre les espèces, mais même au sein d'une même espèce, la gestuelle varie fortement de groupe à groupe. Cela fait de la gestuelle le meilleur candidat à l'évolution primitive du langage. Si nos ancêtres utilisaient les gestes de la même manière flexible, cela a pu être le point de départ de l'évolution de la communication symbolique qui pourrait avoir trouvé son origine dans le domaine gestuel plutôt que vocal (e.g. Corballis 2002).

Bien que nous ayons urgemment besoin d'études plus détaillées, spécialement d'études qui comparent le comportement social et l'intelligence des bonobos avec ceux d'autres grands singes, nous avons néanmoins parcouru un long chemin depuis la première étude de comportement effectuée dans un zoo (Tratz and Heck 1954). Etant donné que le comportement du bonobo est en quelque sorte unique, il a la capacité d'enrichir significativement les scénarios de l'évolution humaine.

# The Bonobo's Adaptive Potential:
# Social Relations under Captive Conditions

Jeroen M.G. Stevens[1,2], Hilde Vervaecke[3], and Linda Van Elsacker[1]

## Introduction

By the end of the 1990s, the reputation of bonobos as a peaceful, egalitarian ape with strong female dominance through female bonding was firmly established (de Waal 1995; de Waal and Lanting 1997; Parish and de Waal 2000; de Waal 2001). Stanford (1998) questioned this reputation, and stated that our knowledge on bonobos lagged behind our knowledge of chimpanzees, because the latter has been studied for a longer span and at more study sites. Knowledge about bonobos stems mainly from captive studies which may not be representative (Stanford 1998). Stanford (1998), Franz (1999) and Hohmann et al. (1999) pointed out that the reported strong female bonds of captive bonobos (Parish 1994, 1996, Parish and de Waal 2000) may be a side effect of life in captivity, similar to chimpanzee females in captivity, wherein similar female bonds occur (de Waal 1982, Baker and Smuts 1994). It certainly cannot be denied that captivity affects behavior, especially in species with fission-fusion systems, such as chimpanzees and bonobos (de Waal 1994). In the wild they form temporary subgroups, "parties," whose composition changes constantly (Van Elsacker et al. 1995). However, in captivity, chimpanzees and bonobos are usually kept in stable groups (but see Fortunato and Berman, this volume), which will certainly influence their social relations. Since the two species, kept under similar conditions, display different behavioral strategies, captive studies can also provide conclusive data on interspecific differences (de Waal 1994).

Moreover, observations made of groups in captivity can yield interesting results because of greater visibility of the study subjects, which can reliably be followed

[1]*University of Antwerp, Department of Biology, Universiteitsplein 1, B 2610 Wilrijk, Belgium*

[2]*Centre for Research and Conservation, Royal Zoological Society of Antwerp, Koningin Astridplein 26, B 2018 Antwerp, Belgium*

[3]*Centre for Ethology and Animal Welfare, KaHoSint-Lieven, Belgium*

T. Furuichi and J. Thompson (eds.), *The Bonobos: Behavior, Ecology, and Conservation*
© Springer 2008

on consecutive days (de Waal 1994). This is especially the case when groups are kept under naturalistic conditions. Under such circumstances, captivity offers an interesting perspective to studying the adaptive potential of a species, which is defined as "the entire range of conditions to which a species can adjust without compromising its health, biological functions (such as reproduction) or major parts of its behavioral repertoire (such as species-typical communication)" (de Waal 1994, p246). A comparison between the behavior of chimpanzees at Arnhem zoo with that of chimpanzees in Tanzania, showed that bonds between males were similar, but "females seem an almost different species in captivity compared to what we know about them living in the wild" (de Waal 1994, p248). While chimpanzees have been kept under naturalistic conditions, including multimale, multifemale groups, since the 1970s, bonobos were for a longtime relatively rare in zoological collections, resulting in very small groups or breeding pairs. Male bonobos were transferred to other zoos when reaching adolescence to avoid inbreeding, while females often remained in the natal group. Only in the 1990s, after field research showed that wild bonobo females migrate and males are philopatric, and that wild communities sometimes contained as many adult males as females (Kano 1992, Hashimoto et al. 2008), did zoos begin to mimic their natural social conditions (Mills et al. 1997). The effect of captivity on relationships between bonobos has not yet been thoroughly studied.

The circumstances under which individuals can display their behavioral repertoire are of particular interest in the light of intraspecific differences. Research with chimpanzees showed a remarkable flexibility, both in captivity (Baker and Smuts 1994, de Waal 1994) and in the wild, where different chimpanzee cultures were documented (Wrangham et al. 1994, Whiten et al. 1999). Chimpanzees can occupy a range of habitats, from dry savannah woodlands, to tropical rain forests, which explains part of the variability (Boesch 2002). While bonobos were long believed to be exclusive inhabitants of dense tropical rain forest, recent research showed that they occupy gallery forests in the southern part of their range (Thompson 2002). Moreover, wild bonobos also showed flexibility, and there are cultural differences between study sites (Hohmann and Fruth 2003a). Hence, the typical distinction between savannah-dwelling chimpanzees and the bonobos from the rain forest became blurred. In addition to comparing bonobo behavior from different field sites, research on captive bonobos and comparisons with data from the wild can shed light on their flexibility. In chimpanzees, a comparison between female relationships at Arnhem Zoo with a colony at Detroit Zoo showed remarkable differences, with competition between females being more expressed in the recently formed colony at Detroit (Baker and Smuts 1994). In bonobos, very little is known about differences between naturalistic groups in captivity, as most studies have focused on single groups with multiple males and females (Vervaecke et al. 1999, 2000a, b, c, Palagi et al. 2004, Paoli et al. 2006) or on multiple groups with one, or at most two, adult males per group (Franz 1999, Parish 1994, 1996).

As a second point of criticism, Stanford (1998) argued that much of the knowledge on social behavior stemmed from only a few captive colonies (Yerkes and San Diego Zoo), which may have biased our knowledge on bonobos. The idea of peaceful, female dominated and egalitarian bonobos may characterize some, but not all zoo groups.

We aim to review our further investigations on the social behavior of several captive groups of bonobos. We examine our earlier published results and provide new additional data about the relationship between dominance, age, and sociosexual behavior of bonobos in captivity. We specifically investigate to what extent the image of bonobos as female-dominated, egalitarian, female-bonded and peaceful is manifest in different captive groups and to describe possible differences among groups. To test some of the current contradictions about dominance and bonding patterns in bonobos, we studied four multimale, multifemale groups, which is the largest study sample of captive bonobos.

# Methods

## *Study Groups and Housing*

We studied four captive groups of bonobos. Although each group contained one or more infants or juveniles, younger than 7 years, these are not included in the analyses. Each group contained at least three males, older than 7 years. Although some of these males are only adolescent (Kano 1992), DNA analyses have shown that each of these adolescent males was able to successfully reproduce (Marvan et al. 2006, P Galbusera unpublished data). Three of the study groups contained at least one adult or adolescent male who that mother reared. Except for one mother-daughter pair at Twycross, all females within groups are unrelated. Furthermore, most of the groups had been stable for at least a few years before our study. Table 1.1 is an overview of all adult and adolescent bonobos, their respective ages, relationships and dominance ranks.

Stevens studied the group at Wuppertal Zoo for 203 hours on 23 days between August and September 1999. It comprised four adult and adolescent males, two adult females, and one juvenile. Female LL was the mother of adolescent male BD and of the juvenile female. LL and LM were raised together, and were joined in 1988 by MT, who sired both BG and BD. In 1996 another female (EJ) joined the group. Haas (1983) described their housing.

Stevens observed the group in Apenheul Primate Park for 490 hours on 74 days between February and May 2001. The group included three adult males, five adult females (older than 8 years old), and three juveniles. All adults were unrelated and had been housed together since March 1998, three years before the study period. Gold (2001) described their housing conditions and group formation.

Stevens studied the group at Twycross Zoo for 263 hours on 34 days in November and December 2001 and in February 2002 for 228 hours on 28 days. The group comprised three males, three females, one juvenile, and one infant. DT was the mother of female KC and male KE. KA was the father of KE. All other group

**Table 1.1** Group composition and individual characteristics of the study groups and animals

| Group | Code | Full Name | Sex | Age | Parents | Rank |
|---|---|---|---|---|---|---|
| Wuppertal | LL | Lisala | F | 19 | Masikini × Catherine | 6 |
| | LM | Lusambo | M | 19 | Masikini × Kombote | 5 |
| | BG | Birogu | M | 10 | Mato × Catherine | 4 |
| | EJ | Eja | F | 9 | Bono × Daniella | 3 |
| | BD | Bondo | M | 8 | Lisala × Catherine | 2 |
| | MT | Mato | M | 36 | Camillo × Margrit | 1 |
| Apenheul | JI | Jill | F | 17 | Bosondjo × Laura | 8 |
| | ZU | Zuani | F | [11] | Wild | 7 |
| | RO | Rosie | F | [11] | Wild | 6 |
| | ML | Molaso | F | [17] | Wild | 5 |
| | HA | Hani | M | 11 | Wild | 4 |
| | LO | Lomela | F | 9 | Bono × Salonga | 3 |
| | MB | Mobikisi | M | [21] | Wild | 2 |
| | MW | Mwindu | M | [17] | Wild | 1 |
| Twycross | DT | Diatou | F | 24 | Masikini × Catherine | 6 |
| | KA | Kakowet II | M | 21 | Kakowet × Linda | 5 |
| | KC | Kichele | F | 12 | Masikini × Diatou | 4 |
| | BY | Banya | F | 11 | ? × Bonnie | 3 |
| | KE | Ke-Ke | M | 7 | Kakowet II × Diatou | 2 |
| | JS | Jasongo | M | 11 | Mato × Lisala | 1 |
| Planckendael | DZ | Dzeeta[1] | F | [27] / − | Wild | 7/− |
| | HE | Hermien[1,2] | F | [21] / [24] | Wild | 6/6 |
| | HO | Hortense[1,2] | F | [21] / [24] | Wild | 5/5 |
| | DE | Desmond[1] | M | [28] / − | Wild | 4/− |
| | RE | Redy[1,2] | M | 9 / 12 | Desmond × Hortense | 3/4 |
| | KO | Kosana[1] | F | [19] / − | Wild | 2/− |
| | KI | Kidogo II [1,2] | M | 16 / 19 | Masikini × Catherine | 1/3 |
| | DJ | Djanoa[2] | F | − / 7 | Santi × Yala | −/2 |
| | VI | Vifijo[2] | M | − / 8 | Kidogo II × Hortense | −/1 |

Age is given in years; numbers between brackets represent estimated ages, following Leus & Van Puijenbroeck (2005). Animals present in Planckendael during the 1999 study period are marked with [1], animals present in the second period in 2002 are marked with [2], their respective ages and ranks are separated with a /. Ranks are taken from Stevens et al. (in press) and are based on the occurrence of "fleeing upon aggression," with the highest rank number given to the most dominant member of the group.

members were unrelated. The group was formed in 1992, when DT and her daughter KC joined males KA and JS. Data from the two periods were pooled, since no changes in group composition occurred. We used matrix correlations to compare behavioral frequencies of the two periods and no significant differences were found.

Stevens observed the group at Planckendael for 190 hours on 24 days in November and December 1999, when the group comprised four adult females, three males, and four infants and juveniles. Except for the mother-son pair HO-RE, all adults and adolescents are unrelated. Stevens studied them again on 73 days for 505 hours between

November 2002 and February 2003, when there were three males three females, and two juveniles. One of the females (DJ) joined the group three months before the onset of the study. Females HE and HO and males RE and KI, were present during the previous study. One male (VI) had reached adolescence by the second period. Apart from the newly introduced female DJ, all other members had been together since 1992 or since they were born into the group. RE and VI were maternal half-brothers and had their mother (HO) in the group. KI was the father of male VI. All other members are unrelated. Stevens et al. (2003) provided more details regarding housing conditions and changes in group composition.

## Behavioral Observations, Categories and Analyses

We used a standardized ethogram, based on those by de Waal (1988) and Vervaecke et al. (2000a). Stevens conducted continuous observations throughout the day, starting in the morning and ending at dusk, when social interactions between the bonobos generally ceased. Frequent night observations at Planckendael revealed that no substantial social interactions occur after nest building or before feeding in the morning. Observations halted only when the bonobos were separated for cage cleaning or management purposes. Between 4 and 8 hours of observations occurred daily.

The observations comprised a combination of focal animal sampling, all occurrence sampling of agonistic, affiliative and sociosexual behaviors and instantaneous scan sampling for proximity. Stevens recorded observations manually and later entered them in the Observer software (Noldus), or entered them directly in the Observer. When social interactions were very frequent, e.g. during feeding bouts, he made video recordings and analyzed them later.

## Dominance Relationships

We determined dominance relationships only on the outcome of decided agonistic interactions, using fleeing upon aggression as a behavioral marker for dominance (Vervaecke et al. 2000a), and analyzed the dominance matrix with MatMan software (de Vries et al. 1993). We calculated Landau's linearity index, corrected for unknown relationships, and tested whether the value of h' differs significantly from the value that is expected under the null hypothesis of random dominance relations (de Vries 1995). When we found significant dominance hierarchies, we reordered the matrices following the I & SI methods, minimizing the number of inconsistencies (I) and the strength of inconsistencies (SI) to reorder the matrix in a manner most consistent with the linear hierarchy (de Vries 1998).

Based on the same marker for dominance, i.e. fleeing upon aggression, we calculated the individual's David's scores (David 1988), a cardinal rank measure which gives a dominance value for each individual, based on the relative numbers of winning and losing conflicts. David's scores have been shown to be more accurate than the index used by Clutton Brock et al. (1979) because 1) they are not disproportionately affected by minor deviations from the main dominance direction within dyads and 2) an individual's rank is independent of interactions in which he was not involved (Gammell et al. 2003). By performing a simple linear regression on individual David's scores, after they have been normalized to control for differences in group size, a measure for the steepness of a dominance hierarchy can be calculated (de Vries et al. 2006). The steepness varies from 0, a complete egalitarian, or shallow hierarchy, to 1, a steep or despotic hierarchy (de Vries et al. 2006).

We briefly reviewed our earlier findings on the linearity and steepness of dominance hierarchies in each of the study groups (Stevens et al. 2007). Furthermore, we correlated dominance with age, testing the idea that dominance and age are not correlated (Vervaecke et al. 2000a, Paoli and Palagi this volume). Patterns of social bonding:

– *Proximity*: every 15 minutes we scored which individuals were within arm's reach of one another (ca. 3 meter, following Furuichi and Ihobe 1994) by means of instantaneous scan sampling (Altmann 1974). From the samples, we took seven random scans for each day to avoid statistical interdependence of the data (Martin and Bateson 1993).
– *Grooming*: we scored grooming bouts by all occurrence sampling (Altmann 1974). In each grooming bout we scored the participation of each partner once. We did not count subsequent switches between the active and passive role as new bouts (Vervaecke et al. 2000b). For intergroup comparisons, we expressed dyadic grooming frequencies as number of bouts per hour.
– *Coalitions*: we scored coalitions per terminology and criteria of de Waal (1978, 1984). A brief overview of our results concerning the direction of support against likely winners or losers is also given.

Previously, we analyzed bonding patterns by lumping data across all groups to look for general trends (Stevens et al. 2006). Here we elaborate on these findings and analyze bonding patterns per group, to look for between groups using matrix comparisons, an approach which has also been used for wild bonobos (White and Burgman 1990). We used a Mantel test (Schnell et al. 1985) to compare each symmetrical matrix of behavioral interactions (spatial association, symmetrical matrices for grooming and for support) with three hypothesis matrices. For each hypothesis, we constructed a matrix, filling in values of 1 for all dyads important for the respective hypothesis, and values of 0 for all the other dyads. *Hypothesis 1*: structure of proximity, grooming or support was caused by preferential female-female associations; *Hypothesis 2*: structure of proximity, grooming or support was caused by preferential association among individual males; *Hypothesis 3*: structure was caused by preferential associations between males and unrelated females. For more details on this approach, see White and Burgman (1990). As we were mainly interested in intersexual bonding between unrelated males and females, we controlled for mother-son dyads by correcting the original data matrix. In the cells containing data on mother-son dyads, we filled in values that would be expected on the basis of the marginal totals of the matrices. We then correlated these adjusted data matrices with each of the hypothesis matrices, via a Mantel test in MatMan (de Vries et al. 1993).

While the Mantel test gives an idea for each group separately whether bonding is caused by preferential female-female, male-female or male-male bonding, we used a Fisher Combination test (Fisher 1954, Sokal and Rohlf 1981,780) to study the effects across groups. Hereto we combined the p-values of all individual Mantel tests. If the null hypotheses are true, the quantity $-2\Sigma lnP$ is expected to be distributed as $\chi 2$ with degrees of freedom= 2 * the number of separate tests and probabilities. Values of $-2\Sigma lnP$ greater than the corresponding $\chi 2$ value allow one to reject the null hypothesis of no effect.

## *Sociosexual Behavior*

We calculated individual sexuality scores for rough comparison with data presented by de Waal (1998, 2001). We used the same definitions and criteria of sociosexual behavior, between all individuals 7 years or older (similar to de Waal's (1998, 2001) adult group).

- *Sex present*: Presenting genital area (penis or anogenital swelling) towards another individual. May or may not be followed by further sexual interactions.
- *Sexual inspection*: Inspecting genital area of another individual by looking at, licking, touching or sniffing it. This category also includes de Waal's (1988) genital massage and oral genital massage.
- *Copulation*: All sexual interactions between mature (> 7 years) heterosexual dyads, which included intromission of the penis and clear thrusting of the pelvis (Furuichi 1997).
- *Non-copulatory mount*: Any sexual interaction involving a) homosexual dyads; b) immature subjects; or c) mature male-female dyads without observations of thrusting or intromission of the penis. Thus this category includes, rump-rump-rubbing, GG-rubbing and any mounting activity.

## Results

## *Dominance Relationships*

### Linearity and Steepness of Dominance Hierarchies

Based on the outcome of decided agonistic interactions, we found a significantly linear dominance hierarchy in each of the study groups (Stevens et al. 2007). Linearity indices varied from 0.86 at Planckendael in 1999 up to the maximum

**Table 1.2** Correlation between rank and age in different groups of bonobos

| Group | Adult group size | Dominance-rank & age | |
|---|---|---|---|
| | N | Kendall tau | $P_r$ |
| Planckendael-1999 | 7 | 0.48 | 0.13 |
| Planckendael-2002 | 6 | 0.69 | 0.05 |
| Wuppertal | 6 | 0.27 | 0.43 |
| Apenheul | 8 | −0.08 | 0.78 |
| Twycross | 6 | 0.82 | 0.02 |

value of 1 in Planckendael in 2002 and in Wuppertal. The individual dominance ranks are provided in Table 1.1. Although females occupied the highest-ranking position in each group, and the lowest-ranking position was always taken by a male, at least one male in every group could dominate at least one female, resulting in non-exclusive female dominance.

We measured the steepness of dominance hierarchies based on the outcome of agonistic interactions (measured by fleeing upon an aggression), and found that groups varied slightly in the steepness of their dominance hierarchy, with steepness values between 0.66 (Apenheul) and 0.81 (Planckendael-1992) (Stevens et al. in press). In general we found that hierarchies between males are steeper than those between females (Stevens et al. 2007).

## Dominance and Age

There is a significant correlation between age and rank only in Twycross zoo and in Planckendael 2002 (Table 1.2), where older bonobos tended to occupy the highest ranking positions in the hierarchy. In all other groups, dominance was not correlated with age.

## *Social Bonding*

### Patterns of Social Bonding

– *Proximity*: Mantel tests showed that females preferred the proximity of other females only in Planckendael-1999 and Apenheul, but the significance level only reached a trend (Table 1.3). In Planckendael-2002, females avoided the company of other females. Female-female preference could not be tested in Wuppertal, because there was only one female-female dyad in the group. When we combined the correlation coefficients of different groups, female preference for other females was not significant (Fisher combination test, p = 0.23). Between unrelated males and females, there are both positive (Planckendael-2002, Wuppertal, and Twycross) and negative (Planckendael-1999 and Apenheul)

**Table 1.3** Results of Mantel's Z correlation tests in which observed patterns of proximity were compared with three hypothetical matrices, assuming 1) bonding between females (f-f), 2) bonding between unrelated males and females (m-f), and 3) bonding between males (m-m)

|  | f-f | m-f | m-m |
| --- | --- | --- | --- |
| Planckendael-1999 | 0.74° | −0.46° | −0.30 |
| Planckendael-2002 | −0.22 | 0.25 | −0.08 |
| Wuppertal | – | 0.41 | −0.41 |
| Apenheul | 0.63° | −0.33 | −0.45* |
| Twycross | 0.13 | 0.08 | −0.24 |
| Fisher combination test | P = 0.23 | P = 0.08 | P = 0.14 |

* $p < 0.05$; °: $0.05 < p < 0.10$.

–: testing for female-female preference was impossible in Wuppertal, because of the low number (n=1) of female dyads.

**Table 1.4** Results of Mantel's Z correlation tests in which observed patterns of grooming were compared with three hypothetical matrices, assuming 1) bonding between females (f-f), 2) bonding between unrelated males and females (m-f), and 3) bonding between males (m-m)

|  | f-f | m f | m-m |
| --- | --- | --- | --- |
| Planckendael-1999 | 0.11 | 0.08 | −0.25 |
| Planckendael-2002 | −0.22 | −0.05 | 0.28 |
| Wuppertal | – | 0.36 | −0.35 |
| Apenheul | 0.04 | 0.15 | −0.31 |
| Twycross | 0.12 | 0.14 | −0.29 |
| Fisher combination | P = 0.99 | P = 0.83 | P = 0.31 |

–: testing for female-female preference was impossible in Wuppertal, because of the low number (n=1) of female dyads.

correlations. The Fisher combination test reached a trend (p = 0.08), indicating that overall, male-female proximity might influence overall group structure. The negative values for male-male proximity indicate that males in all groups tended to avoid proximity to other males, but the effect is only significant in Apenheul. Combining the results of different groups, there is no significant effect (Fisher combination test p = 0.13).

- *Grooming*: In general, grooming relationships between mothers and their adolescent or adult sons were most common, though the difference with grooming between unrelated males and females is not significant (Stevens et al. 2006). Testing against the three hypothesis matrices resulted in no significant effect (Table 1.4). In general, males tended to avoid grooming other males; the relations are always negative, with the exception of Planckendael-2002, but this effect never reached the significance level (Fisher combination test: p = 0.32). Female-female grooming, and grooming between unrelated males and females resulted in both negative and positive correlations, but they did not reach significance (Fisher combination test: female-male p = 0.84; female-females p = 0.99).

**Table 1.5** Results of Mantel's Z correlation tests in which observed patterns of support were compared with three hypothetical matrices, assuming 1) bonding between females (f-f), 2) bonding between unrelated males and females (m-f), and 3) bonding between males (m-m)

|                     | f-f      | m-f       | m-m      |
|---------------------|----------|-----------|----------|
| Planckendael-1999   | 0.67°    | −0.44**   | −0.24    |
| Planckendael-2002   | 0.74**   | −0.39**   | −0.26    |
| Wuppertal           | –        | 0.12      | −0.05    |
| Apenheul            | 0.59*    | −0.42*    | −0.23    |
| Twycross            | 0.37     | −0.20     | −0.21    |
| Fisher combination  | P < 0.01 | P < 0.001 | P = 0.82 |

** p< 0.01; *: p< 0.05; °: 0.05 < p < 0.10.
–: testing for female-female preference was impossible in Wuppertal, because of the low number (n=1) of female dyads.

- *Coalitionary support*: Overall, coalitions between females were significantly more common than coalitions between females and unrelated males, or coalitions between males (Stevens et al. 2006). When analyzed per group, support was significantly more common in female-female dyads than among other dyads in Planckendael-2002 and Apenheul, and there is a positive trend in Planckendael-1999 (Table 1.5). There is no evidence for preferential female-female support in Twycross or Wuppertal. When data from all groups were combined, the effect proved significant (Fisher combination test, p < 0.01), confirming our earlier findings. Comparison of the symmetrical support matrix with the hypothetical matrix for unrelated female-male preference resulted in significantly negative correlations for Planckendael-1999, Planckendael-2002 and Apenheul, suggesting that females and males avoided providing support to members of the other sex. When the results for all groups were combined, this negative effect proved significant (Fisher combination test: p < 0.0001). Male-male relations also resulted in negative correlations, but they never reached statistical significance (Fisher combination test: p = 0.82).

Our further analyses of the use and function of coalitions have shown that females provided significantly more support than males did, but females did not receive more support than males did. Furthermore, males were the usual targets of coalitions (Vervaecke et al. 2000c; unpublished data). Both males and females showed the same marked tendency to support likely winners in conflicts (females and males: 84% in support of likely winners), thus most coalitions were conservative (Chapais 1995). In some cases a lower ranking supporter would opportunistically provide support to a high ranking initial aggressor against an opponent that ranked in between the so-called "bridging alliances." Revolutionary alliances, in which two lower-ranking individuals support one another against a higher-ranking opponent, were extremely rare. This suggests that coalitions in bonobos mainly serve to maintain and reinforce existing dominance hierarchies.

## *Sociosexual Behavior*

We found differences in sociosexual activity between groups. The frequency of sociosexual behavior was remarkably low at Planckendael in 1999, with only 0.13 interactions per hour, a value that lies very close to the value given by de Waal (1998, 2001) for common chimpanzees. In the other bonobo groups, sexual activity was lower than, but close to those observed among the adult group at San Diego (Fig. 1.1).

To explain the low activity at Planckendael in 1999, we looked at the mean age of individuals in the different groups. de Waal's (1988, 1998, 2001) adult group in San Diego was in fact composed of two subgroups. One group comprised one male of 14 years, a 10 year old female, and an adolescent male of 7 years old. The second group included an adult female of 11.5 years, an adolescent male of 8 years old, and an infant. This resulted in a mean individual age of 10 yrs, for the two subgroups combined. In Planckendael, the mean individual age was 20 years.

There is a strong correlation for female individual age to sociosexual activity per hour (Spearman rank rs = −0.63 N = 15, p = 0.01). For males the correlation is slightly weaker but still significant (rs = −0.55 N = 14 p = 0.04, Fig. 1.2). Thus the relatively low frequencies of sexual interactions at Planckendael-1 and Planckendael-2 could be attributed to the presence of several older females, which were less sexually active, though regularly cycling.

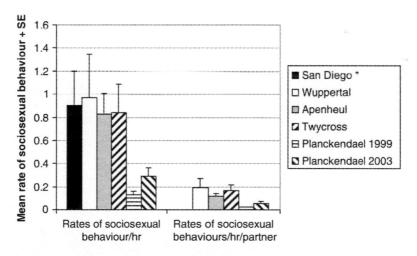

**Fig. 1.1** Mean rates of sociosexual behavior per hour for different groups of bonobo in captivity + SE. Data from San Diego are taken from de Waal (1998; 2001) but were not controlled for the number of partners available.

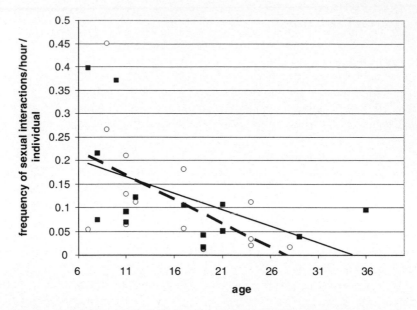

**Fig. 1.2** Correlations between individual age (years) and the frequency of sociosexual interactions per hour and per available partner across all study groups. Males (black squares, black trendline: $r_s = -0.55$, N = 14, p = 0.04) and females (white circles, broken trendline: $r_s = -0.63$, N= 15, p=0.01).

Elsewhere, we showed that mating success (measured as number of copulation bouts, excluding all other forms of sociosexual behavior) between males is not equally distributed. In Apenheul, the highest-ranking male HA obtained the highest mating success (Stevens et al. subm., cf. Paoli and Palagi in press). However, both in Planckendael-2002 and in Twycross, the alpha male did not have the highest overall mating success, with regularly cycling females or with the presumably cycling females when they were in estrus. Although alpha males in each group tried to aggressively monopolize females when they were in estrus by aggressively chasing away lower-ranking males, the younger and lower-ranking males had the highest mating success in Planckendael-1999 and Twycross (Stevens et al. subm.) and also sired offspring (Marvan et al. 2006).

## Discussions

Like wild bonobos (Kano 1992, Kano 1996, Furuichi 1997), in our captive groups, relationships between males are characterized by strongly asymmetric dominance relationships. Dominance hierarchies among captive males are extremely linear, probably because the number of males per group is very small (3–4 males per group), making despotic dominance relationships easier. Both in captivity and in the wild, males can compete over access to females, as is apparent from regular dominance displays and frequencies of male mating harassment (Kano 1992, 1996, Hohmann and Fruth 2003b). Affiliative bonds between males are relatively weak versus those in other dyads; males are rarely near one another and groom each other infrequently. Like wild bonobos (Kuroda 1980, Ihobe 1992, Furuichi and Ihobe 1994, Kano 1992), support between males is very rare, but not completely absent.

Dominance relationships between females are asymmetric, albeit somewhat less overtly expressed than among males (Furuichi 1997; Paoli and Palagi in press, Stevens et al. in press). Rank distances between females are often smaller compared to those between males (Stevens et al. 2007). Competition between females may be less overt, but is not absent, as evident from female mating harassment and abduction of infants, both in the wild and in captivity (Vervaecke and Van Elsacker 2000; Vervaecke et al. 2003; Hohmann and Fruth 2002, 2003b), which confirms the findings for wild bonobos at Lomako (Fruth et al. 1999, Hohmann et al. 1999, Hohmann and Fruth 2002). In two groups, we found a tendency for females to associate preferably with other females, but grooming was not more pronounced between them, which also corresponds to findings by Furuichi & Ihobe (1994) at Wamba, where association between females is pronounced, but grooming is not. Our finding that female-female support is more common than support between the sexes, or support between males, also confirms earlier reports on coalitions (Parish 1994, 1996, Vervaecke et al. 2000 b,c).

While most females could dominate males, female dominance is not complete. In each group, at least one male could dominate one or more females, and females were only able to evoke submission from males in 61% of the conflicts (Stevens et al. 2007). Female dominance was not complete, which we term non-exclusive female dominance (Vervaecke et al. 2000a, Stevens et al. 2007). For wild bonobos, it has been stated that females have about the same rank as males, and that there is a close dominance status between the sexes (Furuichi 1992, 1997, Kano 1992). The term co-dominance, used for wild bonobo males and females (Fruth et al. 1999) and sometimes in captive studies (Paoli et al. 2006, Paoli and Palagi this volume), also suggests that both sexes occupy similar cardinal rank positions, which contradicts our findings.

Bonds between unrelated males and females were equally strong as some female-female bonds, which confirms the findings for bonobos at Wamba and Lomako. Grooming was most frequent between mothers and their sons, but unrelated males also groom females more frequently than female-female grooming. Females support both related and unrelated males in conflict, albeit less frequently than they support other females. Males occasionally provide support to unrelated females. Since females often dominate males, it may pay for unrelated males to invest in long-term friendship relations with dominant females (Furuichi 1989, 1997, Kano 1992, Fruth et al. 1999, Hohmann et al. 1999, Hohmann and Fruth 2002).

The mean number of copulations per hour for each male in our study is 0.19 copulations/hour (range 0.01–0.36), which is similar to that reported for wild bonobos (Takahata et al. 1996: mean of 0.11, range of 0.10–0.20 copulations/hour;

Furuichi and Hashimoto, 2002: mean of 0.13 and 0.18 copulations/hour), thereby dispelling the idea of captive bonobos as being supersexual.

In general, the alleged difference between wild and captive groups of bonobos in term of bonding patterns is less pronounced than in earlier studies. The patterns of social bonding we observed are similar to those in wild bonobos, which confirms the bonobos as a female-centered species with bonds not only between females, but also between males and females instead of female bonded, with primary bonds between females. In contrast to chimpanzees, wherein female relationships seem influenced by captivity and male relations are comparable between the wild and captive conditions (de Waal 1994), we found that bonding patterns among captive bonobos largely resemble those in the wild.

In contrast to affiliative patterns, which may be more intense in captivity, but whose proportional distribution is largely similar to those of wild bonobos, the expression of dominance will undoubtedly be more rigid in captivity, where there are fewer competitors that are nearly always in the vicinity. Similarly, dominance styles may be more despotic compared to those in the wild. Coalitionary support may occur more frequently due to the general higher degree of spatial crowding and the consequent increased potential for occurrence of conflicts.

In contrast to studies by Paoli and colleagues (Paoli et al. 2006, Paoli and Palagi this volume), we consistently found significantly linear dominance hierarchies in all focal groups (Stevens et al. 2007). The difference in results may be due in part to a different behavioral measure. The studies by Paoli and colleagues used a combination of decided agonistic interactions and displace/yielding to measure dominance. Vervaecke et al. (2000a) found that yielding resulted in hierarchies with low linearity and directional consistency indices. This may be partly due to motivations of individuals, which need not concur with dominance ranks. For example, females can retreat to avoid sexually interested lower-ranking males (Vervaecke et al. 2000a). Franz (1999), who also used a combination of fleeing upon aggression and yielding, found significant linear hierarchies nonetheless in the bonobo groups of Stuttgart, Wuppertal and Planckendael. Further, the relation between dominance and age differed strongly among the groups.

Patterns of social bonding differed somewhat between groups, with some of them being more female-bonded than others. For example, in Apenheul and Planckendael-1999, females spent more time in proximity of other females, while this was not the case in the other groups. A combination of the results of all groups failed to confirm traditional assumptions of bonobo grooming patterns: there is no significant female bonding or intra-sexual grooming. Conversely, coalitionary support confirms expectations: female-female support was most common, while support among males was extremely rare.

The most impressive differences between groups are in the rates of sociosexual interactions, most notably at Planckendael, where the frequency of these interactions was much lower than previously reported for bonobos (de Waal 1998, 2001). The age composition of the group may determine the frequency of sexual interactions, because older individuals show a significant decline in sexual activity. There is no consistent pattern regarding male mating success in relation to dominance. The highest-ranking male acquired most copulations in only one of the focal groups. The relationship between male rank and mating success probably is largely influenced by female mate choice: when female choice runs concurrently with male dominance rank (as in Apenheul), dominance predicts mating success. When females show preferences for lower-ranking males, e.g. in Twycross and Planckendael-2002, dominance effects on mating are less clear.

Group dynamic processes may also explain why earlier research found higher degrees of female social bonding in captive groups of bonobos. Parish (1996) conducted the study in San Diego Wild Animal Park during captive group formation. In the wild, when young females migrate to new communities, they look for contacts with resident females (Furuichi 1989, Idani 1991). Later, when they have offspring, the relations with other females weaken, as relationships with their offspring gain importance (Furuichi 1989). Therefore, in newly-formed groups in captivity, females may at first seek contact with other females, while intersexual bonds may take longer to develop. We predict that the importance of female bonding will decrease as groups stabilize. Anecdotic data from 10 years of study at Planckendael support this. When the colony was founded, it comprised three unrelated females, three unrelated males, and one male offspring. Female bonding was more pronounced, with many female-female coalitions directed against the unrelated, lower-ranking males (Vervaecke et al. 2000b). Typically, the females supported each other unconditionally in conflicts with these males. Ten years later, two of the original females have had several offspring. The close bonds between them have weakened, as more conflicts arise between the females and the offspring of their former allies. In these conflicts, support is less unconditional and mothers are only rarely inclined to support their female friends. Instead, they withdraw or make appeasement gestures to both parties of the conflict.

Apart from the differences between bonobo groups in dominance behavior, bonding patterns and sexual behavior, cultural differences have been described for captive bonobo groups. De Waal (1988) described clapping behavior during grooming in the bonobos of San Diego Zoo. Later, Thompson (1994) showed how clapping had spread to other American zoos, where bonobos from San Diego Zoo had been transferred. A bonobo in our study groups, the adult male KA, which was one of the original bonobos studied by de Waal (1988), clapped during grooming. This behavior seemed to be adopted by at least two of the younger bonobos (KE and KC) at Twycross. Similar processes had been reported for other colonies with San Diego Zoo (Parish, cited in de Waal 1994) and seem to indicate social learning as a mechanism of cultural transmission. Apart from clapping behavior, Pika et al. (2005) described two group-specific gestures among young bonobos at Planckendael and Apenheul.

According to de Waal (1994), the capacity to adjust to new conditions is also a good indicator for the study of adaptive potential. He specifically refers to the flexibility of chimpanzees at Arnhem zoo, which coped with crowded winter conditions by increasing friendly grooming behavior to counterbalance an increase in the frequency, but not the intensity, of aggression (Nieuwenhuijsen and de Waal 1982). In Planckendael, we also compared behavior of bonobos in the winter, when they are confined to 600m³ indoor quarters, while in the summer they have access to a

$3000m^2$ island. Van Dyck et al. (2003) found that bonobos at Planckendael groom each other more frequently during winter periods, possibly as a mechanism to cope with the increase in certain types of aggressive behavior during the same study periods (Sannen et al. 2004).

## General Conclusions

In general, our study warns against generalizations derived from studies on a single bonobo group. For instance, the typical bonobo-pattern of female bondedness (Parish 1996) is confirmed by the data on coalitionary support in several groups, but not by the grooming patterns. Further, the dominance related bonding patterns described in the Planckendael group of 1999 (Vervaecke et al. 2000b) no longer persist in the changed group. In some groups, dominant individuals received more grooming or support than subordinates, and in other groups support or grooming was reciprocal (Stevens et al. 2005). However, this variation could be related to variation in dominance steepness, as predicted by biological market theories, wherein one expects reciprocal exchange in groups with a shallow dominance hierarchy, and up-the-hierarchy grooming and interchange of commodities in groups with a more steep dominance gradient (Barrett et al. 1999). Grooming was indeed more reciprocal in groups with a shallow dominance hierarchy, and in relatively steep hierarchies, grooming was not consistently directed at higher-ranking individuals and not interchanged against support or tolerance more frequently (Stevens et al. 2005).

Contrary to the common view derived from single-group studies, we found many variants from the presumed conciliatory, peaceful, and egalitarian bonobo. Bonobos exhibit relatively low conciliatory tendencies. Furthermore, serious aggression occurs. Intersexual aggression is especially common; females gang together against lower-ranking males (Parish 1996). We observed fierce female attacks on lower-ranking males in Planckendael, Twycross, Apenheul, Wuppertal and Frankfurt Zoo. Often they result in the temporary or permanent removal of the target males, though lethal aggression that occurs among chimpanzees (de Waal 1986) is not recorded. The orphan males, which have been hand-reared, have no mother to back them up during conflicts, and may lack social skills to cope with the attacks, are typical scapegoats of redirected aggression and suffer most from violent female attacks. But mothers do not always support their sons in conflicts, and may opt to provide support to the party opposing their own offspring (Stevens et al. subm.). Coalitionary attacks on sons of high-ranking mothers are rarer, but not absent. Furthermore, we also recorded a marked tendency for female bonobos to kidnap or harass offspring of other females (Vervaecke et al. 2003). Although the precise meaning of these interferences is unclear, similar cases of infant abduction in chimpanzees are considered a sign of female competition (Pusey et al. 1997). A wild female bonobo carried another female's newborn offspring, which subsequently died (Hohmann and Fruth 2002).

We did not confirm egalitarianism in dominance relationships in the diverse study groups. Overall, bonobo behavior is so variable and flexible, that studies over longer periods and on multiple groups are a prerequisite to any generalization. Researchers with single groups should reflect cautiously on possible context-related determining factors of observed behavioral patterns. The expectations that have been created by previous bonobo studies should also be put in this perspective in order to observe them without bias.

**Acknowledgements** We are grateful to the directory and keepers of Planckendael Wild Animal Park (Belgium), Apenheul Primate Park (The Netherlands), Wuppertal Zoo (Germany) and Twycross Zoo (United Kingdom) for their help and interest in this study. This research was funded by a Ph.D. grant of the Institution for the Promotion of Innovation through Science and Technology in Flanders (IWT-Vlaanderen: grant number 3340). We thank the Flemish Government for structural support of the CRC of the RZSA.

# References

Altmann J (1974) Observational study of behavior: sampling methods. Behaviour 49: 227–267

Baker KC, Smuts BB (1994) Social relationships of female chimpanzees: diversity between captive social groups. In: Wrangham RW, McGrew WC, de Waal FBM, Heltne PG (eds) Chimpanzee Cultures, Harvard University Press, Cambridge Massachusets and London, pp 227–242

Barrett L, Henzi SP, Weingrill T, Lycett JE, Hill RA (1999) Market forces predict grooming reciprocity in female baboons. Proc Roy Soc London B266:665–670

Boesch C (2002) Behavioural diversity in *Pan*. In: Boesch C, Hohmann G, Marchant LF (eds) Behavioural Diversity in Chimpanzees and Bonobos, Cambridge University Press, Cambridge, pp 1–8

Chapais B (1995) Alliances as a means of competition in primates; evolutionary, developmental and cognitive aspects. Yearbook of Physical Anthropology 38: 115–136

Clutton-Brock T H, Albon S D, Gibson R M, Guinness F E (1979) The logical stag: adaptive aspects of fighting in red deer (*Cervus elaphus L.*). Animal Behaviour, 27, 211–225

David HA (1988) The method of paired comparisons. London: Charles Griffin

Fisher RA (1954) Statistical methods for Research Workers. 12th ed. Oliver & Boyd, Edinburgh

Franz C (1999) Allogrooming behavior and grooming site preferences in captive bonobos (*Pan paniscus*): association with female dominance. International Journal of Primatology 20: 525–546

Fruth B, Hohmann G, McGrew WC (1999) The *Pan* species. In: Dolhinow P, Fuentes A (ed) The Nonhuman Primates. Mayfield Publishing Company, London, pp 64–72

Furuichi T (1989) Social interactions and the life history of female *Pan paniscus* in Wamba, Zaire. International Journal of Primatology 10: 173–197

Furuichi T (1992) The prolonged estrus of females and factors influencing mating in a wild group of bonobos (*Pan paniscus*) in Wamba, Zaire. In: Itoigawa N, Sugiyama Y, Sackett GP, Thompson RK (eds) Topics in Primatology, Vol. 2: Behavior, Ecology and Conservation. Tokyo: University of Tokyo Press. pp 179–190

Furuichi T (1997) Agonistic interactions and matrifocal dominance rank of wild bonobos (*Pan paniscus*) at Wamba. Folia Primatologica 18: 855–875

Furuichi T, Hashimoto C (2002) Why female bonobos have a lower copulation rate during estrus than chimpanzees. In: Boesch C, Hohmann G, Marquardt L (eds) Behavioral diversity of chimpanzees and bonobos. Cambridge Univ Press, New York, pp 156–167

Furuichi T, Ihobe H (1994) Variation in male relationships in bonobos and chimpanzees. Behaviour 130: 212–228

Gammell MP, De Vries H, Jennings DJ, Carlin CM, Hayden TJ (2003) David's score: a more appropriate dominance ranking method than Clutton-Brock et al.'s index. Animal Behaviour 66: 601–605

Gold KC (2001) Group formation in captive bonobos: sex as a bonding strategy. In: The Apes: Challenges for 21st Century, Brookfield Zoo, Brookfield, pp 90–93

Haas G (1983) Neue menschenaffenanlage im Zoo Wuppertal. Der Zoologischer Garten, N.F. 53: 93–101

Hashimoto C, Yasuko T, Hibino E, Mulavwa M, Yangozene K, Furuichi T, Idani G, Takenaka O (in press) Longitudinal structure of a unit-group of bonobos: male philopatry and possible fusion of unit-groups. In: Furuichi T, Thompson J (eds) Bonobos: behavior, ecology, and conservation. Springer, New York, pp 107–119

Hohmann G, Fruth B (2002) Dynamics in social organization of bonobos (Pan paniscus). In: Boesch C, Hohmann,G, Marchant L F (eds) Behavioural Diversity in Chimpanzees and Bonobos. Cambridge, Cambridge University Press. pp 138–150

Hohmann G, Fruth B (2003a) Culture in bonobos? Between-species and within-species variation in behavior. Current Anthopology 44: 563–571

Hohmann G, Fruth B (2003b) Intra-and inter-sexual aggression by bonobos in the context of mating. Behaviour 140: 1389–1413

Hohmann G, Gerloff U, Fruth B (1999) Social bonds and genetic tests: kinship, association and affiliation in a community of bonobos (Pan paniscus). Behaviour 136: 1219–1235

Idani G (1991) Social relationships between immigrant and resident bonobo (Pan paniscus) females at Wamba. Folia Primatol 57:83–95

Ihobe H (1992) Male-male relationships among wild bonobos (Pan paniscus) at Wamba, Republic of Zaïire. Primates 33:163–179

Kano T (1992) The last ape: pygmy chimpanzee behavior and ecology. Stanford California, Stanford University Press

Kano T (1996) Male rank order and copulation rate in a unit-group of bonobos at Wamba, Zadre. In: McGrew WC, Marchant LA, Nishida T (eds) Great Ape Societies. Cambridge: Cambridge University Press. pp 135–145

Kuroda S (1980). Social behavior of pygmy chimpanzees. Primates 21: 181–197

Martin P, Bateson P (1993) Measuring Behaviour: an introductory guide (second edition). Cambridge, Cambridge University Press

Marvan R, Stevens JMG, Roeder AD, Mazura I, Bruford MW, de Ruiter JR (2006) Male dominance rank, mating and reproductive success in captive bonobos. Folia Primatologica 77: 364–376

Mills J, Reinartz G, De Bois H, Van Elsacker L, Van Puijenbroeck B (1997) The care and management of bonobos (Pan paniscus) in captive environments. The Zoological Society of Milwaukee County, Milwaukee

Nieuwenhuijsen K, de Waal FBM (1982) Effects of spatial crowding on social behavior in a chimpanzee colony. Zoo Biol 1:5–28

Palagi E, Paoli T, Tarli SB (2004) Reconciliation and consolation in captive bonobos (Pan paniscus). American Journal of Primatology 62: 15–30

Paoli T, Palagi E, Borgognini, Tarli SMB (2006) Reevaluation of dominance hierarchy in bonobos (Pan paniscus). American Journal of Physical Anthropology xx

Parish AR (1994) Sex and food control in the "uncommon chimpanzee": how bonobo females overcome a phylogenetic legacy of male dominance. Ethology and Sociobiology 15: 157–179

Parish AR (1996) Female relationships in bonobos (Pan paniscus). Human Nature 7: 61–96

Parish AR, de Waal FBM (2000) The other "closest living relative": how bonobos (Pan paniscus) challenge traditional assumptions about females, dominance, intra- and inter-sexual interactions, and hominid evolution. Annals of the New York Academy of Sciences 907: 97–113

Pika S, Liebal K, Tomasello M (2005) Gestural communication in subadult bonobos (*Pan panis-cus*): repertoire and use. Amer J Primatol 65:39–61

Pusey A, Williams J, Goodall J (1997) The influence of dominance rank on the reproductive success of female chimpanzees. Science 277:828–831

Sannen A, van Elsacker L, Eens M (2004) Effect of spatial crowding on aggressive behavior in a bonobo colony. Zoo Biol 23:383–395

Schnell G, Watt D, Douglas M (1985) Statistical comparison of proximity matrices: applications in animal behaviour. Animal Behaviour 33, 239–253

Sokal RR, Rohlf FJ (1981) Biometry, 2nd edition. Freeman, San Francisco

Stanford CB(1998) The social behavior of chimpanzees and bonobos. Current Anthropol 39: 399–420

Stevens J, Vervaecke H, Melens W, Huyghe M, De Ridder P, Van Elsacker L (2003) Much ado about bonobos: ten years of management and research at Planckendael Wild Animal Park, Belgium. In: Gilbert TC, (ed) Proceedings of the Fifth Annual Symposium on Zoo Research, 7–8 July 2003, Marwell, Marwell Zoological Park, pp114–125

Stevens JMG, Vervaecke H, De Vries H, Van Elsacker L (2005) The influence of steepness of dominance relations on reciprocity and interchange in captive bonobos (*Pan paniscus*). Behaviour 142: 941–960

Stevens JMG, Vervaecke H, De Vries H, Van Elsacker L (2006) Social structures in *Pan panis-cus*: Testing the female bonding hypothesis. Primates 47: 210–217

Stevens JMG, Vervaecke H, De Vries H, Van Elsacker L (2007) Sex differences in steepness of dominance hierarchies in captive groups of bonobos International Journal of Primatology DOI 10.1007/s10764-007-9186-9

Takahata Y, Ihobe H, Idani G (1996) Comparing copulations of chimpanzees and bonobos: do females exhibit proceptivity or receptivity. In: McGrew WC, Marchant LF, Nishida T (eds) Great ape societies. Cambridge Univ Press, Cambridge

Thompson JA Myers (1994) Cultural diversity in the behavior of *Pan*. In: Quiatt D, Itani J (eds) Hominid culture in primate perspective. Univ Press of Colorado, Niwot, pp 95–115

Thompson JA Myers (2002) Bonobos of the Lukuru Wildlife Research Station. In: Boesch C, Hohmann G, Marchant LF. (eds) Behavioural Diversity in Chimpanzees and Bonobos, Cambridge University Press, Cambridge, pp 61–70

Van Dyck S, Stevens J, Meuleman B, Van Elsacker L (2003) Effects of the change in accommodation and group composition on the affinitive behaviour of a captive bonobo group (*Pan paniscus*). In: Gilbert TC (ed) Proceedings of the Fifth Annual Symposium on Zoo Research, 7–8 July 2003, Marwell Zoological Park, Marwell, pp 205–210

Van Elsacker L, Claes G, Melens W, Struyf K, Vervaecke H, Walraven V (1993) New outdoor exhibit for a bonobo group at Planckendael: design and introduction procedures. Bonobo Tidings: 35–47

Van Elsacker L, Vervaecke H, Verheyen RF (1995) A review of terminology on aggregation patterns in bonobos (*Pan paniscus*). International Journal of Primatology 16: 37–52

Vervaecke H, Van Elsacker L (2000) Sexual competition in a group of captive bonobos (*Pan paniscus*). Primates 41: 109–115

Vervaecke H, de Vries H, Van Elsacker L (1999) An experimental evaluation of the consistency of competitive ability and agonistic dominance in different social contexts in captive bonobos. Behaviour 136, 423–442

Vervaecke H, de Vries, H, Van Elsacker L (2000a) Dominance and its behavioral measures in a captive group of bonobos (*Pan paniscus*). International Journal of Primatology 21: 47–68

Vervaecke H, de Vries H, Van Elsacker L (2000b) The pivotal role of rank in grooming and support behavior in a captive group of bonobos (*Pan paniscus*). Behaviour 137: 1463–1485

Vervaecke H, de Vries H, Van Elsacker L (2000c) Function and distribution of coalitions in captive bonobos (*Pan paniscus*). Primates 41: 249–265

Vervaecke H, Stevens J, Van Elsacker L (2003) Interfering with others: female-female reproductive competition in *Pan paniscus*. In: Jones CB (ed) Sexual Selection and Reproductive Competition in primates: new perspectives and directions, American Society of Primatologists, pp 1235–1246 de Vries H (1995) An improved test of linearity in dominance hierarchies containing unknown or tied relationships. Animal Behaviour 50: 1375–1389

de Vries H (1998) Finding a dominance order most consistent with a linear hierarchy: a new procedure and review. Animal Behaviour 55: 827–843

de Vries H, Netto WJ, Hanegraaf PLH (1993) Matman: a program for the analysis of sociometric matrices and behavioural transition matrices. Behaviour 125: 157–175

de Vries H, Stevens JMG, Vervaecke H (2006) Measuring the steepness of a dominance hierarchy based on normalised David's score and the dyadic dominance index corrected for chance. Animal Behaviour 585–592

de Waal FBM (1978) Exploitative and familiarity dependent support strategies in a colony of semi-free living chimpanzees. Behaviour 66: 268–312

de Waal, FBM (1982) Chimpanzee Politics: power and sex among apes. London, Jonathan Cape Ltd

de Waal, FBM (1984) Sex differences in the formation of coalitions among chimpanzees. Ethology and Sociobiology 5: 239–255

de Waal FBM (1986) The brutal elimination of a rival among captive male chimpanzees. Ethology and Sociobiology 7: 237–251

de Waal FBM (1988) The communicative repertoire of captive bonobos (*Pan paniscus*) compared to that of chimpanzees. Behaviour 106: 183–251

de Waal FBM (1994) Chimpanzee's adaptive potential. In: Wrangham RW, McGrew WC, de Waal FBM, Heltne PG. (eds) Chimpanzee Cultures, Harvard University Press, Cambridge Massachusets and London, pp 243–260

de Waal FBM (1995) Bonobo sex and society. Scientific American 272: 58–64

de Waal FBM (1998) Comment on CB Stanford The social behavior of chimpanzees and bonobos. Current Anthropology 39: 407–408.

de Waal FBM (2001) Apes from Venus: bonobos and human social evolution. In: de Waal FBM, (ed) Tree of Origin: What Primate Behavior Can Tell Us about Human Social Evolution, Harvard University Press, Cambridge, Massachusetts & London, England pp 41–68

de Waal F, Lanting FL (1997) Bonobo, The Forgotten Ape. Berkeley and Los Angeles: University of California Press

White FJ, Burgman MA (1990) Social organization of the pygmy chimpanzee (*Pan paniscus*): Multivariate analysis of intracommunity association. American Journal of Physical Anthropology 83: 193–201

Whiten A, Goodall J, McGrew WC, Nishida T, Reynolds V, Sugiyama Y, Tutin CEG, Wrangham RW, Boesch C (1999) Cultures in chimpanzees. Nature 399: 682–685

Wrangham RW, McGrew WC, de Waal FBM, Heltne, PG (1994) Chimpanzee Cultures. Harvard University Press, Cambridge Massachusets and London

# What Does Agonistic Dominance Imply in Bonobos?

Tommaso Paoli[1] and Elisabetta Palagi[1]

## Introduction

### Hierarchy in Bonobos: An Up-to-date Review

Social dominance is a relevant factor in the study of animal behavior, primatology in particular (Bernstein 1981, Walters and Seyfarth 1987, Newton-Fisher 2004). Social dominance is determined by repeated interactions between pairs of individuals, thus dyadic interactions are important in shaping the nature of the relationship (Hinde 1976). Given that social dominance allows each individual to resolve intragroup contests without engaging in energetically expensive, risky, agonistic interactions, the dominant individual (one with the higher probability of winning any contest) generally acquires the contested resource with only a minimum cost of time and energy, while the subordinate individual (one with the lower probability of winning) avoids wasting both time and energy in a contest that it is likely to lose anyway (Newton-Fisher 2004). Therefore, both individuals avoid potential injuries, which are expected to be greater for the subordinate. This view of dominance is generally based on agonistic interactions and is more precisely defined as agonistic dominance (Bernstein 1981, Walters and Seyfarth 1987, Drews 1993, Mason 1993).

On the other hand, dominance style refers to the pattern of expressed asymmetry in agonistic relationships (de Waal 1989, de Waal and Luttrell 1989): it refers to how dominants treat subordinates and vice versa (de Waal 1996). Many studies have revealed dominance style in chimpanzees (*Pan troglodytes*), in which males are fairly linearly ranked, whereas females generally are not (Wittig and Boesch 2003). Our knowledge of dominance style in *Pan paniscus*, however, is still controversial (Hohmann and Fruth 2003, Paoli et al. 2006a).

[1]*Department of Ethology, Ecology and Evolution, Anthropology Unit, University of Pisa, Via Roma 79, 56011 Calci, Pisa, Italy*

T. Furuichi and J. Thompson (eds.), *The Bonobos: Behavior, Ecology, and Conservation*　　　39

In wild bonobos, dominance ranks have been consistently recorded among males (Furuichi 1997, Furuichi and Ihobe 1994, Kano 1992), but are generally not so clear among females (Kano 1992). In captive bonobos, Franz (1999) and Vervaecke et al. (2000a) described a linear hierarchy with results drawn from both sexes taken together. Specifically, Vervaecke et al. (2000a) showed the occurrence of a linear hierarchy in the bonobo colony of Planckendael (Belgium). Franz (1999) also reported linearity of hierarchy in the Stuttgart and the Wuppertal bonobo groups. De Vries et al. (2006), Stevens et al. (2005b), and Stevens and Vervaecke (this book) showed that the steepness of the bonobo dominance hierarchy fluctuates slightly in different groups. In addition, Stevens and Vervaecke (this book) suggested that in bonobos, dominance relationships between males and between females can be semidespotic. Paoli et al. (2006a) showed that, in a group of unrelated adult bonobos (Apenheul Primate Park, the Netherlands), there was unclear non-linear hierarchy in one study period whereas there was a fairly clear hierarchy in another period, though it just fell to reach statistical linearity. Thus, the dominance style of bonobos may be loose and differentially expressed in diverse groups and/or even in the same group with shifting conditions.

Another peculiarity of bonobos is that they show no formal sign of subordinance, unlike chimpanzees' pant-grunting and bobbing (Kano 1992, Furuich, 1992, Furuichi and Ihobe 1994, Wrangham 1999). In fact, the meaning of pant-grunting in *Pan paniscus* (de Waal 1988, Bermejo and Omedes 1999) remains ambiguous, and in some bonobo groups it is rare (Furuichi and Ihobe 1994, Palagi, 2006). Further, de Waal (1987) and Hohmann and Fruth (2000) hypothesized that genito-genital rubbing signals dominance, but recent data from the Apenheul colony showed no overall asymmetry in performance or invitation to this behavior (Paoli et al. 2006b). Vervaecke et al. (2000a) suggested that even peering (Kano 1992) expresses subordinance, but it is surely not ritualized and appears to be highly polyvalent (Furuichi 1989, Ihobe 1991, Stevens et al. 2005a).

Contrary to the evident male-oriented chimpanzee society, bonobo male bonds are definitely weak (Kano 1992, Parish 1994, White 1996, Fruth et al. 1999, Palagi et al. 2004). In addition, females often dominate males: in fact, even though the adult female is generally slightly physically smaller than the adult male, she is either co-dominant or has a moderate dominance advantage over her male counterpart (Kano 1992, Furuichi 1997, Vervaecke et al.,, 2000a). In agreement with this view, as reported by Vervaecke et al. (2000a), the alpha position in bonobo colonies is often occupied by a female. As stated by Wrangham (1999), the relative lack of interest of male bonobos in high status may be partly a consequence of a system that unites concealed ovulation (Paoli et al. 2006b) with multiple mating, thus reducing the benefits of being a high-ranking male (Kano 1992, Furuichi, 1997, Vervaecke et al. 2000a).

Thus the literature on the dominance style of bonobos is often contradictory and sometimes incomplete. Further research is needed to enhance the understanding of this subject.

## *Filling the Gap: Additional Investigations*

We aim to extend the understanding of the bonobo dominance style. The emerging picture is that *Pan paniscus* shows a flexible and complex society in which agonistic dominance exists, though with variable linearity. Thus, if agonistic dominance occurs in bonobos, what is its meaning? In the attempt to clarify what agonistic dominance implies in bonobos, we focus on some important traits of their social behavior, trying to relate them to the observed rank in two different study periods. We use new data and a review of our published findings to clarify some major aspects about bonobo dominance that have not been adequately described:

1. Linearity and steepness of hierarchy

    - How does linearity vary along with shifting group conditions?
    - How does steepness vary according to changes in linearity of hierarchy and group composition?

2. Individual attributes

    - Does sex influence the dominance rank?
    - Does rank correlate with age and body mass?

3. Social and sexual interactions

    - Does rank correlate with:

    i. Grooming exchange
    ii. Food-sharing exchange
    iii. Peering exchange
    iv. Frequency of genito-genital-rubbing (GG-rubbing) and GG-rubbing invitation exchange

    - Does rank determine any asymmetry in the pattern of performance of GG-rubbing (mounter and mountee roles)?
    - Do males benefit from higher rank in copulatory rate?
    - Post-conflict behavior
    - Does rank influence reconciliation and consolation levels?

## Methods: Study Groups, Data Collection and Analysis

We collected behavioral data during two observation sessions (July-October 2000 and April-July 2002) on the group of *Pan paniscus* housed at the Apenheul Primate Park (Apeldoorn, The Netherlands), first established in 1998. The composition of the colony varied over the time (Table 2.1). Details on the study group and the methods used for i) collecting data on agonistic dominance, ii) testing the linearity of hierarchy, and iii) determining the rank using David's scores are described in

**Table 2.1** The colony of *Pan paniscus* in the Apenheul Primate Park (Apeldoorn, The Netherlands). Individuals marked with an * died after the first session of observations (July-October 2000). All the bonobos from Democratic Republic of Congo (DRC) were previously housed in a Rescue Center and came from different collection sites

| Subject | Sex | Class | Date of Birth | Origin, Arrival Date |
|---------|-----|-------|---------------|----------------------|
| H, Hani* | M | Adult | 1989, wild | DRC, 1998 |
| MB, Mobikisi | M | Adult | 1981, wild | Antwerp, 1996 |
| MW, Mwindu | M | Adult | 1985, wild | DRC, 1998 |
| J, Jill | F | Adult | 1985, captivity | San Diego, 1997 |
| R, Rosie* | F | Adult | 1989, wild | DRC, 1998 |
| MO, Molaso | F | Adult | 1985, wild | DRC, 1998 |
| Z, Zuani | F | Adult | 1990, wild | DRC, 1998 |
| LO, Lomela | F | Adult | 1992, captivity | Frankfurt, 1998 |
| LI, Liboso | F | Juvenile | 1997, captivity, Zuani's daughter | DRC, 1998 |
| T, Tarishi | M | Infant | 1998, captivity, Jill's son | Apenheul |
| K, Kumbuka | F | Infant | 1999, captivity, Molaso's daughter | Apenheul |

Paoli et al. (2006a). Tables 2.2 and 2.3 report the frequency of aggressions and displacements for each study period and the calculated rank (David's score).

The steepness of hierarchy is a measure which can vary between 0 (a complete egalitarian, or shallow hierarchy) and 1 (a steep or despotic hierarchy) and is independent from the number of individuals, thus useful for comparing different conditions. It is defined as the absolute slope of the straight line fitted to the normalized David's scores (calculated on the basis of a dyadic dominance index corrected for chance) plotted against the subjects' ranks (de Vries et al. 2006). While the linearity depends on the number of established binary dominance relationships and the degree of transitivity in these relationships (Appleby 1983), the steepness measures the degree to which individuals differ from each other in winning dominance encounters. Linearity and steepness are complementary measures to characterize a dominance hierarchy. To obtain a steepness measure that varies between 0 and 1, it is necessary to convert David's scores into normalized David's scores (NDS) to control for differences in group size, as suggested by de Vries et al. (2006). The use of NDS allows one to obtain steepness values which are independent from the number of individuals characterizing a social group.

We took into account behavioral data collected via scan sampling at 5 minute intervals, and focal animal sampling (Altmann 1974) in both observation periods. We collected data on grooming by scan observations (session 1: 352 h, session 2: 356 h) whereas we collected data on food-sharing, peering, GG-rubbing and copulations via focal animal sampling (session 1: 41 h *per* individual, session 2: 57 h *per* individual).

To evaluate the exchange of social interactions we used ratios calculated as logarithm [(performed +1) / (received +1)] per individual, thus obtaining an index that is positive when the individual gives more than it receives and negative when it gives less.

When trying to relate conciliatory and consolatory levels to the observed rank, given that with a break-up approach (considering the two periods separately), post-conflict interactions were insufficient for a proper evaluation, we used the following method:

**Table 2.2** Frequency of aggressions and displacements. Observed during 450 hours. Rate/hour = 0.67. Unknown relationships 3.6%, one-way relationships 50%, two-way relationships 46.4%. Weak and non-significant improved index of linearity (Matman, 10,000 permutations): h' = 0.428, P = 0.252 one-tailed; Directional Consistency Index DC = 0.63

| Actor | Recipient | | | | | | | | Total | David's score |
|---|---|---|---|---|---|---|---|---|---|---|
| | JILL | HANI | MOBIKISI | ZUANI | LOMELA | ROSIE | MOLASO | MWINDU | | |
| JILL | – | 41 | 22 | 2 | 34 | 5 | 6 | 15 | 125 | 17.44 |
| HANI | 28 | – | 7 | 0 | 8 | 2 | 0 | 25 | 70 | 4.79 |
| MOBIKISI | 1 | 1 | – | 9 | 6 | 1 | 3 | 12 | 33 | 0.05 |
| ZUANI | 1 | 1 | 1 | – | 3 | 1 | 1 | 0 | 8 | –2.08 |
| LOMELA | 0 | 4 | 10 | 12 | – | 1 | 3 | 6 | 36 | –3.22 |
| ROSIE | 0 | 0 | 1 | 0 | 2 | – | 1 | 0 | 4 | –3.51 |
| MOLASO | 0 | 1 | 0 | 0 | 15 | 0 | – | 1 | 17 | –3.84 |
| MWINDU | 0 | 0 | 5 | 1 | 1 | 0 | 0 | – | 7 | –10.98 |
| Total | 30 | 48 | 46 | 24 | 69 | 10 | 14 | 59 | 300 | |

1) We determined NDS as suggested by de Vries et al. (2006).
2) We then calculated the mean value for the NDS between the two periods for each individual (individuals present only in the first period held the value of the first period).
3) We determined the group mean for all individual mean NDS.
4) The animals showing a value of mean NDS over the group mean were high-ranked (J, H, MB); the others were low-ranked (LO, Z, R, MO, MW).
5) We compared the levels of reconciliation (corrected conciliatory tendency, CCT) and consolation (triadic contact tendency, TCT) via the Mann-Whitney test for two independent samples, using CCT and TCT levels published in Palagi et al. (2004). CCTs and TCTs are percentage values.

We used the Spearman test to evaluate the correlation between rank and exchanged social interactions in the whole group and then separately in females in both study periods. Unfortunately, the correlations could not be carried out for males because there were only three adult individuals.

When comparing the GG-rubbing frequency in each dyad with the rank distance, we evaluated the latter via the absolute value of the difference in David's scores between the two individuals of the dyad. We then used MatMan's row-wise correlation tool with 10,000 permutations and a two-tailed test.

# Results

## *Linearity of Hierarchy*

During the first period of observations, the matrix of aggressions and displacements (Table 2.2) showed a weak and non-significant linearity index (h' = 0.428, p = 0.252, one-tailed) and a directional consistency index of 0.63. In one dyad (Mwindu-Rosie), no interactions occurred; therefore the percentage of unknown relationships was 3.6%.

During the second period of observations, the matrix (Table 2.3) showed a fairly high linearity index (h' = 0.91) just failing to reach statistical significance (p = 0.055, one-tailed). The directional consistency index was 0.88, but in four dyads no interactions occurred; therefore the percentage of unknown relationships was 26.7%.

## *Steepness of Hierarchy*

In the first study period (Table 2.2), the steepness value is 0.378 (Fig. 2.1a). When we tested the observed steepness against the null hypothesis of random wins for all pairs of individuals (randomization test procedure with 2000 repetitions), we

**Table 2.3** Frequency of aggressions and displacements. Observed during 516 hours. Rate/hour = 0.48. Unknown relationships 26.7%, one-way relationships 33.3%, two-way relationships 40%. Almost-significant improved index of linearity (Matman, 10,000 permutations): h' = 0.91, P = 0.055 one-tailed; Directional Consistency Index DC = 0.88

| Actor | Recipient | | | | | | Total | David's score |
|---|---|---|---|---|---|---|---|---|
| | MOBIKISI | JILL | MOLASO | LOMELA | ZUANI | MWINDU | | |
| MOBIKISI | – | 10 | 3 | 5 | 2 | 71 | 91 | 6.32 |
| JILL | 5 | – | 2 | 50 | 6 | 69 | 132 | 4.48 |
| MOLASO | 1 | 0 | – | 0 | 0 | 0 | 1 | –1.07 |
| LOMELA | 1 | 0 | 0 | – | 5 | 8 | 14 | –1.21 |
| ZUANI | 0 | 5 | 0 | 1 | – | 0 | 6 | –1.75 |
| MWINDU | 2 | 0 | 0 | 0 | 0 | – | 2 | –6.78 |
| Total | 9 | 15 | 5 | 56 | 13 | 148 | 246 | |

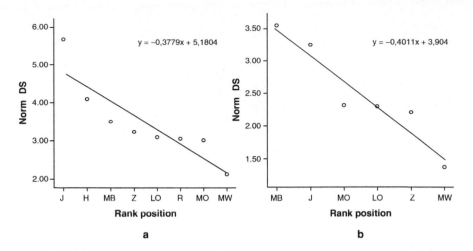

**Fig. 2.1** Steepness of hierarchy in the first (a) and the second (b) study periods. Norm DS = Normalized David's scores.

obtained a non-significant value (P = 0.07). That means the hierarchy cannot be defined as steep.

In the second period (Table 2.3), the steepness value is 0.401 (Fig. 2.1b). When we tested the observed steepness against the null hypothesis of random wins for all pairs of individuals (randomization test procedure with 2000 repetitions), we obtained a significant value (P = 0.026), allowing us to label the hierarchy as steep.

## *Individual Attributes*

### Rank and Gender

David's scores of males and females obtained from the first study period are not statistically different (Mann-Whitney test: U = 7, $n_1$ = 3, $n_2$ = 5, n.s., two-tailed). The Mann-Whitney test is not applicable to compare David's scores of males and females from the second study period due to small sample size. Nevertheless, the ranks of males and females seem to be comparable in the group.

### Rank and Age/body Mass

Individual values of David's score of both sexes taken together are not significantly correlated with age or body mass in either observation period (age, first period: rs = 0.025, n = 8, n.s., two-tailed; second period: rs = 0.58, n = 6, n.s., two-tailed; body mass, first period: rs = 0.307, n = 8, n.s., two-tailed; second period: rs = 0.319, n = 6, n.s., two-tailed; Paoli et al., 2006a). When considering only females, we obtained the same results (age, first period: rs = 0.462, n = 5, n.s., two-tailed; second period: rs = 0.105, n = 4, n.s., two-tailed; body mass, first period: rs = 0.224, n = 5, n.s., two-tailed; second period: rs = 0.211, n = 4, n.s., two-tailed). Thus, individual attributes do not noticeably influence rank.

## Social and Sexual Interactions

### Exchanged Social Interactions

The grooming ratio is not correlated with rank in either study period (first period $r_s$ = 0, n = 8, n.s., two-tailed; second period: $r_s$ = 0.43, n = 6, n.s., two-tailed). We obtained the same result for food sharing ratio and rank (first period $r_s$ = 0.24, n = 8, n.s., two-tailed; second period: $r_s$ = −0.6, n = 6, n.s., two-tailed). Similarly, rank is not correlated with the ratio of peering in either period (first period $r_s$ = 0.167, n = 8, n.s., two-tailed; second period: $r_s$ = −0.086, n = 6, n.s., two-tailed). Even the ratio of invitation to GG-rubbing is not correlated with rank (first period $r_s$ = −0.4, n = 8, n.s., two-tailed; second period: $r_s$ = 0.2, n = 6, n.s., two-tailed) and the result is the same for comparisons of the GG-rubbing frequency in each dyad with the observed rank distance (Matman's row-wise correlation, first period: Kr = 13, $tau_{rw}$ = 0.44, n.s., two-tailed; second period: Kr = −3, $tau_{rw}$ = −0.26, n.s., two-tailed). Accordingly, rank-related asymmetries in social interactions are not apparent in the bonobo group.

### Rank and Postural Pattern in GG-rubbing

Paoli et al. (2006b) reported the absence of any overall asymmetry in the pattern of performance of GG-rubbing among all female dyads (mother-daughter pairs excluded, immature individuals included), though in some dyads there was asymmetry. Specifically, even top-ranking females (J and Z, tables 2.2 and 2.3) performed GG-rubbing as mounter or mountee in relation to different partners (J: 433 bouts as mounter, 522 as mountee: Z: 127 as mounter, 272 as mountee).

### Copulations and Male Rank

In the first study period, copulations performed by males during focal observations were not equally distributed ($\chi^2$ = 29.7, df = 2, p < 0.001), Hani (the alpha male according to David's scores) having the highest frequency (n = 70), followed by Mwindu (n = 34) and Mobikisi (n = 22) (Paoli et al., 2006a). Conversely, in the

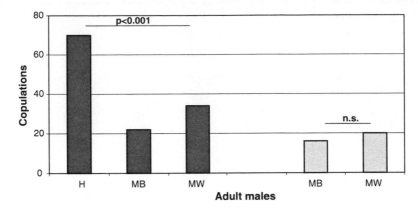

**Fig. 2.2** Frequency of male copulations in the first (black bars) and the second (grey bars) study periods.

second study period, copulations performed by the two males are comparable ($\chi^2$ = 0.44, df = 1, n.s.; Mwindu: n = 20; Mobikisi: n = 16) (see Fig. 2.2).

**Postconflict Behaviors**

We found no clear influence of rank on reconciliation in adults: CCT levels in high-ranking and low-ranking individuals are not statistically different (mean CCT = 14.7% ±12.2% S.E. for high-ranking individuals, CCT = 30.8% ±23.0% S.E. for low-ranking individuals; Mann-Whitney test: U = 5.5, $n_1$ = 3, $n_2$ = 5, n.s.), though low-ranking subjects showed a higher mean CCT. Even for consolatory levels among adults, there is no statistical difference between high- and low-ranking individuals (mean TCT = 20.8% ±7.2% S.E. for high-ranking individuals, TCT = 20.5% ±10.8% S.E. for low-ranking individuals; Mann-Whitney test: U = 7, $n_1$ = 3, $n_2$ = 5, n.s.) (see Fig. 2.3).

## Discussion

Complementing the study by Paoli et al. (2006a) on the hierarchy of the Apenheul bonobos, we expanded the overall analysis based on new results. First, the suggested ill-defined hierarchy characterizing the first study period (Table 2.2) has been confirmed by an insignificant steepness value. Conversely, the almost-significant linearity of the hierarchy characterizing the second study period is accompanied by a significant steepness value. Therefore, this additional investigation at the steepness level confirms that an overall change has occurred in the hierarchy of the Apenheul bonobos across the two periods. The deaths of the two adults (Hani and

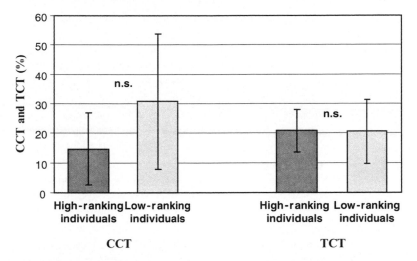

**Fig. 2.3** Reconciliation (CCT) and consolation (TCT) levels ± SD in the Apenheul adult bonobos as a function of rank.

Rosie) after the first study session, and the effects of their relationships upon other group members may account for the different results of the two periods. They belonged to the middle/low-ranking class, which according to Furuichi (personal communication), generally shows unclear and non-linear relationships: a decrease in middle/low-ranking individuals reduced the number of uncertain relationships, thereby increasing linearity and steepness in the second study period.

Considering individual attributes, there is no correlation between age and rank in the group or in females during either study period, in contrast to findings by Vervaecke et al. (2000a). Moreover, no correlation between body mass and rank was observed in the group or in females during either study period, which is in line with findings by Vervaecke et al. (2000a). Tests for sexual influences on dominance rank via David's scores of males and females reveal no statistical difference, in both the first and second study periods. Thus, it appears that in Apenheul bonobos, being male or female is not an effective predictor of likely social status.

Considering exchanged social interactions, Vervaecke et al. (2000b) described the occurrence of up-hierarchy grooming in agreement with Seyfarth's model (1980). We used a different approach to assess the occurrence of correlation between rank (David's scores) and grooming exchange. In fact, we correlated individual ranks with the grooming exchange index calculated for each individual, whereas Vervaecke et al. (2000b) employed matrix correlations and an arbitrary assignment of ranks (from high = 6, to low = 1) to the six individuals of the Planckendael colony, thus creating a ranking method which seems less accurate in comparison to David's scores. Given our definition of the exchange index, a ratio of performed over received, a negative correlation between rank and grooming ratios might be expected if subordinate individuals groom more than they are groomed. There is no correlation in either study period. Our results do not agree with those by Vervaecke et al. (2000b), though we underscore that we used a different approach. Another possible explanation for the difference is that the Planckendael bonobos always have been described as strictly linearly ranked (Vervaecke et al. 2000b), whereas the Apenheul bonobos are not. Thus, it may be that the steeper and more linear the hierarchy, the higher the chance to observe a correlation between rank and grooming exchange. However, in our second study period, with an almost-significant linearity of hierarchy and significant steepness, there was no clear correlation, and moreover, the observed tendency (evaluated by the positive Spearman's $r_s$) suggested a possible, though non-significant, positive correlation (i.e., down-hierarchy grooming) instead of a negative one. Stevens et al. (2005b) also reported the absence of up-hierarchy grooming in the Apenheul colony, which is in line with our findings.

There is no correlation between rank and food-sharing exchange, in either study period, although there are opposite overall tendencies per the sign of $r_s$. Thus, in the Apenheul bonobo group, high- and low-ranking individuals seem to share and receive food from others to the same extent. The other findings on exchanged social interactions are in line with this framework: there is no correlation between rank and peering exchange, rank and ratio of invitation to GG-rubbing or between GG-rubbing frequency in each dyad and the observed rank distance. All these results indicate a hierarchy-independent distribution of the exchange of social interactions in the Apenheul group.

Further, GG-rubbing, besides not being correlated with rank, showed no asymmetry in the role of performers according to rank (Paoli et al. 2006b). Furuichi (1989) reported similar results, whereas de Waal (1987) and Hohmann and Fruth (2000) described asymmetries in initiation and performance of genital contacts, with high-ranking females more often the mounter (top position) than the mountee. However, the life history of a social group and individual temperaments may influence the patterns of a given behavior to a great extent.

Although we could not test for a correlation between male rank and copulatory frequency (cf. chimpanzees: Newton-Fisher 2004, bonobos: Kano 1996), it is remarkable that copulations were not equally distributed among males in the first study period, with the highest ranking male (Hani) accredited for about 51.5% of the total copulations. Conversely, in the second study period, copulations were equally distributed among males, with the new highest ranking male (Mobikisi) accredited for 44.4% of them. Previous studies on captive and wild populations have indicated various contradictory results. In wild bonobos, Kano (1996) reported a positive relationship between dominance and copulation. Conversely, Gerloff et al. (1999) and Furuichi and Hashimoto (2004) reported the absence of such an effect, illustrating that high-ranking males do not necessarily have the highest copulation rates. Stevens et al. (2001) reported that in some captive groups, males do not monopolize copulations, and even where an unequal distribution of copulations among males occurred, the alpha male did not perform the majority of copulations (Marvan et al. 2006). Our data do not clarify the mixed evidence on the subject.

However, we suggest that the high female rank observed in the Apenheul and other bonobo groups may imply that the correlation between dominance rank and copulatory frequency of males may be disturbed by the mate choice of females (Furuichi 1992, Kano 1996, Fruth et al. 1999), which could be inferred from the long-lasting and frequent maximum swelling characterizing bonobo females (Paoli et al. 2006b). This interpretation is also consistent with the absence of any information on sexual coercion by bonobo males: high-ranking females may choose their mating partners with few or no objections from other males. In addition, the distinctive temperament of each male may play a primary role in determining his attitude to exert monopolization of females, even if the male is a high-ranking individual. For example, in the Apenheul group, Mobikisi was the second-ranking individual in the first study period and the top-ranking one in the second session, but he always showed the lowest copulation frequency.

Another relevant aspect characterizing bonobo sociality is their post-conflict behavior: they reconcile and console to a great extent (de Waal 1987, Palagi et al. 2004). Friendship, evaluated by contact sitting and grooming frequencies, positively affected the level of reconciliation (Palagi et al. 2004), thus supporting the "good relationships hypothesis" (Aureli et al. 1989), and consolation levels were comparable among adult males and females. Our results suggest that bonobo post-conflict behavior is not noticeably affected by rank: high- and low-ranking individuals do not show significantly different rates of reconciliation and consolation. Nevertheless, the mean value of reconciliation in high-ranking individuals is lower than that of low-ranking ones. This finding might be interpreted, even with caution, as the effect of the nature of conflicts among high-ranking animals. In fact, a high-ranking individual is generally the victim of aggression by another high-ranking animal (a higher-ranking individual is more likely to be an aggressor (Tables 2.2 and 2.3). Thus, closely-ranked animals may be more interested in trying to outrank each other than in repairing a relationship put at risk by the aggression between them. In fact, the shifting of the hierarchy is a never-ending process that is probably more evident at the top. In addition, good relationships (Palagi et al. 2004) are more important than rank asymmetries in determining the level of reconciliation. The fact that comparable TCT levels occurred in high- and low-ranking bonobos suggests that social status implies no privilege in receiving reassurance gestures by third-parties. De Waal and Aureli (1996) stated that consolatory affiliations seem to be more common in egalitarian than in despotic societies, i.e., the "social constraints hypothesis." Egalitarian is not a proper term for bonobos, which should be defined as tolerant. Nevertheless, bonobo society is surely not despotic. Thus it is not surprising to observe high consolatory levels in the species. Our finding of no consolatory asymmetry related to rank fits perfectly into the overall scenario on the dominance style of bonobos: in a condition where the social structure is flexible (Hohmann and Fruth 2002, 2003), tolerant and loose, consolation is probably not offered up in the hierarchy as an appeasement gesture, e.g. to a high-ranking victim, but instead is more likely driven by other complex mechanisms such as empathy.

## Conclusions

It is difficult to determine what agonistic dominance implies in bonobos: we did not single out any clear benefit of being a high-ranking individual in terms of asymmetries in social interactions. Even the evidence for a positive relationship between male rank and copulations was mixed when comparing the results of the two study periods and reviewing the literature. Parish (1994, 1996) showed that dominance ranks fit perfectly with feeding priority in a captive group of bonobos: specifically she illustrated that all adult females had priority over the sole adult male. We have no datum on feeding priority from our study group, but the result by Parish seems compatible with data from wild groups, wherein females are rarely attacked by males and enjoy feeding priority (Wrangham 1993, Furuichi 1997, Furuichi and Hashimoto 2002). Even if it is problematic to draw generalizations, we can suggest that a likely primary benefit of being a high-ranking individual is the priority of access to food resources, and that a high social status is generally observed in females, including the Apenheul group. This, along with the availability of large food patches and feed-as-you-go foraging characterizing bonobos (Wrangham 2000), may provide a basis for the occurrence of mixed-sex parties regardless of the female swelling phase (Furuichi 1997, Gerloff et al. 1999). In fact, in this framework where females have high status and feeding priority, their costs for group-living are reduced and their reproductive success is probably increased (Mulavwa et al. this volume).

In addition, an emerging relevant aspect is that bonobos are characterized by a flexible society with constantly shifting relationships in both the wild and captivity (Hohmann and Fruth 2002, Stevens and Vervaecke this book). Yet, given that detailed data on dominance are relatively scarce in wild groups, we strongly encourage further investigations in the field to document more thoroughly the bonobo social system, including ecological data, e.g. changing food availability/ quality, in the overall scenario. This may help the understanding of the relationships among social status, feeding priority and reproductive strategies in this species.

**Acknowledgements** We thank the Apenheul Primate Park (The Netherlands): the Director Leobert E.M. de Boer, the General Curator Frank Rietkerk and the bonobo keepers for allowing and facilitating this work; Benedetta Balsotti for collaboration in data collection; Han de Vries for statistical advice; Takeshi Furuichi, Frans de Waal and Richard Wrangham for previous enlightening discussion; Giada Cordoni for useful comments.

## References

Altmann J (1974) Observational study of behaviour sampling methods. Behaviour 49: 227–265
Appleby MC (1983) The probability of linearity in hierarchies. Anim Behav 31: 600–608
Aureli F, van Schaik CP, Van Hoof JARAM (1989) Functional aspects of reconciliation among captive long-tailed macaques (*Macaca fascicularis*). Am J Primatol 19: 39–51

Bermejo M, Omedes A (1999) Preliminary vocal repertoire and vocal communication of wild bonobos (*Pan paniscus*) at Lilungu (Democratic Republic of Congo). Folia Primatol 70: 328–357

Bernstein IS (1981) Dominance: the baby and the bathwater. Behav Brain Sci 4: 419–458

Drews C (1993) The concept and definition of dominance in animal behaviour. Behaviour 125: 283–313

Franz C (1999) Allogrooming behavior and grooming site preferences in captive bonobos (*Pan paniscus*): association with female dominance. Int J Primatol 20: 525–546

Fruth B, Hohmann G, McGrew WC (1999) The *Pan* species. In: Dolhinov P and Fuentes A (eds) The non human primates. California: Mayfield Publishing Company. pp 64–72

Furuichi T (1989) Social interactions and the life history of female *Pan paniscus* in Wamba, Zaire. Int J Primatol 10: 173–197

Furuichi T (1992) Dominance relations among wild bonobos (*Pan paniscus*) at Wamba, Zaire. Paper presented at the 14th Congress of the International Primatological Society, Strasbourg

Furuichi T (1997) Agonistic interactions and matrifocal dominance rank of wild bonobos (*Pan paniscus*) at Wamba. Int J Primatol 18: 855–875

Furuichi T, Hashimoto C (2002) Why female bonobos have a lower copulation rate during estrus than chimpanzees. In: Boesch C, Hohmann G, Marchant L (eds) Behavioral Diversity of Chimpanzees and Bonobos. Cambridge University Press, Cambridge, pp 156–167

Furuichi T, Hashimoto C (2004) Sex differences in copulation attempts in wild bonobos at Wamba. Primates 45: 59–62

Furuichi T, Ihobe H (1994) Variation in male relationships in bonobos and chimpanzees. Behaviour 130: 211–228

Gerloff U, Hartung B, Fruth B, Hohmann G, Tautz D (1999) Intracommunity relationships, dispersal pattern and paternity success in a wild living community of bonobos (*Pan paniscus*) determined from DNA analysis of faecal samples. Proc R Soc Lond 266: 1189–1195

Hinde RA (1976) Interactions, relationships and social structure. Man 11: 1–17

Hohmann G, Fruth B (2000) Use and function of genital contacts among female bonobos. Anim Behav 60: 107–120

Hohmann G, Fruth B (2002) Dynamics in social organizsation of bonobos (*Pan paniscus*). In: Boesch C, Hohmann G, Marchant L, (eds) Behavioural Diversity in Chimpanzees and Bonobos. Cambridge University Press, Cambridge, pp 138–150

Hohmann G, Fruth B (2003) Intra- and inter-sexual aggression by bonobos in the context of mating. Behaviour 140: 1389–1413

Ihobe H (1991) Male relationships of pygmy chimpanzees of Wamba, Republic of Zaire. In: Ehara A et al. (eds) Primatology Today: The Proceedings of the XIIIth Congress of the International Primatological Society. Amsterdam: Elsevier. pp 231–234

Kano T (1992) The Last Ape. Palo Alto: Stanford University Press.

Kano T (1996) Male rank order and copulation rate in a unit-group of bonobos at Wamba, Zaire. In: McGrew WC, Marchant LF, Nishida T (eds) Great Ape Societies. Cambridge. University Press, Cambridge, pp 135–155

Marvan R., Stevens JMG, Roeder AD, Mazura I, Bruford MW, de Ruiter JR (2006) Male dominance rank, mating and reproductive success in captive bonobos (*Pan paniscus*) Folia Primatol 77: 364–376

Mason WA (1993) The nature of social conflict: a psycho-ethological perspective. In: Mason WA, Mendoza SP (eds) Primate social conflict. State University of New York Press, Albany, pp 13–47

Newton-Fisher NE (2004) Hierarchy and social status in Budongo chimpanzees. Primates 45: 81–87

Palagi E (2006) Social play in bonobos (*Pan paniscus*) and chimpanzees (*Pan troglodytes*): implications for natural social systems and inter-individual relationships. Am J Phys Anthropol 129: 418–426

Palagi E, Paoli T, Borgognini Tarli SM (2004) Reconciliation and consolation in captive bonobos (*Pan paniscus*). Am J Primatol 62: 15–30

54 T. Paoli and E. Palagi

Paoli T, Palagi E, Borgognini Tarli SM (2006a) Reevaluation of dominance hierarchy in bonobos (*Pan paniscus*). Am J Phys Anthropol 130: 116–122

Paoli T, Palagi E, Borgognini Tarli SM (2006b) Perineal swelling, intermenstrual cycle, and female sexual behavior in bonobos (*Pan paniscus*). Am J Primatol 68: 333–347

Parish AR (1994) Sex and food control in the "uncommon chimpanzee": how bonobo females overcome a phylogenetic legacy of male dominance. Ethol Sociobiol 15: 157–179

Parish AR (1996) Female relationships in bonobos (*Pan paniscus*): evidence for bonding, cooperation, and female dominance in a male-philopatric species. Human Nat 7: 61–69

Seyfarth RM (1980) The distribution of grooming and related behaviors among adult female vervet monkeys. Anim. Behav. 28: 798–813

Stevens JMG, Vervaecke H, van Elsacker L (2001) Sexual strategies in *Pan paniscus*: implications of female dominance. Primate Report Special Issue 60: 42–43

Stevens JMG, Vervaecke H, de Vries H, van Elsacker L (2005a) Peering is not a formal indicator of subordination in bonobos (*Pan paniscus*). Am J Primatol 65: 255–267

Stevens JMG, Vervaecke H, de Vries H, van Elsacker L (2005b) The influence of the steepness of dominance hierarchies on reciprocity and interchange in captive groups of bonobos (*Pan paniscus*).- Behaviour 142: 941–960

Vervaecke H, de Vries H, van Elsacker L (2000a) Dominance and its behavioral measures in a captive group of bonobos (*Pan paniscus*). Int J Primatol 21: 47–68

Vervaecke H, de Vries H, van Elsacker L (2000b) The pivotal role of rank in grooming and support behaviour in captive bonobos (*Pan paniscus*). Behaviour 137: 1463–1485

de Vries H, Stevens JMG, Vervaecke H (2006) Measuring and testing the steepness of dominance hierarchies. Anim Behav 71: 585–592

de Waal FBM (1987) Tension regulation and nonreproductive functions of sex in captive bonobos (*Pan paniscus*). Nat Geographic Research 3: 318–335

de Waal FBM (1988) The communicative repertoire of captive bonobos (*Pan paniscus*), compared to that of chimpanzees. Behaviour 106: 183–251

de Waal FBM (1989) Dominance 'style' and primate social organisation. In: Staden V, Foley RA (eds) Comparative Socioecology. Oxford: Blackwell. pp 243–263

de Waal FBM (1996) Conflict as negotiation. In: McGrew WC, Marchant LF, Nishida T (eds) Great Ape Societies. Cambridge: Cambridge University Press. pp 159–172

de Waal FBM, Aureli F (1996) Consolation, reconciliation, and a possible cognitive difference between macaques and chimpanzees. In: Russon AE, Bard KA, Parker ST (eds) Reaching into thought : The minds of the great apes. Cambridge University Press, Cambridge, pp. 80–110

de Waal FBM, Luttrell LM (1989) Toward a comparative socioecology of the genus *Macaca*: different dominance styles in rhesus and stumptail monkeys. Am J Primatol 19: 83–109

Walters JR, Seyfarth RM (1987) Conflict and cooperation. In: Smuts BB, Cheney DL, Seyfarth RM, Wrangham RW, and Struhsaker TT, editors. Primate societies. University of Chicago Press,Chicago, pp 306–317

White FJ (1996) Comparative socio-ecology of *Pan paniscus*. In: McGrew WC, Marchant LF, Nishida T (eds) Great Ape Societies. Cambridge University Press ,Cambridge, pp 29–41

Wittig RM, Boesch C (2003) Food Competition and Linear Dominance Hierarchy among Female Chimpanzees of the Taï National Park. Int J Primatol 24: 847–867

Wrangham RW (1993) The evolution of sexuality in chimpanzees and bonobos. Human Nature 4: 47–79

Wrangham RW (1999) The evolution of coalitionary killing. Yearbook Phys Anthropol 42: 1–30

Wrangham RW (2000) Why are male chimpanzees more gregarious than mothers? A scramble competition hypothesis. In: Kappeler P (ed) Male Primates. Cambridge University Press, Cambridge, pp 248–258

Wrangham RW, Peterson D (1996) Demonic Males, New York: Houghton Mifflin.

# Social Play in Bonobos:
# Not Only an Immature Matter

**Elisabetta Palagi[1] and Tommaso Paoli[1]**

## Introduction

### Defining Play: A Hard Work

In the *Confessions* (Book XI, Chapter XIV), St. Augustine pondered the meaning of time. His answer is, "Quid ergo est tempus? Si nemo ex me quaerit, scio: si quaerenti explicare velim, nescio" [What is time then? If no one asks me, I know. If I want to explain it to those who ask, I no longer know]. Perhaps St. Augustine's frustration is felt most by ethologists when dealing with play.

We have no doubt what play is, and we have few uncertainties when it comes to understanding when our cats, dogs, and children play. Yet when we check the extensive literature devoted to play, we discover there are as many definitions of play as there are authors who studied it (Fagen 1981, Martin and Caro 1985, Power 2000). The difficulty in finding an objective definition derives from the fact that we cannot describe a distinctive characteristic of play; we can only state that play lacks certain characteristics that are typical of serious behavior.

Some authors attempted to define play as a functionless behavior (Bierens de Haan 1952), but the notion that play has no obvious benefit involves a subjective interpretation on the part of the observer (Martin and Caro 1985). Play probably has many benefits (both delayed and immediate), but they are not easily detectable. Play can be defined as all activity that appears to an observer to have no obvious immediate benefits for the performer, but which involves motor patterns typical of functional contexts, such as agonistic, anti-predatory, and mating behavior (Martin

[1]*Department of Ethology, Ecology and Evolution, Anthropology Unit, University of Pisa, Via Roma 79, 56011 Calci, Pisa, Italy*

T. Furuichi and J. Thompson (eds.), *The Bonobos: Behavior, Ecology, and Conservation*
© Springer 2008

and Caro 1985, Pellis and Pellis 1996, Bekoff 2001). The difference between play-ful and serious contexts is not in the actual behavioral patterns performed, but in how they are performed (Pellis and Pellis 1998). In fact, compared to behaviors in serious functional contexts, behavioral patterns during play are often, exaggerated, reordered, incomplete, brief, repeated, varied in sequence, and inhibited (Burghardt 2005). However, Bekoff and Allen (1998) suggest that the study of play should proceed on the basis of an intuitive understanding of play. Accordingly, while fol-lowing the above definition to some extent, it should be kept in mind that no defini-tion should include or exclude any specific pattern from the category of play.

## What is Play for? A Brief Overview

As a subject matter, play challenges psychologists to discover its consequences on behavioral development, anthropologists to identify its role in the evolution of social and cognitive skills, and evolutionary biologists to search for the functions of an apparently nonfunctional behavior. Assuming that play is a functional behav-ior, is it also an adaptive behavior? Gould and Vrba (1982) suggested that the term adaptive can be used only for a phenotypic character that promotes the fitness of the organism and performs the function for which it was selected. For the traits that produce benefits for their present role, but were not originally selected for that role, the authors suggested the term ex-aptation. In this paper, when we use the terms benefit and function of play, we refer to its possible current utility (effects) and not to its historical genesis.

Its ubiquitous nature and diversity suggest that play may assume multiple adap-tive roles depending on the species, age, sex of the players, and context in which it occurs (Loizos 1967, Jolly 1985, Nunes et al. 2004).

Play behavior has an important role in the ontogeny of many primate species (Martin and Caro 1985, Fagen 1993). A research carried out on the timing of play in juvenile vervets supports the theories that play evolved to influence neural selec-tion during early brain development (Fairbanks 2000). In the neural selection model, she suggests that new forms of play have evolved to promote the acquisition of uniquely human linguistic and cognitive capacities via neural selection.

Further functions of mammalian and, specifically, nonhuman primate play have been the topic of a number of reviews in the past two decades (Fagen 1981, 1993, Power 2000, Burghardt 2005). Social play may have an important role in establish-ing and maintaining relationships among animals likely to interact with each other in the future (e.g., social skill hypothesis; Baldwin and Baldwin 1974, Bekoff 1974, Fagen 1981, Holmes 1994, 1995, Maestripieri and Ross 2004, Palagi 2006). A variety of effects on physical and motor development have also been hypothe-sized to derive from social play (motor training hypothesis; Byers and Walker 1995, Byers 1998). For example, locomotor-rotational play might have immediate benefits to young animals, such as providing important physical exercise that develops endurance, control of body movements, and perceptual-motor integration (Power 2000, Palagi and Paoli 2007). Further immediate benefits include testing social roles and improving skills in communication (Špinka et al. 2001, Dugatkin and Bekoff 2003, Palagi et al. 2004, 2006, Burghardt 2005). Such immediate benefits seem to be particularly evident in adult play, a phenomenon that is very often neglected in this research field (Pellis and Iwaniuk 1999, Palagi 2006).

Whatever the functions of social play are, it is one of the most sophisticated types of social communication (Fagen 1981, 1993). In fact, during a session, a constant fine-tuning of the sequences of the patterns has to be maintained in order to prevent play from escalating into real aggressions (Byers 1998, Power 2000). In addition, for play sessions to take place, a recognition of those stimuli that might appear ambiguous is needed, e.g. intentions of other animals. The ability to interpret such ambiguous features of social signaling could represent a central issue in the evolution of behavioral flexibility and intelligence in primates (Fagen 1981, 1993, Bekoff 1995, Pellis and Pellis 1996). Accordingly, investigating if primates are able to finely adjust such activity, according to shifting social contexts and to the availability of different playmates, could provide information on their social competence and help to hypothesize some possible roles of social play.

## *Focusing on Adult Play in Bonobos*

Observers have described play in captive and wild *Pan paniscus* only by anecdotic reports: no systematic or empirical datum was available. The few reports on bonobo play are often a side-effect of studies on sexual behavior (Hashimoto 1997). Yet, *Pan paniscus* is an extremely playful species (de Waal 1989, Kano 1992). We believe that bonobos, with their tolerant society, peculiar social structure, flexible inter-individual relationships, and playful tendency, represent a good model species to test empirically many emerging hypotheses on adult play behavior, the multiple functions of which are not well understood.

When adults engage in social play, they most often do so with immature partners. However, this phenomenon is difficult to interpret, as the playful interactions are generally initiated by the immature playmates. Therefore, the most convincing data to fully understand the inclination of adults to engage in social play are those involving adult-adult dyads or those during which adults trigger the session.

We aimed to thoroughly investigate modalities and frequency distribution of adult social play in a captive group of bonobos (Apenheul Primate Park, the Netherlands), testing some predictions on the possible functions of their play.

As a first step, we present data on the variation of play behavior according to the age-class combinations of the playmates. We also examine whether differences in play modality occur in such combinations of playmates. Secondly, we test the possibility that play behavior can be used to prevent conflict escalation by comparing data on the use of such behavior in the periods around feeding time, and in a controlled condition to evaluate if play is used as a mechanism of celebration.

The last step focuses on a comparison of data collected on the two *Pan* species in order to assess whether, despite their phylogenetic closeness and similar social structure (fission-fusion society), chimpanzees (*Pan troglodytes*) and bonobos show differences in adult play behavior.

## Methods

### *The Bonobo Colony and Data Collection on Play*

We collected behavioral data during four observation seasons, for a total of ca. 12 months, from a group of *Pan paniscus* housed in the Apenheul Primate Park (Apeldoorn, The Netherlands). We observed the subjects during 6-hr sessions, encompassing both the morning and the afternoon. The composition of the colony changed over time: during session 1 (July–October 2000) there were 11 individuals (five mature females, three mature males, and three infants). In session 2 (April–July 2002), there were nine individuals (four mature females, two mature males, one juvenile, and two infants). In session 3 (September–December 2002), there were seven individuals (three mature females, two mature males, one juvenile, and one infant), and in session 4 (March–June 2003), there were eight individuals (one more newborn; Table 2.1 in Paoli and Palagi 2008).

The wild-caught bonobos came from a rescue center in the Democratic Republic of Congo. They were collected from different sites in different periods, and therefore we can be fairly confident that they were unrelated. The group was first established in 1998 from some wild individuals for conservation purposes. They lived in an enclosure with both an indoor and an outdoor facility (about $230\,m^2$ and $5000\,m^2$, respectively). Because they were not always within sight in the indoor facilities, we stopped the observations when more than one animal was out of sight. Both facilities had everything necessary to allow the movements of bonobos in all three dimensions; they could move from the indoor to the outdoor enclosure after the first feeding session (at 0845 hr) and obtain food (milk enriched by vitamins and proteins, monkey chow, vegetables, and fruit) four times a day at about 0845, 1230, 1430, and 1630 hr.

By focal animal sampling (Altmann 1974), we collected play interactions in the four observation periods. We followed each focal animal for about 41 h, 56 h, 52 h, and 52 h during each of the four observation sessions respectively. To evaluate play intensity (rough and gentle play sessions), we used the parameters applied by Palagi (2006) to record locomotor-rotational (L) play, and for contact (C) play we followed the description of Wilson and Kleiman (1974) and Burghardt (2005). Complete descriptions of the recorded play patterns and their categories are in Table 3.1.

**Table 3.1** - Play behavioral patterns recorded during the observation sessions

| | Initials | Definition<br>Gentle Play Patterns |
|---|---|---|
| Airplane | AIR | An adult lies on its back and raises an infant up with its hands and feet |
| Grab Gentle | GRG | An individual massages gently another |
| Play Push | PPS | A bonobo pushes a playmate either with its hands or feet |
| Play Bite | PBIT | An individual gently bites a playmate |
| Play Recovering a Thing | PRCO* | An individual chases a playmate and attempts to grab an object carried by it |
| Play Slap | PSL | An individual slaps any part of a playmate's body |
| Tickle | TK | The partner's body is contacted either with the mouth or with the hands |
| | | Rough Play Patterns |
| Pirouetting | PIRO* | One or more individuals together turn, somersault or roll over either on the ground or on vertical supports |
| Acrobatic Play | ACP* | One (solitary play) or more bonobos (social play) climb, jump and dangle from supports of the environment (e.g. branches) |
| Play Run | PRUN* | A bonobo runs alone (solitary play) or chases a play partner (social play) |
| Play Stamping | PST | An individual jumps on a play partner with its feet |
| Rough and Tumble | RT | Two bonobos (or more) grasp, slap, and bite each other. This pattern is typical of immature individuals |
| Play Brusque Rush | PBR | An individual jumps with its four limbs on a playmate |
| Play Retrieve | PRE | An individual holds a playmate to avoid its flight |
| | | Other |
| Play Invitation | PI | A bonobo approaches a possible play partner, pats it and then goes away. This display is used to start a play session |
| Play Face | PF | Playful facial display: the mouth is opened with only the lower teeth exposed |
| Full Play Face | FPF | Playful facial display: the mouth is opened with the upper and lower teeth exposed |

The patterns marked by an * belong to the locomotor play category (L play).

Via scan sampling (Altmann 1974), we recorded the frequency of play and grooming interactions, noting the individual identity of interactants using a 5-min interval between consecutive scans. We scored a total of 4224 scans, corresponding to 380 h collected during session 1.

## The Chimpanzee Colony

In order to compare data on play behavior between the two *Pan* species, we selected a group of *Pan troglodytes* at the ZooParc de Beauval (St. Aignan sur Cher, France). The colony, the largest group (n=19) of chimpanzees in France, comprised 10 adults (2 males and 8 females), 4 juveniles, and 5 infants. The animals lived in an enclosure with both indoor and outdoor facilities (ca. 200 m$^2$ and 2,000 m$^2$, respectively) similar to the one at Apenheul. The group had access to food at 9:00 AM, 2:00 PM, and 4:30 PM. We observed them over a 6-hr period, encompassing both morning and afternoon. We followed them via the procedure used for bonobos (Palagi et al. 2004, Palagi 2006). By focal animal sampling, we collected ca. 44 h per subject; via scan sampling, we recorded 4,128 scans corresponding to 344 h. Because play behavior varies according to age (Fagen 1981, 1993, Caine 1986, Pusey 1990, Mendoza-Granados and Sommer 1995, Dolhinow 1999, Palagi et al. 2002), we selected the eight youngest subjects (2 males and 6 females) of the 10 adult chimpanzees as focal individuals, in order to have adult chimpanzees and bonobos of comparable ages.

## Definition of the Periods around Food and Control

To investigate whether or not the presence of food affected play and grooming behavior, we defined four different periods by preliminary observations: prefeeding (the last 25 minutes before food provisioning), feeding (the 25 min block starting from food provisioning), postfeeding (the 25 minutes after feeding as defined earlier), and control (the time block from 1000 h to 1200 h, the farthest from feeding times, when individuals showed high activity levels, and a sufficiently long time span to represent a control condition). The parameter for delimiting the 3 periods linked to feeding activity was the usual time span necessary for complete food consumption, i.e., 25 min.

We followed 2 feeding times per day (1245 and 1630 h). Feeding times were highly predictable, with an imprecision of only *ca.* ± 5 min. We also defined cofeeding as the frequency with which a certain dyad sits in contact, i.e. contact sitting during feeding, to establish if there is a correlation between grooming/play level in the prefeeding time and cofeeding.

## Statistics

We employed nonparametric statistical tests for analyses at the individual level, (Lehner 1996, Siegel and Castellan 1988, Zar 1999). We used exact tests according to the threshold values suggested by Mundry and Fischer (1998).

For analyses at the dyadic level, we employed randomization tests with a number of 10,000 shuffles to avoid errors due to non-independence of the data (Manly 1997). We used the software Resampling Procedures 1.3 by David C. Howell (freeware).

For the correlation analyses, we used MatMan's row-wise correlation tool with 10,000 permutations.

We set the level of significance at 5%. All the analyses are 2-tailed. We performed statistical analyses via SPSS 12.0.

# Results

## *Play Frequency and Modalities in Adult Bonobos*

As a first step we compared the levels of play performed by adults with other adults and with immature individuals (excluding mother-offspring play from the analysis). The analysis showed no significant difference (Wilcoxon Exact test T = 6, ties = 0, n = 8, p = 0.11).

In the first observation session, adults directed play invitations (PINV) more frequently to each other than to immature subjects (Palagi 2006). In the following observation sessions (specifically, sessions 3 and 4), we also recorded the frequency of successful and unsuccessful play invitations ($PINV_{succ}$ and $PINV_{unsucc}$). We compared the results obtained by the following ratio

$$(PINV_{succ} - PINV_{unsucc})/(PINV_{succ} + PINV_{unsucc}) \quad (3.1)$$

in order to assess whether the successful PINVs performed by adults toward other adults and toward immature individuals differed significantly. There proved to be no significant difference (Wilcoxon Exact test T = 10, n = 6, ties = 0, p = 0.99). Both age-matched and immature subjects responded with a similar motivation to the play invitations performed by adults (Fig. 3.1).

Then we carried out a comparison of the play modalities according to the age of playmates. In order to evaluate such a difference, we compared the following ratio

$$(C\,play - L\,play)/(C\,play + L\,play) \quad (3.2)$$

between adult-adult and adult-immature combinations.

We found that adults used higher levels of contact play (C play) during play sessions with immature subjects, whereas during adult-adult sessions they preferred the locomotor modality (L play) (Wilcoxon Exact test T = 1, n = 8, ties = 0, p = 0.016) (Fig. 3.2).

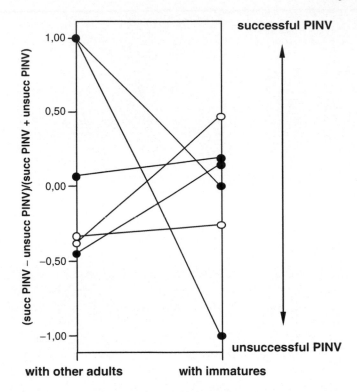

**Fig. 3.1** Proportion of successful play invitations showed by adult bonobos with other adults or immatures. A line with open dots represents an adult male, and a line with closed dots represents an adult female.

As for the occurrence of sexual contacts per play session, adult-adult dyads used sexual patterns more frequently than adult-immature dyads (Wilcoxon Exact test $T = 1$, $n = 8$, $p = 0.016$). The use of sexual patterns per play session is significantly more frequent in males than in females (Exact Mann Whitney $U = 0$, $n_m = 3$, $n_f = 5$, $p = 0.036$) (Fig. 3.3).

Play frequency among adults did not follow a random distribution in relation to the partner's sex (randomization ANOVA, one-way: $F = 8.837$, $p = 0.007$) (Palagi 2006). The frequency of play among bonobo females was higher than in male-female dyads ($t = 3.501$, $n_{ff} = 10$, $n_{mf} = 15$, $p = 0.001$). Males, on the other hand, never played together. Here, we provide an additional analysis which showed a difference in play modality according to the sex of the players. In order to evaluate the difference we compared the ratio (3.2) between female-female and male-female dyads. The randomization test for two independent samples showed that C play occurred with significantly higher frequency among female-female compared to male-female dyads ($t = 3.841$, $n_{ff} = 10$, $n_{mf} = 15$, $p = 0.0007$) (Fig. 3.4).

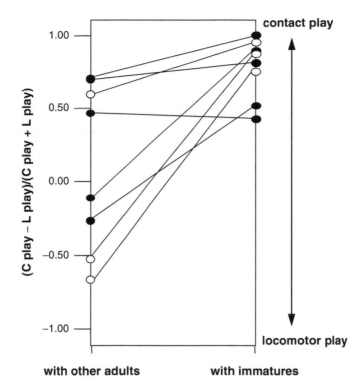

**Fig. 3.2** Proportion of contact play performed by adult bonobos with other adults or immatures. A line with open dots represents an adult male, and a line with closed dots represents an adult female.

## *Social Play as a Tension Reduction Tactic*

Palagi et al. (2006) analyzed the distribution of grooming and social play across the four conditions: prefeeding, feeding, postfeeding, and control. Neither social activity among adults followed a random distribution (for grooming, Exact Friedman $\chi^2_r = 19.95$, n = 8, df = 3, p = 0.001; for social play, Exact Friedman $\chi^2_r = 14.19$, n = 8, df = 3, p = 0.006).

Considering grooming interactions, post hoc tests revealed a significant increase during prefeeding compared to feeding (q = 3.68, p < 0.01, n = 8) and postfeeding (q = 4.25, p < 0.05, n = 8). Adult dyads showed no correlation between the frequency of grooming performed during the prefeeding period and the frequency of cofeeding during the feeding period ($Kr = 35$, $tau_{rw} = 0.22$, n.s.). Moreover, the frequency of grooming in the control period was not correlated with the cofeeding ($Kr = 37$, $tau_{rw} = 0.23$, n.s.).

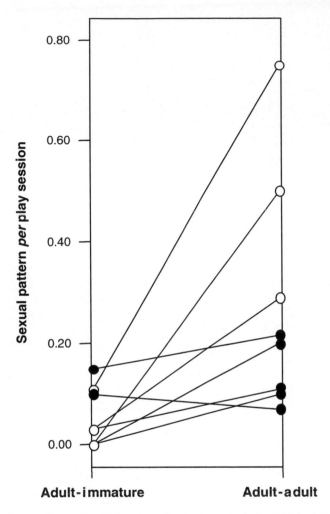

**Fig. 3.3** Proportion of sexual pattern performed per play session by adult bonobos with other adults or immatures. A line with open dots represents an adult male, and a line with closed dots represents an adult female.

Post hoc analysis on adult play frequencies revealed a significant peak level of such behavior during the prefeeding condition compared to any other possible condition (prefeeding vs feeding: q = 3.5, p < 0.01, n = 8; prefeeding vs postfeeding: q = 2.37, p < 0.01, n = 8; prefeeding and control: q = 3.20, p < 0.01, n = 8) (Fig. 3.5).

In the matrices of the play performed during the prefeeding period with that of cofeeding, there is a positive correlation ($Kr = 52$, $tau_{rw} = 0.42$, p = 0.007). We could not test for a correlation between the frequency of play in the control period and the rates of cofeeding due to the few dyads that played in the control period.

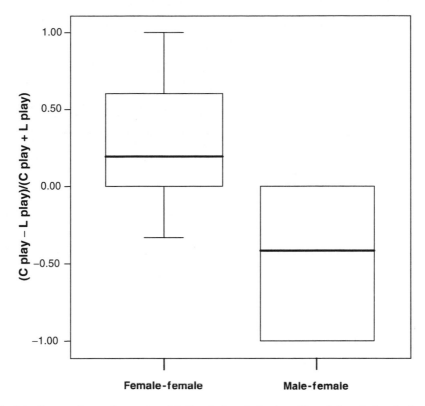

**Fig. 3.4** Contact play preference in adult female-female dyads and by adult male-female dyads. Solid horizontal lines indicate medians, boxes correspond to inter-quartile range, thin horizontal lines indicate range of observed values.

## *Play and Play Signals in Bonobos and Chimpanzees: A Comparative Approach*

Palagi (2006) attempted the first direct comparison of play distribution and modality in *Pan paniscus* and *Pan troglodytes*. The comparison of the overall social play performed by the adults of the two species showed no significant difference (Mann-Whitney U = 52, $n_b$ = 8, $n_c$ = 8, n.s). In contrast, the frequency of play interactions among adult bonobos is significantly higher than that in chimpanzees (Mann-Whitney U = 9, $n_b$ = 8, $n_c$ = 8, p = 0.007).

The frequency of PINV performed and received by adults also differs significantly in the two species, with adult bonobos showing higher frequencies (Mann-Whitney U = 0, $n_b$ = 8, $n_c$ = 8, p < 0.001; Fig. 3.6).

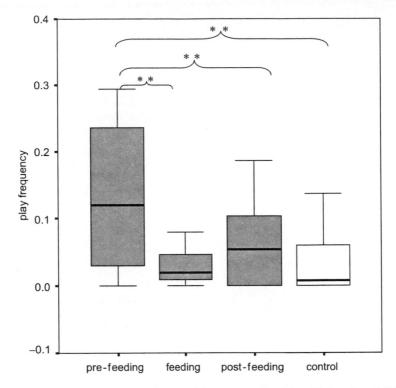

**Fig. 3.5** Play distribution across the four conditions among the eight adult bonobos. Solid horizontal lines indicate medians, length of the gray boxes corresponds to inter-quartile range, thin horizontal lines indicate range of observed values.

We carried out an additional analysis on the choice of playmates according to the age of the chimpanzee players. The adults showed a strong preference for immature subjects as playmates (excluding mother-offspring play from the analysis) versus age-matched fellows (Wilcoxon Exact test T = 0, n = 8, p = 0.008). This result contrasts with data for bonobos, which showed no preference.

Adult bonobos performed play facial displays significantly more often than adult chimpanzees (play face *plus* full play face: Exact Mann-Whitney U = 13, $n_b$ = 8, $n_c$ = 8, p = 0.04). Considering the two variants of the play facial displays separately, we found no significant difference in the use of play faces (Exact Mann-Whitney U = 30, $n_b$ = 8, $n_c$ = 8, n.s.), but adult bonobos had a significantly higher frequency of the full play face (Exact Mann-Whitney U = 10.5, $n_b$ = 8, $n_c$ = 8, p = 0.018).

In bonobos, there is a difference in play distribution according to the sex combinations, with females playing mainly with each other. In adult chimpanzees, there is no significant difference with regard to sex of playmates (randomization ANOVA, one-way: F = 0.124, n.s.).

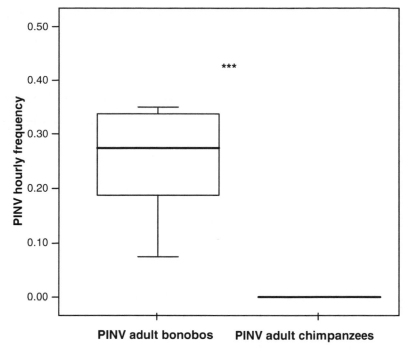

**Fig. 3.6** Hourly frequency of Play Invitations (PINV) performed by adult bonobos (Apenheul) and chimpanzees (Beauval). Solid horizontal lines indicate medians, length of the gray boxes corresponds to inter-quartile range, thin horizontal lines indicate range of observed values.

## Discussion

### Play Modality and Distribution in Adult Bonobos

Each phase of the animal life span is peculiar. However, no sudden change occurs and each step gradually shifts to the next. Animals never stop changing throughout their lives, both physically and socially, and adults are no exception. Play has an important role in the assembly of adult behavior, however, it can continue into adulthood as well. Because adult play is probably linked to neoteny (Bekoff 1974), the choice of age-mates as play-partners by adults could represent an empirical confirmation of the neoteny hypothesis. Describing playful behavior in bonobos, Enomoto (1990) provided an example of play between an adult male and an adult female. He suggested that adult-adult sessions may reflect the neotenous nature of the species. Indeed, neotenic traits in bonobos have been reported for some features of their morphology (Coolidge 1933) and behavior (Dahl 1986, Kuroda 1980). The comparable levels of adult-adult and adult-immature play sessions we recorded support the paradigm of a neoteny of play in bonobos.

We also found differences in the modality of play sessions according to age combinations. During adult-adult play sessions, the locomotor modality (L play) was preferred over the contact one (C play). In contrast to the amount of debate concerning the multiple functions of C play, there is significantly more consensus regarding L play (Power 2000). Because most physical training effects are not permanent, but disappear quickly after exercise stops (Byers 1998), the L play may provide general exercise for adults and allow an evaluation of the rapidity of the playmates without running into the typical risks of contact play sessions, e.g., damaging aggression or injury (Fagen 1981). The hypothesis on reduced amount of risk seems to be supported by the extremely low frequency of play signals during the L play sessions (Palagi and Paoli 2007).

Kuroda (1980 p. 186) reported that in Wamba bonobos, "play between a male and female occurred before and/or after copulations." Our findings support his anecdotical report; in fact, we found higher levels of sexual interactions 1) during adult-adult play versus adult-immature play and 2) in adult male-female dyads versus female-female dyads. Accordingly, play may have a role in courtship.

Palagi (2006) found that the frequency of social play among bonobo females is higher than in any other sex combinations, and our study demonstrated a difference in the distribution of C play according to the sex of the adult players, with females showing higher levels of C play with each other. C play, which is also called rough and tumble or play fighting, involves cooperation and reciprocity, and it is one of the most sophisticated and complex forms of social interaction. During C play sessions, the playmates have to trust each other to maintain the rules of the game (Bekoff 2001, Pellis et al. 2005). In primates, strong affinitive relationships between unrelated females are rare (Silk et al. 2003). They are associated with grouping patterns in which females do not migrate from their natal groups; hence they are female philopatric (Pusey and Packer 1986). Bonobo females are an exception to the general trend. In fact, like female chimpanzees they migrate from their natal groups (female dispersal, Furuichi, 1989; Hashimoto et al., this volume) but, dissimilarly, they form strong coalitions and alliances with other unrelated females (Doran et al. 2002). Bonobo female play patterns are similar to those of spotted hyenas (*Crocuta crocuta*), in which adult females have high levels of social play (Fagen 1981, Burghardt 1999). Spotted hyenas are characterized by a fission-fusion society, with female dominance and male dispersal. If social play is a biological adaptation, the occurrence of high levels of play behavior in bonobo and spotted hyena females may have arisen from evolutionary convergence. Because play patterns reflect the social organization of a species (Cheney 1978, Miller and Nadler 1981, Zucker et al. 1986, Watts and Pusey 1993, Maestripieri and Ross 2004), females of both species might use play in the same functional way (Palagi 2006).

Each social play session may be viewed as a conflict of interest because each animal might have its own preferred play manner as a result of age, sex, dominance, and individuality (Konner 1975). However, if two or more individuals play together for long periods of time, then cooperation has to occur much more frequently than conflict of interest (Fagen 1993, Bekoff 2001). In fact, failure to negotiate and cooperate prevents animals from continuing to play together and can lead to a decline in honest interaction and cooperation in other behavioral contexts (Dugatkin and Bekoff 2003). Accordingly, we can hypothesize that play among female bonobos serves to maintain social bonds, to test the weakness of play partners, and therefore, to gain social advantage (Pellis et al. 1993, Paquette 1994).

## Social Play as a Tension Reduction Tactic

Henzi and Barrett (1999), Loizos (1967), Merrick (1977), and Goodall (1968) emphasized the similarity shared by social play and grooming. In fact, both behaviors entail close physical contact for long periods and have an important role in social cohesion. Grooming maintains social stability by reducing tension and providing appeasement during contexts characterized by conflict of interest (Aureli et al. 2002, Merrick 1977, Schino et al. 1988). Social primates use diverse strategies to mitigate tension and to prevent conflict escalation such as communicative displays, dominance relationships, and greeting gestures (Preuschoft and van Schaik 2000, Whitham and Maestripieri 2003).

In bonobos, the presence of food increased the rate of sexual interactions (de Waal 1987, 1992, Kitamura 1989). Particularly during feeding time, bonobos selectively used non-reproductive sexual interactions such as mounting and genito-genital rubbing, whereas copulations did not increase (Paoli et al. 2007). De Waal (1992) and Paoli et al. (2007) interpreted these findings as an appeasement mechanism of tension regulation.

In the Apenheul bonobos, the frequency of adult-adult play was significantly higher during prefeeding than in any other condition, and there was a positive correlation between play performed during that period and cofeeding (Palagi et al. 2006). It seems that anticipatory play increases the level of tolerance around food. The correlation also may be an independent consequence of other variables that favor friendly social relationships, so the dyads playing together often also feed in contact. However, the extremely low frequency of play recorded in the control condition does not seem to support this hypothesis. The positive correlation between play recorded in the prefeeding period and cofeeding suggests that bonobos anticipate the forthcoming tension associated with feeding by increasing their rates of social play, thus making the foraging peaceful (Palagi et al. 2006). This finding supports several anecdotic reports from captive and wild bonobos. De Waal (1989) described play behavior in a context of social tension during the joining of two different groups hosted at the San Diego zoo. Enomoto (1990) reported a play bout in the wild between an adult male and an older female when the two individuals were aggressively competing over food. A possible function for social play might be to protect the members of a social group from escalating such competitive interactions. Our hypothesis is supported by an experimental study comparing the ability of bonobos and chimpanzees to cooperatively solve a food-retrieval problem (Hare et al. 2007). Hare et al. (2007) found that bonobos were more tolerant of cofeeding than chimpanzees and that during cofeeding tests, only bonobos exhibited sociosexual behavior, and they played more often.

Chimpanzees (*Pan troglodytes*) cope with competitive tendencies via grooming among adults (de Waal 1992, Koyama and Dunbar 1996) and via playing between adults and unrelated immature individuals (Palagi et al., 2004). The mechanisms seem to alleviate the tension that rises before feeding time (celebration, de Waal 1992). Probably, the tolerant society of bonobos, with its less differentiated roles between the sexes (de Waal 2001, Kano 1980, Kuroda 1979, 1980, Paoli and Palagi 2008), may facilitate adult-adult play during situations of high tension. The distribution of play among adult bonobos matches that observed for grooming in adult chimpanzees (Koyama and Dunbar 1996, Palagi et al. 2004). The use of play during stressful social situations (prefeeding) is particularly interesting given the commonly held view that stress suppresses play: here play appears to regulate stress.

Grooming seems not to function for bonobos as a tension reduction strategy in the feeding context, in fact, the grooming frequency in the prefeeding period, though higher than that in the feeding condition, failed to reach statistical significance versus rates recorded in the control period. Moreover, for adult dyads there is no correlation between grooming rates (prefeeding) and cofeeding. This finding is similar to that obtained by de Waal (1987) in the San Diego group. In conclusion, bonobos seem to cope with social tension by two diverse tactics. Play has a role in tension regulation at a short-term level (anticipation and celebration), whereas sociosexuality seems to work at an immediate level (appeasement and reassurance) once the stressful situation has occurred (Furuichi and Ihobe 1994, Palagi et al. 2006, Paoli et al. 2007).

## Adult Play in Pan Species in Comparative Perspective

In 1999, Pellis and Iwaniuk suggested that adult social play occurs to a higher extent in species characterized by fluid societies. When individuals meet each other periodically, adult play may have a role in social assessment. In rhesus monkeys, play among adults is used to manipulate some social situations (Brueggeman 1978). Bonobos and chimpanzees are characterized both by a fission-fusion society and male philopatry, yet they show striking differences in play frequency and modality among adults. Accordingly, the type of social system does not seem to be a sufficient condition by which adult-adult play may have been favored selectively in bonobos.

Play among adults may be important when relationships are not codified and structured according to rank rules. In this view, the observed differences in the level of bidirectional conflicts (more frequent in bonobos than in chimpanzees) and the complete absence of formal submissive displays in bonobos suggest that they have a more tolerant and flexible society than that of chimpanzees (Palagi 2006, Paoli and Palagi 2008).

The playful temperament of adult bonobos is not a side-effect of the presence of immature individuals, because adults show no preference according to the age of the playmate. Conversely, chimpanzee adults selectively choose immatures as play partners.

A play session can occur only if the playmates are capable of some degree of metacommunication, e.g. exchanging the signals that carry the message: this is play (Hayaki 1985). An agreement to play rather than to fight, mate, or engage in any serious pattern, can be negotiated in different ways. Some play markers have evolved into signals for the beginning of a play session or to maintain a play mood or both (Bekoff 2001). In primates, the signals are represented by facial displays. Playful facial displays, including play face and full play face (Palagi 2006), were significantly more frequent in bonobos than in chimpanzees, whereas a separate analysis for play face and full play face showed that in bonobos, facial display with greater tooth exposure is preferred.

These findings may be interpreted as a consequence of the occurrence of more rough play than gentle play in bonobos (Palagi 2006). Play-fighting can be risky (Bekoff 1995, Bekoff and Allen 1998), and a friendly appeasing signal could modulate the interaction (Pellis and Pellis 1996). For example, when the roughness of a playful interaction escalates and one of the two playmates manifests weakness, the dominance relationship inside that dyad may be reversed. Conversely, if one of the playmates counter-escalates the encounter by increasing his/her vigor, then the other may use a play signal to cope with the tense situation (retroactive function hypothesis; Pellis and Pellis 1996). The selective use of play signals by bonobos in rough vs. gentle play sessions supports this hypothesis. An alternative hypothesis is that given the higher degree of uncertainty in rank relationships among adult bonobos, it is necessary to signal more clearly and frequently that it is only play, probably due to the lack of role-reversal, which becomes useless within a given dyad characterized by uncertain rank relationships.

Social play can be viewed as a balance between cooperation and competition. Konner (1975) stated that a social play session is characterized by a conflict of interest, during which each playmate might show a peculiar play mood in relation to age, sex, dominance, and temperament. Fagen (1993) and Bekoff (2001) underlined that when two or more playmates engage in long play sessions, cooperation is expected to overcome the conflict of interest. A failure to negotiate and to cooperate deprives animals of the possibility to continue playing together and leads to a decrease in cooperation and fair interaction in other behavioral contexts.

**Acknowledgements** We thank Takeshi Furuichi and Jo Thompson for inviting us to contribute this volume. Thanks go also to the Director Francoise Delord, veterinarian Romain Potier, the chimpanzee keepers of the ZooParc de Beauval (France), and Director Leobert E.M. de Boer, General Curator Frank Rietkerk, and the bonobo keepers of the Apenheul Primate Park (The Netherlands) for allowing and facilitating this work; Giada Cordoni for useful discussion and comments.

# References

Altmann J (1974) Observational study of behaviour sampling methods. Behaviour 49:227–265
Aureli F, Cords M, van Schaik CP (2002) Conflict resolution following aggression in gregarious animals: a predictive framework. Anim Behav 63:1–19

Baldwin JD, Baldwin JI (1974) Exploration and social play in squirrel monkeys (*Saimiri*). Am. Zool. 14:303–315

Bekoff M (1974) Social play and play-soliciting by infant canids. Am Zool 14:323–340

Bekoff M (1995) Play signals as punctuation: the structure of social play in canids. Behaviour 132:419–429

Bekoff M (2001) Social play behaviour. Cooperation, fairness, trust, and the evolution of morality. J Consc Stud 8:81–90

Bekoff M, Allen C (1998) Intentional communication and social play: how and why animals negotiate and agree to play. In: Bekoff M, Byers JA (eds) Animal play. Evolutionary, comparative, and ecological perspectives. Cambridge University Press, Cambridge, pp 97–114

Bierans de Haan JA (1952) The play of a young solitary chimpanzee. Behaviour 4:144–156

Brueggeman JA (1978) The function of adult play in free-living *Macaca mulatta*. In: Smith EO (ed) Social Play in Primates, Academic Press, New York, pp 169–192

Burghardt GM (1999) Play. In: Greenberg G, Haraway MM (eds) Comparative psychology: a handbook. Garland Publishing Co, New York, pp 725–735

Burghardt GM (2005) The genesis of animal play. Testing the limits. MIT Press, Cambridge, MA

Byers JA (1998) Biological effects of locomotor play: general or specific? In: Bekoff M, Byers JA (eds) Animal play: evolutionary, comparative, and ecological perspectives. Cambridge University Press, Cambridge, pp 205–220

Byers JA, Walker C (1995) Refining the motor training hypothesis for the evolution of play. Am Natur 146:25–40

Caine NG (1986) Behavior during puberty and adolescence. In: Mitchell G, Erwin J (eds) Comparative Primate Biology, Vol. 2, Part A. Behavior, Conservation, and Ecology. Liss, New York, pp 327–362

Cheney DL (1978) The play partners of immature baboons. Anim Behav 26:1038–1050

Coolidge HJ (1933) *Pan paniscus*: pigmy chimpanzee from south of the Congo river. Am J Phys Anthropol 17:1–57

Dahl JF (1986) Cyclic perineal swelling during the intermenstrual intervals of captive female pigmy chimpanzees (*Pan paniscus*). J Hum Evol 15:369–385

Dolhinow P (1999) Play: a critical process in the developmental system. In: Dolhinow P, Fuentes A (eds) The nonhuman primates. Mayfield Publishing Company, Mountain View, California, pp 231–236

Doran DM, Jungers WL, Sugiyama Y, Fleagle JG, Heesy CP (2002) Multivariate and phylogenetic approaches to understanding chimpanzee and bonobo behavioural diversity. In: Boesch C, Hohmann G, Marchant LF (eds) Behavioural diversity in chimpanzees and bonobos. Cambridge University Press, Cambridge, pp 14–34

Dugatkin LA, Bekoff M (2003) Play and the evolution of fairness: a game theory model. Behav Process 60:09–14

Enomoto T (1990) Social play and sexual behavior of the bonobo (*Pan paniscus*) with special reference to flexibility. Primates 31:469–480

Fagen R (1981) Animal play behavior. Oxford University Press, New York

Fagen R (1993) Primate juvenile and primate play. In: Pereira ME, Fairbanks LA (eds) Juvenile primates. Oxford University Press, Oxford, pp 182–196

Fairbanks LA (2000) The developmental timing of primate play. A neural selection model. In: Parker ST, Langer J, McKinney ML (eds) Biology, brains, and behavior. The evolution of human development. School of American Research Press, Santa Fe, pp 211–219

Furuichi T (1989) Social interactions and the life history of female *Pan paniscus* in Wamba, Zaire. Int J Primatol 10:173–197

Furuichi T, Ihobe H (1994) Variation in male relationships in bonobos and chimpanzees. Behaviour 130:211–228

Goodall J van Lawick (1968) The behaviour of free-living chimpanzees in the Gombe Stream Reserve. Anim Behav Monogr 1:161–311

Gould SJ, Vrba ES (1982) Exaptation: a missing term in the science of form. Paleobiology 8: 4–15

Hare B, Melis AP, Woods V, Hastings S, Wrangham R (2007) Tolerance allows bonobos to out-perform chimpanzees on a cooperative task. Curr Biol pp 17, 619–623

Hashimoto C (1997) Context and development of sexual behavior of wild bonobos (*Pan paniscus*) at Wamba, Zaire. Int J Primatol 18:1–21

Hayaki H (1985) Social play of juvenile and adolescent chimpanzees in the Mahale Mountains National Park, Tanzania. Primates 26:343–360

Henzi SP, Barrett L (1999) The value of grooming to female primates. Primates 40:47–59

Holmes WG (1994) The development of littermate preferences in juvenile Belding's ground squirrels. Anim Behav 48:1071–108

Holmes WG (1995) The ontogeny of littermate preferences in juvenile golden-mantled ground squirrels: effects of rearing and relatedness. Anim Behav 50:309–322

Jolly A (1985) The Evolution of Primate Behavior. Macmillan Publishing Company, New York

Kano T (1980) Social behavior of wild pygmy chimpanzees (*Pan paniscus*) of Wamba: a prelimi-nary report. Human Evol 9:243–260

Kano T (1992) The last ape: pygmy chimpanzee behavior and ecology. Stanford Univ Press, Stanford

Kitamura K (1989) Genito-genital contacts in the pygmy chimpanzee (*Pan paniscus*). African Study Monogr 10:49–67

Konner MJ (1975) Relations among infants and juveniles in comparative perspective. In: Lewis M, Rosemblum LA (eds) Friendship and Peer Relations. Wiley, New York, pp 99–124

Koyama NF, Dunbar RIM (1996) Anticipation of conflict by chimpanzees. Primates 37:79–86

Kuroda S (1979) Grouping of the pigmy chimpanzees. Primates 20:161–183

Kuroda S (1980) Social behavior of the pigmy chimpanzee. Primates 21:181–197

Lehner PN (1996) Handbook of Ethological Methods. Cambridge University Press, Cambridge

Loizos C (1967) Play behaviour in higher primates: a review. In: Morris D (ed) Primate ethology. Anchor Books, Chicago, IL, pp 226–282

Maestripieri D, Ross SR (2004) Sex differences in play among western lowland gorilla (*Gorilla gorilla gorilla*) infants: Implications for adult behavior and social structure. Am J Phys Anthropol 123:52–61

Manly BFJ (1997) Randomization, bootstrap and Montecarlo methods in biology. Chapman and Hall, London

Martin P, Caro TM (1985) On the functions of play and its role in behavioral development. Adv Stud Behav 15:59–103

Mendoza-Granados D, Sommer V (1995) Play in chimpanzees of the Arnhem zoo: self serving compromises. Primates 36:57–68

Merrick NJ (1977) Social grooming and play behavior of a captive group of chimpanzees. Primates 18:215–224

Miller LC, Nadler RD (1981) Mother-infant relations and infant development in captive chimpan-zees and orang-utans. Int J Primatol 2:247–261

Mundry R, Fischer J (1998) Use of statistical programs for nonparametric tests of small samples often leads to incorrect P-values: examples from Animal Behaviour. Anim Behav 56:256–259

Nunes S, Muecke EM, Sanchez Z, Hoffmaier RR, Lancaster LT (2004) Play behavior and motor development in juvenile Belding's ground squirrels (*Spermophilus beldingi*). Behav Ecol Sociobiol 56:97–105

Palagi E (2006) Social play in bonobos (*Pan paniscus*) and chimpanzees (*Pan troglodytes*): impli-cations for natural social systems and interindividual relationships. Am J Phys Anthropol 129:418–426

Palagi E, Paoli T (2007) Play in adult bonobos (*Pan paniscus*): Modality and potential meaning. Am J Phys Anthropol 134:219–225

Palagi E, Gregorace A, Borgognini Tarli SM (2002). Development of olfactory behavior in cap-tive ring-tailed lemurs. Int J Primatol 23:587–599

Palagi E, Cordoni G, Borgognini Tarli SM (2004) Immediate and delayed benefits of play behav-iour: new evidence from chimpanzees (*Pan troglodytes*). Ethology 110:949–962

Palagi E, Paoli T, Borgognini S (2006) Short-term benefits of play behavior and conflict prevention in *Pan paniscus*. Int J Primatol 27:1257–1270

Paoli T, Palagi E (2008) What does agonistic dominance imply in bonobos? In: Furuichi T, Thompson J (eds) The bonobos: behavior, ecology, and conservation. Springer, New York, pp 39–54

Paoli T, Tacconi G, Borgognini Tarli S, Palagi E (2007) Influence of feeding and short-term crowding on the sexual repertoire of captive bonobos (*Pan paniscus*). Ann Zoo Fenn 44:81–88

Paquette D (1994) Fighting and playfighting in captive adolescent chimpanzees. Aggress Behav 20:49–65

Pellis SM, Pellis VC (1996) On knowing it's only play: the role of play signals in play fighting. Aggr Viol Behav 1:249–268

Pellis SM, Pellis VC (1998) The structure-function interface in the analysis of play fighting. In: Bekoff M, Byers JA (eds) Animal play – evolutionary, comparative, and ecological perspectives. Cambridge University Press, Cambridge, pp 115–140

Pellis SM, Iwaniuk AN (1999) The problem of adult play-fighting: a comparative analysis of play and courtship in primates. Ethology 105:783–806

Pellis SM, Pellis VC, McKenna MM (1993) Some subordinates are more equal than others: play fighting amongst adult subordinate rats. Aggr Behav 19:385–393

Pellis SM, Pellis VC, Foroud A (2005) Play fighting. Aggression, affiliation, and the development of Nuanced Social Skills. In: Tremblay RE, Hartup WW, Archer J (eds) Developmental origins of aggression. Guilford Press, New York, pp 47–62

Power TG (2000) Play and exploration in children and humans. Erlbaum, Mahwah, NJ

Preuschoft S, van Schaik CP (2000) Dominance and communication. In: Aureli F, de Waal FBM (eds) Natural Conflict Resolution. University of California Press, Berkeley, pp 77–105

Pusey AE (1990) Behavioural changes at adolescence in chimpanzees. Behaviour 115:203–246

Pusey AE, Packer C (1986) Dispersal and philopatry. In: Smuts BB, Cheney DL, Seyfarth RM, Wrangham RW, Struhsaker TT (eds) Primate societies. University of Chicago Press, Chicago, pp 183–204

Schino G, Scucchi S, Maestripieri D, Turillazzi PG (1988) Allogrooming as a tension-reduction mechanism: a behavioral approach. Am J Prim 16:43–50

Siegel S, Castellan NJJ (1988) Non parametric statistics for the behavioral sciences. McGraw Hill New York

Silk JB, Alberts SC, Altmann J (2003) Social bonds of female baboons enhance infant survival. Science 302:1231–1234

Špinka M, Newberry RC, Bekoff M (2001) Mammalian play: training for the unexpected. Quart Rev Biol 76:141–167

de Waal FBM (1987) Tension regulation and nonreproductive functions of sex in captive bonobos (*Pan paniscus*). Nat. Geograph. Res. 3:318–335

de Waal FBM (1989) Peacemaking among primates. Harvard University Press, Cambridge, MA

de Waal FBM (1992) Appeasement, celebration, and food sharing in the two *Pan* species. In: Nishida T, McGrew WC, Marler P, Pickford M (eds) Topics in Primatology, Vol. 1: Human Origins. Tokyo University Press, Tokyo, pp 37–50

de Waal FBM (2001) Apes from Venus: bonobos and human social evolution. In: de Waal FBM (ed) Tree of origin: what primate behavior can tell us about human social evolution. Harvard University Press, Cambridge, MA, pp 41–68

Watts DP, Pusey AE (1993) Behavior of juvenile and adolescent great apes. In: Pereira ME, Fairbanks LA (eds) Juvenile Primates - Life History, Development, and Behavior. Oxford University Press, New York, pp 148–167

Whitham JC, Maestripieri D (2003) Primate rituals: the function of greetings between male Guinea baboons. Ethology 109:847–859

Wilson SC, Kleiman DG (1974) Eliciting play: a comparative study. Am Zool 14:341–370

Zar JH (1999) Biostatistical Analysis. Prentice Hall, Upper Saddle River, New Jersey

Zucker EL, Dennon MB, Puleo SG, Maple TL (1986) Play profiles of captive adult orangutans: a developmental perspective. Devel Psychobiol 19:315–326

# Gestures and Multimodal Signaling in Bonobos

Amy S. Pollick[1], Annette Jeneson[2], and Frans B. M. de Waal[3]

## Introduction

Studies on bonobos have come a long way in the last several decades. Our understanding of this remarkable ape's ecology, sexual behavior, dominance style, and conservation issues is constantly evolving. We know a great deal about the bonobo's vocal repertoire, as described by de Waal (1988), Hohmann and Fruth (1994), and Bermejo and Omedes (1999). Facial expressions have not been studied nearly as fully, perhaps because doing so requires close observation. For this and other forms of visual communication, captive studies remain invaluable. They allow for the observation of detailed social behavior at close range, as well as the observation of complex social interactions in their entirety.

One of the most interesting and least studied forms of social communication in apes is gesture. We see all four species of great ape – bonobo, chimpanzee, gorilla, and orangutan – using their hands to communicate, but gestures, as with facial expressions, are very difficult to study in the wild. Most studies of gesture concern human-trained ones, such as American Sign Language taught to a handful of individuals (Patterson 1979, Gardner et al. 1989, Miles 1990). Additionally, we know next to nothing about how natural gestures work in concert with other communicative signals.

Although there have been some advances, we still know relatively little about the evolutionary history of language (Christiansen and Kirby 2003). An understanding of this complex issue must be grounded in a range of disciplines, including linguistics,

[1] Association for Psychological Science, 1010 Vermont Ave. NW, Suite 1100, Washington, DC 20007 United States

[2] Department of Psychology, University of California, United States

[3] Living Links, Yerkes National Primate Research Center, Emory University, United States

psychology, neuroscience, philosophy, archaeology, and primatology. An important way to further our knowledge is through the comparative study of closely related primate species (Marler 1976, Cheney and Seyfarth 1990). Though nonhuman primate vocalizations have long been the focus of language evolution theories, gesture also has much to contribute.

We will first review the general tenets of the gestural origins of language theory, after which we will briefly summarize the aspects of human gesture that relate to its probable evolution. We will also review what is known about ape gestures and how they may or may not fit the theory of gestural origins of language (Corballis 2002). We will then provide detailed descriptions of bonobo gestures, emphasizing their flexibility relative to other communicative signals. We will review multimodal communication and describe how gestures function within a multimodal scheme in bonobos. The flexible nature of gestures as compared to other communicative signals will provide food for thought for the role that gestural communication may have played in the evolution of human language (Corballis 2002).

We observed four groups of captive apes: one bonobo group at the San Diego Zoo and one at the San Diego Wild Animal Park, both in California, and two separate chimpanzee groups at the Yerkes National Primate Research Center in Lawrenceville, Georgia. All of these apes live in social groups in primarily outdoor settings, and each group contained a mix of sexes and ages. We recorded data onto videotape, which we used for subsequent analysis (with the exception of a small subset of focal data), and considered only social interactions that were initiated by a communicative signal (Pollick 2006, Pollick and de Waal 2007).

## Gestural Origins of Language Theory

Historically, primate communication research has focused on the vocal modality, usually with the exclusion of other forms of communicative signals. This focus is probably a reflection of prevalent theories of language evolution at the time, emphasizing the vocal trajectory as the evolutionary origin of language (Marler 1965, Cheney and Seyfarth 1990). Though theories of human communication have long underscored the interplay of different modalities, it is only recently that theoretical debates about the evolutionary history of language have opened up to the possibility that other communicative behaviors conceivably evolved along with or perhaps even earlier than spoken language. Several decades ago, Hewes (1973) proposed a gestural origin of language theory, which Corballis (1999, 2002, 2003) has further developed. Corballis argues that there are several convincing pieces of evidence for why gesture may have been the original medium for evolving language in our hominid ancestors, which can be summarized as follows: 1) the advantage of manual communication in the hunting-and-gathering phase in early hominid society (silent communication to coordinate hunts); 2) paleoarchaeological evidence suggesting that the early hominid brain was "language ready" before the vocal apparatus was ready to produce complex speech; 3) the observation that apes use manual gestures in a more controlled manner than they do their voices; and 4) the fact that gesture use is lateralized in the Broca's area homologue in great apes.

Corballis thus argues that all of these elements add up to a plausible scenario in which gesture assumed the burden for the burgeoning linguistic capacity that was spilling from our fast developing neocortex, until the vocal tract further developed and human society came to rely predominantly on speech as its means of language. The theory has not been fully embraced by human gesture and linguistic researchers, who doubt the strength of the evidence Corballis calls upon (Jackendoff 2002, Pollick and de Waal 2004, McNeill et al. 2005). The importance of this theory, however, may not lay so much in prioritizing gesture over speech, but in the attention called to their co-evolution. A compromise that seems reasonable is that gestures and vocalizations may have been incorporated into a multimodal communication strategy. While we rely heavily on speech to convey the majority of linguistic information, this multimodal communication strategy is nevertheless evident in humans today.

## Gesturing as an Integral Part of Human Communication

Human gesture has been studied for over 60 years (Efron 1941, Kendon 1972, 1980, McNeill 1992, Goldin-Meadow and Wagner 2005), and we know some of the ways in which gesture facilitates and enhances vocal communication as well as cognitive and symbolic processes. Gesture produced while speaking can enhance information transfer and supplement the meaning of the linguistic signal (McNeill 1992). Though not normally produced without speech in hearing people, gesture can assume linguistic properties when users are prevented from talking (Goldin-Meadow 2001), even in children raised in linguistically poor environments (Goldin-Meadow and Mylander 1984). When a person is having trouble expressing a thought through speech, simultaneous gesturing may facilitate lexical retrieval (Morrel-Samuels and Krauss 1992), and even provide a kind of cognitive arena in which to think when speech does not provide the appropriate means of expression (Goldin-Meadow et al. 2001).

While some specific human gestures are universal, many are culture-specific. But we also show so-called "beat" gestures, which simply emphasize the flow of speech (McNeill 1992). We habitually gesture in the presence of speech, often in precise synchrony with speech (McNeill 1985). We even gesture in the absence of a visible audience, as we do when talking on the phone (Morris 1977, 1994), or communicating with blind individuals (Iverson and Goldin-Meadow 1998). The fact that gestures accompany speech even in situations in which its communicative value seems null, emphasizes its automaticity and encourages investigation into the possible evolution of this ubiquitous behavior.

What is even more remarkable is that some human gestures occur without learning from others. The ethologist Eibl-Eibesfeldt, who followed the expressive behavior of a congenitally deaf and blind girl named Sabine, observed her stretching her hand and pushing it back, palms facing outwards, in a gesture of rejection (Eibl-Eibesfeldt 1973), without her ever having observed such a gesture. Comparative theorists who view the difference between human language and other forms of communication as one of degree only argue that human linguistic capacity expanded from abilities already present in other animals, particularly closely related species. If this hypothesis is correct, and if gesturing is integral to human communication, we should expect to find certain precursors of this communication strategy in nonhuman primates.

## Ape Gesturing

As pointed out by de Waal (2003), free hand gestures are virtually limited to the Hominoidea. This is not a mere quantitative difference with monkeys, but a qualitative one. Facial expressions and vocalizations are common means of communication in all primates and many other animals, but with the exception of a single gesture in a single species, monkeys lack ritualized hand gestures. Macaques may slap the ground with a hand when threatening another, or reach back to their partner during a sexual mount, but these are the limits of their manual communication. Contacts with a substrate or partner function as a signal, but involve more than the hand. In contrast, bonobos wave at each other, shake their wrists when impatient,

**Fig. 4.1** Bonobo reaching out his arm in a gesture (Photograph by Frans B.M. de Waal.).

beg for food with an open hand held out, flex their fingers towards themselves when inviting contact, move an arm over a subordinate in a dominance-gesture, and so on. They even gesture with their feet (de Waal 1988).

Like facial expressions, many free hand gestures of apes are ritualized, that is, they are stereotypical, exaggerated, and tied to specific contexts. The begging gesture, which is also universal in humans, most likely derives from a cupped hand held under the mouth of a food possessor. The origin of this gesture is visible in the only known ritualized monkey gesture, which is hand-cupping by capuchins (*Cebus apella*). If one monkey possesses food, another will reach out a hand and hold it under the possessor's chin so as to catch dropping morsels. This seems an instrumental act, but the same gesture can also be given from a distance – for example, when two capuchins are separated by mesh and one is consuming food (de Waal 1997). In those instances, the gesture is used as a distant signal, divorced from its instrumental function, similar to the way all of the great apes use gestures. An important difference remains, however, in that apes have generalized the meaning of the begging gesture to apply to a variety of situations, whereas in capuchins, the gesture appears to be entirely food-specific.

Apes and humans gesture more with the right hand than the left hand (Annett 1985, Hopkins and Morris 1993, Hopkins and de Waal 1995). Since the right hand is left-brain controlled, this means that ape gestures share the same lateralization as human language. The highly flexible use of ritualized hand gestures, their recent appearance on the evolutionary scene (compared with other means of communication), and their culture-dependency in both humans and apes have implications for the role that gestural communication may have played in the evolution of human language (e.g. Corballis 2002).

There has been a resurgence of interest in natural gestural behavior in various ape species. The great apes use their hands extensively in daily life: in play, sharing food, getting one's attention, and grooming (Goodall 1968, van Hooff 1973, de Waal 1988, Kano 1992, Tanner and Byrne 1999, Tanner 2004). Studies of ape behavior in the wild included some attention to gestures (Goodall 1968, Kuroda, 1984, Kano 1992, Veà and Sabater-Pi 1998), but more abundant are observational studies of naturally occurring gestures in captive chimpanzees (Ladygina-Kohts 1935, van Hooff 1973, Plooij 1978, 1984, Tomasello et al. 1985, 1989, 1994, 1997), bonobos (Savage-Rumbaugh et al. 1977, de Waal 1988, Pika et al. 2005) and gorillas (Tanner and Byrne 1996, 1999, Pika et al. 2003, Tanner 2004).

Researchers have used gesture in apes as an experimental tool with which to ask questions about imitation (Custance et al. 1995), intentionality and perspective taking (Hopkins and Leavens 1998), linguistic ability (Gardner et al. 1989), and laterality (Cantalupo and Hopkins 2001). In these cases, gestures were usually either taught as part of an artificial system or generated using food or other desirable objects. While these experiments have been able to explore questions about theory of mind and intentionality, no current research truly asks questions about how gesture is used to mediate social life in a naturalistic environment.

# A Word About Definitions

In reviewing the literature on gestural communication in primates, it soon becomes obvious that researchers use the term gesture in different ways. Prior behavioral studies (Goodall 1986, Plooij 1978), and none of the studies by Tomasello and colleagues, comprehensively defined gesture: investigators selected the contents of their ethograms on broadly described gestures. Some studies included facial expressions, body postures, or even locomotion patterns in their gesture definitions, e.g., back offer, belly offer, lip-lock, genital offer, spit-at, or swagger (Tomasello et al. 1997). Researchers who took care to define gesture, such as Savage-Rumbaugh et al. (1977), restricted the definition to movements of the hands or upper forelimbs, and generally did not include body postures or general body movement, even if directed at another individual. Tanner and Byrne's (1999) working definition is the most precise, and served as the model for our own studies: they defined gesture as all discrete, nonlocomotor limb and head movements that appear to be communicative, and the movement should be intentionally directed toward another individual. For a tactile interaction to be considered a gesture, it needs to involve a transformation of purposive behavior, so that it is no longer mechanically effective and communicates a specific desire, intent, or feeling (Bretherton and Bates 1979, Goldin-Meadow and Mylander 1984, Gomez 1990). Pollick (2006) provided an extensive working definition of gesture.

It is crucial that the study of gestures in apes is restricted to the limbs. This is not only so in relation to theories about the evolution of language, but also because the detection of manual activity in monkeys has been shown to be neurologically distinct from general body movements (Perrett et al. 1985). In humans, the neural space that houses language (Broca's area) is also active during the observance and performance of manual gestures, but not other body movements (Rizzolatti et al. 1996). Hence, a sharp distinction needs to be drawn between brachiomanual gestures and any other nonvocal bodily-based forms of communication.

# Manual Gestures in Bonobos

Given the above restriction to the study of gestures in apes, we were able to finely discriminate against many different kinds of manual gestures in the San Diego Zoo and Wild Animal Park bonobos (Table 4.1).

For example, when stretching the arm and hand out in a gesture, the palm can face upwards, downwards, or to the side (a distinction made for chimpanzees by van Hooff 1972). However, we did not observe the three being used interchangeably with respect to social context: the reach out side gesture was more often made in food contexts, reach out up was made typically when requesting a grooming session, and reach out down was often produced in play. Of 32 different manual gestures observed, the bent wrist gesture was rarely produced, and when it was, it

**Table 4.1** Gesture usage in bonobos and chimpanzees

| BONOBOS | | CHIMPANZEES | |
|---|---|---|---|
| Gesture | Percentage of Total gestures | Gesture | Percentage of total gestures |
| gentle touch | 40.4 | gentle touch | 25.9 |
| reach out down | 10.2 | bent wrist | 11.1 |
| reach out up | 9 | arm raise | 9.7 |
| arm raise | 6.8 | throw aimed | 9.2 |
| hard touch | 5.2 | reach out down | 7.9 |
| Pat | 5.1 | throw hold | 5.8 |
| foot/leg | 3.1 | hard touch | 3.9 |
| Dab | 3 | beg hand | 3.4 |
| reach out side | 2.9 | reach out up | 3.4 |
| slap ground | 2.9 | dab | 2.2 |
| poke | 1.4 | reach out side | 2.2 |
| shake wrist | 1.4 | rap knuckles | 2.1 |
| swing | 1.3 | slap ground | 1.8 |
| hunchover | 1 | shake wrist | 1.6 |
| rap knuckles | 1 | foot/leg | 1 |
| clap | 0.8 | pat | 1 |
| flap | 0.7 | armwave | 0.9 |
| bent wrist | 0.5 | hand/mouth | 0.9 |
| slapstomp | 0.5 | swing | 0.9 |
| throw aimed | 0.5 | flap | 0.7 |
| armwave | 0.4 | poke | 0.7 |
| beg hand | 0.4 | beckon | 0.6 |
| hand to hand | 0.4 | flail | 0.6 |
| beckon | 0.1 | hunchover | 0.6 |
| finger/mouth | 0.1 | clap | 0.4 |
| flail | 0.1 | finger flex | 0.3 |
| hand lead | 0.1 | point | 0.3 |
| point | 0.1 | slapstomp | 0.3 |
| stomp | 0.1 | stomp | 0.3 |
| | | clasp self | 0.1 |

Total number of gestures: 763 Total number of gestures: 673.
Both bonobos and chimpanzees used the *gentle touch* gesture more than any other, but contextual usage varied (see Pollick 2006).

was never in an agonistic situation. This is in stark contrast with chimpanzees, which often use this gesture to ask for or provide appeasement (Goodall 1968; Figure 4.2).

Another contrast with chimpanzees lay in the tactile nature of some gestures: bonobos use more gestures that involve touching (albeit not forcefully), such as gentle touch and pat: tactile gestures comprised 55.8% of the all observed gestures in bonobos and 34.6% of those observed in chimpanzees. Perhaps bonobos, being generally less aggressive and dominance-oriented, are more tolerant of communicative touching than chimpanzees (Table 4.2).

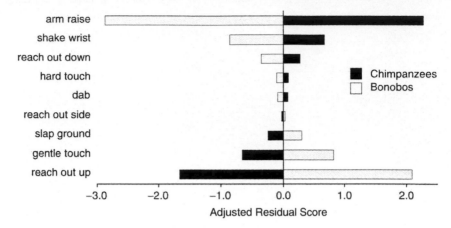

**Fig. 4.2** Some manual gestures compared between bonobos and chimpanzees.

**Table 4.2** Gestures observed in bonobos and chimpanzees

| Gesture | Description |
|---|---|
| arm raise | one or both arms raised, which initially hang more or less down, forwards with usually a quick, jerky movement; fingers are flexed slightly and palm of hand may be oriented towards the other individual and upwards, or away from the other individual and downwards; arms stop rising at horizontal position, and hand may swing further upwards; recipient is never struck |
| armwave | rising to a bipedal position while facing another individual and either swinging arms in front of torso or raising one or both arms rapidly into the air (not as part of a swagger/bluff display) |
| beckon | one or both arms raised forward and upward sweepingly and stiffly with the elbows more extended than in the arm raise; hands are hanging down limply with finger flexes usually; movement is held at end of upward swing while individual stares at recipient |
| beg with hand | placing one or both hands around or under lips, or chin and lips, of recipient that has food in mouth; or touching the hand by the mouth of individual containing the food |
| bent wrist | flexing the wrist while holding the back or side of hand out towards another individual; contact possible |
| clap hands/feet[*] | flat palms of hands are brought into contact with each other either in vertical or horizontal position; can be repetitive |
| clasp self[*] | arms are crossed in front of torso, with hands curled and usually slapped on individual's arms, repeated two or three times in succession |
| dab | touching approaching or stationery individual with back of flexed fingers where after touching hand is withdrawn immediately; sequence can be repeated a number of times in quick succession |
| finger flex | palm can be up or down, and wrist is not bent; fingers move rapidly back and forth |

(continued)

**Table 4.2** (continued)

| Gesture | Description |
| --- | --- |
| finger/hand in mouth | putting a finger or hand into another individual's mouth |
| flail* | arms and hands are completely raised above head and are shaken in rapid succession (usually in tantrum or approach); repetitive |
| flap | one arm and hand raised and makes a downward slapping movement of the hand in direction of another individual – no forceful contact with substrate (ground, wall, etc.) |
| foot/leg gesture | any extension of leg or foot towards another individual |
| gentle touch | any sort of contact made with hand (front or back) or fingertips with another individual, without appreciable force |
| hand lead | taking the hand of another individual and bringing it into contact with his own body, but without sufficient force to move recipient's entire body |
| hard touch | any sort of contact made with hand (front or back) or fingertips with another individual, without appreciable force, but the actual contact itself is more forceful than a simple laying of the hand on another's body |
| hunchover* | one arm is swept over back of another individual, but there is no hugging or extended contact (less than two seconds) |
| pat | rapidly repeatedly contacting another individual with flattened palm surface of hand; not in play; repetitive |
| point* | either whole hand or one or more digits directed to recipient, another individual, or object |
| poke | pushing one more fingertips with sudden movement onto body part of another individual; repetitive |
| rap knuckles* | knuckles of one or both hands are rapped on ground or wall or object while looking at recipient; repetitive |
| reach out down∧ | holding a hand toward another individual by extending the arm, wrist, and hand in more or less horizontal position, and stretching the fingers while palm is facing downwards; other individual is not touched |
| reach out side | same as reach out down except the palm of the hand is directed sideways |
| reach out up∧ | same as reach out down except that the palm of the hand is directed upwards |
| shake wrist | shaking the hand vigorously with flexible wrist towards another individual; repetitive |
| slap ground* | flattened palm of hand is forcefully brought into contact with ground in front of self or on an inanimate fixture such as a wall or net |
| slapstomp | simultaneous slap ground and stomp |
| stomp∧ | hitting an object or ground with sole(s) of foot (feet); can be done with both feet in quick alternation |
| swing | arm is swung in an underhanded arch; can involve contact |
| throw aimed | over or underarm throw of object, including loose dirt, in forward direction while looking at target; not in play |
| throw hold | arm is raised above head, as if in a throw, but movement not carried out for at least two seconds (if at all) |

All labels and descriptions, except where noted, are based on Plooij (1984).

Those marked with * are descriptions based on personal observations and are similarly described in Nishida et al. (1999).

Those marked with ∧ are based on Plooij's (1984) descriptions but are labeled differently.

# Gestures as a Suitable Candidate for Language Evolution

Characteristics that we share with apes but not monkeys likely evolved recently. Hence, they may have provided a basis for the development of even more unique patterns found only in humans (de Waal 2003). In this context, the difference in gesture usage between apes and monkeys is highly relevant, and becomes even more intriguing if we consider that apes appear to possess greater control over the production of gestures versus other signals (Preuschoft and Chivers 1993, Wiesendanger 1999). This hypothesis is supported by several observations, and the case of cultural transmission of gestures is one example. Just as there are cultural variations of gestures in humans, population-specific communicative behaviors are also known to exist in chimpanzees, such as leaf-clipping (Nishida 1980) and handclasp grooming (McGrew and Tutin 1978, de Waal and Seres 1997, Bonnie and de Waal 2005). In chimpanzees and all other great apes species, manual gestures are more culture-specific than facial expressions, which tend to be relatively invariant. The tendency of cultural communication patterns to be nonfacial and nonvocal is probably due to the ape's limited control over face and voice. In humans too, facial expressions seem universal (Ekman 1972), whereas many gestures vary by culture (Kendon 1995).

The fact that apes appear to have greater cortical control over limb movements than vocalizations, is further supported by observations that while efforts to teach chimpanzees to modify their vocalizations have failed dismally (Hayes 1952), apes can learn to employ American Sign Language in a referential manner (Gardner et al. 1989). In fact, each of the species of great ape has been taught to communicate using visual and manual signals. Both chimpanzees and bonobos have learned to use a keyboard containing symbols, which they point to in sequence to deliver messages. Kanzi, a bonobo, spontaneously added gestures to this repertoire (Savage-Rumbaugh et al., 1998).

Greater control over gestures than other signals is also suggested by observations of deception, in which apes may use their hands to modify a facial expression (de Waal, 1982) or a vocalization. Goodall (1986) reported how a chimpanzee attempted to muffle his excited pant-hoot, signaling the discovery of food, by covering his mouth with his hand, presumably in an attempt to keep the food to himself. Finally, monkeys also seem to have great difficulty producing vocal signals in the absence of a triggering situation (Goodall 1986). This is no doubt why in so-called ape language studies, the forelimbs have proven a more promising candidate for intentional communication.

These observations in conjunction with one another support the gestural hypothesis about human language evolution, which is further bolstered by the theory that the early human brain was capable of producing language before the vocal chords (Lieberman et al. 1972), the early appearance of gestural communication in human infants (Petitto and Marentette 1991), and the right-hand (hence left-brain) bias of both human and ape gestures.

# Towards a More Flexible Communication Strategy: Contextually Defined Meaning

Whereas monkeys possess a rich repertoire of communicative signals, some with a demonstrable degree of referentiality (e.g., Seyfarth et al. 1980, Gouzoules et al. 1984, Zuberbühler 2000), they are by-and-large fixed signals with regards to emotional and/or social context. Ape gestures seem quite different: a single gesture may communicate entirely different needs or intentions depending on the social context in which it is used. Unlike the majority of facial expressions and vocalizations, manual gestures are more flexible (Tomasello et al. 1985, 1989, 1994, 1997) in the sense that they can be divorced from highly arousing contexts.

Because many gestures do not seem tied to a specific social situation, there is a great deal of equipotentiality in these communicative signals, and we don't really understand how they acquire meaning (in the absence of other discrete signals such as facial expressions and vocalizations). In the case of apes, for example, the begging gesture has absolutely no meaning unless one can deduce its referent from the context. For instance, a chimpanzee stretching out an open hand toward a third party during a fight signals a need for support, whereas the same gesture towards a possessor of food likely signals a desire for a share (de Waal and van Hooff 1981).

Given this distinction, we set out to test the hypothesis that gestures are less tightly tied to behavioral contexts than facial or vocal signals. Calculating the percentage that each communicative signal, be it gesture, facial expression, or vocalization, occurred in the context in which it was produced with the highest frequency, Pollick and de Waal (2007) found that, as a group, gestures showed far looser contextual associations than facial or vocal signals. Gestures also showed far greater contextual variation than facial and vocal displays both between bonobos and chimpanzees, and between groups within each species. Thus, knowing the usage of a facial/vocal display in one species allows one to predict how it will be used the other species, whereas knowing the usage of a gesture in one species does not allow one to predict how the other species uses it, and sometimes not even how other members of the same species use it in other groups. For example, the facial expression of silent bared teeth and the vocalization scream were almost always produced in agonistic contexts in both ape species, yet the arm raise gesture was used mostly in play in bonobos, but in chimpanzees it was used mostly to solicit grooming.

This suggests that the meaning of, for example, a gentle touch is informed by other signals as well as by the situation, and that individuals need to interpret these manual actions in light of the behavioral context (Goodall 1968, de Waal and van Hooff 1981). The flexibility of this class of signals suggests that gestural communication may have been one through which symbolic meaning was acquired in our hominid ancestors, alongside referential vocalizations (Corballis 2002, Pollick and de Waal 2004).

## Multimodal Communication

The production and perception of communicative signals such as vocalizations, gestures, and facial expressions generally do not occur in isolation, but instead occur more often in combinations.

Different modes of communicative signals such as facial expression, gesture, body posture, head movement, touch, and vocalization often work together in a multimodal strategy that is common in humans and other animals. It may be that gestures combined with other signals have different effects than either have on their own. Researchers are becoming increasingly aware that a deeper understanding of the evolution of communication must be based on comparative studies of vocal as well as other communicative abilities, but also of how the signals work in concert to convey information. Multimodal communication may have been the springboard for the evolution of the almost infinite flexibility of human language (Rizzolatti and Arbib 1998, Corballis 2002).

Although the bulk of the animal signaling data concentrates on signals sent via a single sensory modality, multimodal signaling is quite common. Researchers have long understood the importance of multimodal signals (Møller and Pomiankowski 1993, Johnstone and Grafen 1993, Partan and Marler 1999), and the majority of the data has been collected just over the past decade or so. Multimodal signaling occurs across taxa, from snapping shrimp to spiders to birds, and in many different contexts, though ones involving courtship and mating are the best documented (Pollick 2003).

There are many ways to characterize multimodal signaling, from simply documenting which modalities are involved, to describing intricate temporal patterns of the signals. Of course, bonobos employ a battery of communicative signals, including head movement, posture, and gaze, among others. Here, we talk only about three of the more distinguishable and easily observed signals: manual gestures, facial expressions, and vocalizations. How the patterns differ from those of chimpanzees serves as interesting contrast.

Both facial and vocal signals were equally likely to occur in the bonobo combinations, whereas vocalizations were much more prevalent in chimpanzee combinations: 50% of bonobos and 66% of chimpanzees. Chimpanzees aren't necessarily more vocal than bonobos; the discrepancy is likely the result of more combinations in agonistic situations in chimpanzees, which usually involve much vocalizing. Within a combination, the facial or vocal signal tended to occur first, just before the gesture, which was also true of chimpanzees (cf. van Hooff 1973). It may be that the facial/vocal signals are more uninhibited, highly arousing, and tied to specific contexts, and perhaps the subsequent gesture informs or emphasizes the meaning of the first signal in a more cognitive or deliberate manner.

A multimodal signaling strategy can serve a variety of functions, including redundancy, amplification, and modulation (Partan and Marler 1999). Whatever the exact function of multimodal communication, it is clear that ape gestural flexibility, combined with their graded facial/vocal signal system (Parr et al. 2005), may be

**Fig. 4.3** Multimodal communication: a bonobo gestures and vocalizes simultaneously (Photograph by Frans B.M. de Waal).

advantageous over the more stereotyped signals of monkeys in that it allows for greater communicative complexity. One specific benefit of multimodal signaling may be its effectiveness in altering the recipient's behavior. Pollick and de Waal (2007) found that combinations of gestures and facial or vocal signals in bonobos were significantly more effective in getting the recipient to respond (defined as any change in overt behavior shown within 10 seconds of the signal). Although chimpanzees produce more combinations than bonobos, they seem to be less effective in getting the receiver to respond. Possibly, the relative scarcity of combinations in bonobos renders them more salient and more likely to affect behavior.

## Directions for Future Research

We have given a broad overview of bonobo and chimpanzee gesturing, but much remains to be studied, starting with multimodal signaling. We discussed data on facial and vocal signals accompanying gestures, but there are other possible combinations; for example, a gesture could not only be accompanied by a vocalization, it could also be followed by a full-body bow or a head nod. These other kinesic movements have been shown to modulate meaning in humans (Kita 2003). It remains to be seen, however, if those signals are as meaningfully or consistently combined with gesture or other signals, in ways that affect receiver behavior.

Another way to expand on the data discussed here involves the nature of combinations during an interaction. There was a clear temporal pattern in the combinations data in this study, but the timing of combinations may be very different during an ongoing interaction, especially when repetitions or sequences are considered. The latter factors may have a different impact on the receiver's response and the nature of the entire interaction. We employed a narrow lens with which to view combinatorial signaling (looking only at signals which initiated an interaction), so alternate or more encompassing methods need to be tried. If signals during ongoing interactions are studied, alternate criteria for discerning responses would also be necessary. Perhaps instead of looking at the immediate behavior subsequent to the signal, the larger behavioral state change could be analyzed over a broader timeframe, such as 10 minutes following the signal.

Bonobos have individual, complex social relationships with one another, and their behavioral flexibility allows for very different ways of interacting with and signaling within particular dyads. For example, a high-ranking female may have a better chance of getting a male to react to her with a single gesture than a low-ranking female does with a combination, or an adolescent male is more likely to react to his mother than he is to an unrelated female. So it is not about the efficacy of the signals themselves so much as it may be about the relationship between the two participants, since so little of their gestural communication is stereotyped. Ideally, individual baseline rates of signaling and responding need to be established for every possible pair of social partners being studied, which would then be used to compare combinations in each dyad. This way, on an individual or dyadic basis, true differences in response can be determined.

Variations on what is considered effectiveness need to be explored. Because there were relatively few combinations on which to conduct analysis, all responses had to be pooled into a dichotomous "response/no response" category. These responses ranged from positive reactions, such as engaging in sex or sharing food, to negative ones, such as direct aggression or fleeing (or more broadly, appropriate versus inappropriate to context). These data can speak only to how responsive individuals are to different signal strategies, not necessarily how effective the strategies are at communicating a specific message. More data and analyses are needed on the quality of the response and how it varies as a function of the type of gesture and combination used; in this way, we will get closer to the issue of meaning.

Acoustic analysis comparing vocalizations produced alone and those produced in combinations might reveal other aspects of efficacy. Recent work on chimpanzee vocalizations (Slocombe and Zuberbühler 2005) has revealed a level of complexity and possible referentiality heretofore unexplored in apes, and much remains to be investigated in these signals within combinatorial strategies.

Finally, there is a great need for experimental work on the perception and classification of gesture. Description of how signals work behaviorally and how they visibly affect receivers' actions is a necessary component of our understanding of communication, but there is a growing awareness that animals may not perceive, and more importantly, may not categorize their signals the way we do (Evans 1997). Studies have shown that chimpanzees classify their own communicative signals (largely facial expressions) and can do so according to their emotional meaning (Parr 2001). It is completely unknown whether great apes categorize their gestures in the same way. It would also be interesting to see if they associate particular gestures with social contexts, such that if presented with a choice between a bent wrist and a reach out down after viewing a fight, would they always pick the bent wrist, or if supplementation with another signal affects that choice.

## Conclusions

Manual gestures play a significant role in bonobo communication. The flexible nature of these gestures is highlighted by the fact that facial/vocal signals correlated to a much higher degree with regard to contextual usage than did gestures. This flexibility is all the more striking when we consider the fact that apes gesture and monkeys do not. Monkeys possess a rich repertoire of communicative signals, some of which have been demonstrated to contain the seeds of referentiality (Seyfarth et al. 1980, Zuberbühler 2000), but they are, generally, signals bound to specific emotional and/or social contexts (with the exception of deceptive use). Manual gestures, on the other hand, have been repeatedly shown to be flexible signals that can be divorced from highly arousing contexts. Thus, there is a great deal of equipotentiality in gestural signals, and how they acquire meaning (in the absence of other discrete signals such as facial expressions and vocalizations) remains to be investigated.

Few studies have attempted to observe multimodal signaling in ape communication, and the data discussed here examined how gestures are combined with facial and vocal signals. In this study, facial expressions and vocalizations constituted the other half of a combination at equal rates, while bonobos combined their gestures with vocalizations less often than with facial expressions than did chimpanzees. This may be due to the overall prevalence of vocal activity in chimpanzees (de Waal 1988), but it may also concern the issue of control. If bonobos can better regulate their vocal output and divorce them from highly arousing contexts, it is not necessarily the case that this will happen at the start of social interactions. We observed much vocal "chattering" among the bonobos, by which we mean vocalizing (to each other or to humans) to garner attention in the absence of excitement, low-intensity vocalizing in the presence of food, but not in a chorus-like manner the way chimpanzees do, low-intensity vocalizing when traveling, and dialogue-like vocalizing in alarm situations. This greatly contrasted with the vocal output of chimpanzees, which was mostly restricted to highly charged situations such as aggression or food anticipation. Thus, greater control over this modality may not necessarily translate to greater production in general or greater use in initiating social interactions.

When combinations occurred in both species, the two signals tended to overlap rather than occur separately in time. Combinations did occur in highly charged situations, such as fights and reconciliations, but this was not always the case. The overlap, therefore, was not necessarily due to lack of inhibition during emotionally charged situations. It may be that, as has been theorized for humans, there is a common cognitive underpinning for both signals (McNeill 1992). In humans, this underpinning, what McNeill calls the "growth point," represents the initial form of a thought that is eventually expressed in two modalities. The growth point's defining characteristic is the tight co-expression of gesture and speech. While this pattern was not consistently observed in the apes, there were several notable instances in which the strokes of a repetitive gesture were closely matched by pauses in the accompanying vocalization. This tight synchrony is possible evidence of a shared neural space from which symbolic communication evolved (Cantalupo and Hopkins 2001, Corballis 2002).

Across all signals and contexts, combinations of gestures and facial/vocal signals were more effective at eliciting a response than gestures alone. This supports the bulk of the multimodal signaling literature across animal taxa (Møller and Pomiankowski 1993, Partan 2002), in that multimodal signaling has a differential effect, whether that effect is simply a response at all, an enhanced response, or a compound response (Partan and Marler 1999). This was true only for the bonobos, however, and not the chimpanzees, which is interesting given that combinations occur less frequently in the bonobos. It may be that the relative scarcity of combinations renders them more salient and more likely to affect behavior, whereas the relative ubiquity of combinations in chimpanzees is associated with a lower rate of response. This held true even when combinations were broken down into specific contexts.

Environmental noise, however, can affect the efficacy of signaling, and it often exerts evolutionary pressure on the signals themselves (Brown and Waser 1988). For example, if a species typically lives in a heavily forested area and in a fission-fusion society, communication will evolve to overcome these barriers and rely more heavily on vocal rather than visual signals (Brown and Waser 1988). There are some differences in the wild habitats of bonobos and chimpanzees: bonobos live in more humid forest while chimpanzees occupy a drier forest (Kano 1992), but these differences do not seem to dramatically differ with respect to visual or vocal barriers. One intriguing difference, however, lies in the fact that chimpanzees spend considerably more time foraging for food than bonobos do (the latter species' environment is more abundant in fruit; Kano 1992), and thus may spend more time communicating about food. This greatly contrasts with how signals are used in captive apes, and remains an important difference in comparing wild and captive ape communication.

Kano (1992) suggested that the characteristics of bonobos are more original and closer to those of the common ancestor, having retained a larger number of ancestral genes due to a slower rate of selection. The habitat of bonobos likely resembles our shared ancestor's; thus, Kano suggests, we should look to them as a model of the physical and behavioral characteristics of the common ancestor of the African great apes and humans. From that perspective, the bonobos' more flexible gestural repertoire and greater responsiveness to combinatorial signaling may be characteristic of the communicative repertoire of the early hominoid lineage, and perhaps of our direct ancestors as well.

# References

Annett M (1985) Left, right, hand and brain: the right shift theory. Erlbaum Associates, London

Bermejo M, Omedes A (1999) Preliminary vocal repertoire and vocal communication of wild bonobos (*Pan paniscus*) at Lilungu (Democratic Republic of Congo). Folia Primatol 70: 328–357

Bonnie KE, de Waal FBM (2005) Affiliation promotes the transmission of a social custom: hand-clasp grooming among captive chimpanzees. Primates 47: 27–34

Bretherton I, Bates E (1979) The emergence of intentional communication. In: Uzgiris IC (ed) Social interactions and communication during infancy. Jossey-Bass, San Francisco, pp 81–100

Brown CH, Waser PM (1988) Environmental influences on the structure of primate vocalizations. In: Todt D, Goedeking P, Symmes D (eds) Primate vocal communication. Springer Verlag, Berlin, pp 51–66

Cantalupo C, Hopkins WD (2001) Asymmetric Broca's area in great apes: a region of the ape brain is uncannily similar to one linked with speech in humans. Nature 414: 505

Cheney DL, Seyfarth RM (1990) How monkeys see the world. University of Chicago Press, Chicago

Christiansen MH, Kirby S (2003) Language evolution: the hardest problem in science? In: Christiansen MH, Kirby S (eds) Language evolution. Oxford University Press, Oxford, pp 1–15

Corballis MC (1999) The gestural origins of language. American Scientist 87: 38–145

Corballis MC (2002) From hand to mouth: the origins of language. Princeton University Press, Princeton NJ

Corballis MC (2003) From mouth to hand: gesture, speech, and the evolution of right-handedness. Behav Brain Sci 26: 199–260

Custance DM, Whiten A, Bard KA (1995) Can young chimpanzees (*Pan troglodytes*) imitate arbitrary actions – Hayes and Hayes (1952) revisited. Behavior 132: 837–859

Efron D (1941) Gesture, race, and culture. Mouton, The Hague

Eibl-Eibesfeldt I (1973) The expressive behaviour of the dead-and-blind born. In: Cranach M, Vine I (eds) Social communication and movement. Academic Press, New York, pp 163–194

Ekman P (1972) Universals and cultural differences in facial expressions of emotion. In: Cole J (ed) Nebraska Symposium on motivation, 1971. University of Nebraska Press, Lincoln NE, pp 207–283

Evans CS (1997) Referential signals. In: Owings, D, Beecher, M, Thompson N (eds) Perspectives in ethology. Plenum, New York, pp 99–143

Gardner RA, Gardner BT, van Cantfort TE (1989) Teaching sign language to chimpanzees. State University of New York Press, Albany NY

Goldin-Meadow S (2001) Giving the mind a hand: the role of gesture in cognitive change. In: McClelland JL, Siegler RS (eds) Mechanisms of cognitive development: behavioral and neural perspectives. Erlbaum, Mahwah NJ, pp 5–31

Goldin-Meadow S, Mylander C (1984) Gestural communication in deaf children: the effects and noneffects of parental input on early language development. Monographs of the Society for Research in Child Development 49: 207

Goldin-Meadow S, Wagner SM (2005) How our hands help us learn. Trends Cognit Sci 9: 234–241

Goldin-Meadow S, Nusbaum H, Kelly SD, Wagner S (2001) Explaining math: gesturing lightens the load. Psychol Sci 12: 516–522

Gomez J (1990) The emergence of intentional communication as a problem-solving strategy in the gorilla. In: Parker ST, Gibson KR (eds) "Language" and intelligence in monkeys and apes. Cambridge University Press, Cambridge UK, pp 333–355

Goodall J (1986) The chimpanzees of Gombe: patterns of behavior. The Belknap Press of Harvard University, Cambridge MA

Goodall J, van Lawick (1968) The behaviour of free-living chimpanzees in the Gombe Stream Reserve. Anim Behav Monographs 1: 161–311

Gouzoules H, Gouzoules S, Marler P (1984) Rhesus monkey (Macaca mulatta) screams – representational signaling in the recruitment of agonistic aid. Anim Behav 32: 182–193

Hayes C (1952) The ape in our house. Gollancz, London

Hewes GW (1973) Primate communication and the gestural origins of language. Curr Anthropol 14: 5–24

Hohmann G, Fruth B (1994) Structure and use of distance calls in wild bonobos (pan paniscus). Int J Primatol 15: 767–782

van Hooff JARAM (1972) A comparative approach to the phylogeny of laughter and smiling. In: Hinde RA (ed) Non-verbal communication. Cambridge University Press, Cambridge, pp 209–241

van Hooff JARAM (1973) A structural analysis of the social behavior of a semi-captive group of chimpanzees. In: von Cranach M, Vine I (eds) Social communication and movement. Academic Press, London & New York, pp 75–162

Hopkins WD, Leavens DA (1998) Hand use and gestural communication in chimpanzees (*Pan troglodytes*). J Comp Psychol 112: 95–99

Hopkins WD, Morris RD (1993) Handedness in great apes: a review of findings. Int J Primat 14: 1–25

Hopkins WD, de Waal FBM (1995) Behavioral laterality in captive bonobos (*Pan paniscus*): replication and extension. Int J Primat 16: 261–276

Iverson JM, Goldin-Meadow S (1998) Why people gesture when they speak. Nature 396: 228

Jackendoff R (2002) Commentary in Eakin E, some language experts think humans spoke first with gestures. New York Times, Section B p 7, col 5

Johnstone RA, Grafen A (1993) Dishonesty and the handicap principle. Animal Behaviour 46: 759–764

Kano T (1992) The last ape: pygmy chimpanzee behavior and ecology. Stanford University Press, Stanford CA

Kendon A (1972) Some relationships between body motion and speech. In: Siegman AW, Pope B (eds) Studies in dyadic communication. Pergamon Press, New York, pp 177–210

Kendon A (1980) Gesticulation and speech: two aspects of the process. In: Key MR (ed) The relation between verbal and nonverbal communication. Mouton, The Hague, pp 207–228

Kendon A (1995) Gestures as illocutionary and discourse structure markers in Southern Italian conversation. Journal of Pragmatics 23: 247–279

Kita S (2003) Pointing: where language, culture, and cognition meet. Lawrence Erlbaum, Mahwah, NJ

Kuroda S (1984) Rocking gesture as communicative behavior in the wild pygmy chimpanzees in Wamba, central Zaire. J Ethnology 2: 127–137

Ladygina-Kohts NN (2002/1935) Infant chimpanzee and human child. Oxford University Press, New York

Lieberman P, Crelin ES, Klatt DH (1972) Phonetic ability and related anatomy of the newborn and adult human, Neanderthal man, and the chimpanzee. Amer Anthropol 74: 287–307

Marler P (1965) Communication in monkeys and apes. In: DeVore I (ed) Primate behavior: field studies of monkeys and apes. Holt, Rinehart and Winston, New York, pp 544–584

Marler P (1976) Social organization, communication, and graded signals: the chimpanzee and the gorilla. In: Bateson PP, Hinde RA (eds) Growing points in ethology. Cambridge University Press, London, pp 239–279

McGrew WC, Tutin CEG (1978) Evidence for a social custom in wild chimpanzees? Man 13: 234–251

McNeill D (1985) So you think gestures are nonverbal? Psychol Rev 92: 350–371

McNeill D (1992) Hand and mind: what gestures reveal about thought. University of Chicago Press, Chicago

McNeill D, Bertenthal B, Cole J, Gallagher S (2005) Gesture-first, but no gestures? Behav Brain Sci 28: 138–139

Miles HL (1990) The cognitive foundations for reference in a signing orangutan. In: Parker ST, Gibson KR (eds) "Language" and intelligence in monkeys and apes: comparative developmental perspectives. Cambridge University Press, New York, pp 511–539

Møller AP, Pomiankowski A (1993) Why have birds got multiple sexual ornaments? Behav Ecol Sociobiol 32: 167–176

Morrel-Samuels P, Krauss R (1992) Word familiarity predicts temporal asynchrony of hand gestures and speech. J Exp Psychol Learn Mem Cognit 18: 615–622

Morris D (1977) Manwatching. Harry N Abrams Inc, New York

Morris D (1994) The human animal: the language of the body. In: The human animal, Discovery Channel Productions

Nishida T (1980) The leaf-clipping display: a newly discovered expressive gesture in wild chimpanzees. J Hum Evol 9: 117–128

Parr LA (2001) Cognitive and physiological markers of emotional awareness in chimpanzees (*Pan troglodytes*). Anim Cognit 4: 223–229

Parr LA, Cohen M, de Waal FBM (2005) Influence of social context on the use of blended and graded facial displays in chimpanzees. Int J Primat 26: 73–103

Partan SR (2002) Single and multichannel signal composition: facial expressions and vocalizations of rhesus macaques (Macaca mulatta). Behaviour 139: 993–1027

Partan SR, Marler P (1999) Communication goes multimodal. Science 283: 1272–1273

Patterson F (1979) Linguistic capabilities of a lowland gorilla. In: Schiefelbusch RL, Hollis J (eds) Language intervention from ape to child. University Park Press, Baltimore, p 325–356

Perrett D, Smith PAJ, Mistlin AJ, Chitty AJ, Head AS, Potter DD, Broenniman R, Milner AP, Jeeves MA (1985) Visual analysis of body movements by neurons in the temporal cortex of the macaque monkey. Behav Brain Res 16: 153–170

Petitto LA, Marentette P (1991) Babbling in the manual mode: evidence for the ontogeny of language. Science 251:1483–1496

Pika S, Liebal K, Tomasello M (2003) Gestural communication in young gorillas (Gorilla gorilla): gestural repertoire, learning, and use. Am J of Primat 60: 95–111

Pika S, Liebal K, Tomasello M (2005) Gestural communication in subadult bonobos (*Pan paniscus*): repertoire and use. Am J of Primat 65: 39–61

Plooij FX (1978) Some basic traits of language in wild chimpanzees. In: Lock A (ed) Action, gesture and symbol: the emergence of language. Academic Press, London & New York, pp 111–131

Plooij FX (1984) The behavioral development of free-living chimpanzee babies and infants. Ablex, Norwood NJ

Pollick AS (2003) Signalling theory and multimodal communication: a theoretical and empirical review. Unpublished manuscript

Pollick AS (2006). Gestures and multimodal signaling in bonobos and chimpanzees. Ph.D. Dissertation, Emory University. ISBN 0-542-61523-1

Pollick AS, de Waal FBM (2004) Gestural gab: review of "from hand to mouth: the origins of language". Am J Psychol 117: 124–129

Pollick AS, de Waal FBM (2007) Ape gestures and language evolution. Proc Nat Acad Sci 104:8184–8189

Preuschoft H, Chivers DJ (1993) Hands of primates. Springer-Verlag/Wien, Slovenia

Rizzolatti G, Arbib MA (1998) Language within our grasp. Trends Neurosci 21: 188–194

Rizzolatti G, Fadiga L, Matelli M, Bettinardi V, Perani D, Fazio F (1996) Localization of grasp representations in humans by positron emission tomography: 1. Observation versus execution. Exp Brain Res 111: 246–252

Savage-Rumbaugh ES, Wilkerson BJ, Bakeman R (1977) Spontaneous gestural communication among conspecifics in the pygmy chimpanzee (Pan paniscus). In: Bourne GH (ed) Progress in ape research. Academic Press, New York, pp 97–116

Savage-Rumbaugh S, Shanker S, Taylor TJ (1998) Apes, language, and the human mind. Oxford University Press, New York

Seyfarth RM, Cheney DL, Marler P (1980) Monkey responses to three different alarm calls: evidence for predator classification and semantic communication. Science 210: 801–803

Slocombe KE, Zuberbühler K (2005) Functionally referential communication in a chimpanzee. Curr Biol 19: 1779–1784

Tanner JE (2004) Gestural phrases and gestural exchanges by a pair of zoo-living lowland gorillas. Gesture 4: 1–24

Tanner JE, Byrne RW (1996) Representation of action through iconic gesture in a captive lowland gorilla. Curr Anthropol 37: 162–173

Tanner JE, Byrne RW (1999) The development of spontaneous gestural communication in a group of zoo-living lowland gorillas. In: Parker ST, Mitchell RW, Miles HL (eds) The mentalities of gorillas and orangutans: comparative perspectives. Cambridge University Press, New York, pp 211–239

Tomasello M, Call J (1997) Primate cognition. Oxford University Press, New York

Tomasello M, George BL, Kruger AC, Farrar MJ, Evans A (1985) The development of gestural communication in young chimpanzees. J Hum Evol 14: 175–186

Tomasello M, Gust D, Frost GT (1989) A longitudinal investigation of gestural communication in young chimpanzees. Primates 30: 35–50

Tomasello M, Call J, Nagell K, Olguin R, Carpenter M (1994) The learning and use of gestural signals by young chimpanzees: a trans-generational study. Primates 35: 137–154

Tomasello M, Call J, Warren J, Frost JT, Carpenter M, Nagell K (1997) The ontogeny of chimpanzee gestural signals: a comparison across groups and generations. Evol Comm 1: 223–59

Veà JJ, Sabater-Pi J (1998) Spontaneous pointing behaviour in the wild pygmy chimpanzee (Pan paniscus). Folia Primatol 69: 289–290

de Waal FBM (1982) Chimpanzee politics: power and sex among apes. Jonathan Cape, London

de Waal FBM (1988) The communicative repertoire of captive bonobos (Pan paniscus) compared to that of chimpanzees. Behaviour 106: 183–251

de Waal FBM (1997) Attitudinal reciprocity in food sharing among brown capuchin monkeys. Anim Behav 60: 253–261

de Waal FBM (2003) Darwin's legacy and the study of primate visual communication. In: Ekman P, Campos JJ, Davidson RJ, de Waal FBM (eds) Emotions inside out: 130 years after darwin's the expression of the emotions in man and animals. New York Academy of Sciences, New York, pp 7–31

de Waal FBM, van Hooff JARAM (1981) Side-directed communication and agonistic interactions in chimpanzees. Behaviour 77: 164–198

de Waal FBM, Seres M (1997) Propagation of handclasp grooming among captive chimpanzees. Am J of Primat 43: 339–346

Wiesendanger M (1999) Manual dexterity and the making of tools – an introduction from an evolutionary perspective. Exp Brain Res 128: 1–5

Zuberbühler K (2000) Referential labeling in Diana monkeys. Behavior 59: 917–927

# Part II
# Ecological Study Section

# Foreword to Ecological Study Section

Richard Wrangham[1]

Bonobos and chimpanzees look so similar that when Robert Yerkes studied a juvenile of each species in the 1920s, he was famously unaware that they belonged to different taxa. Science's early failure to recognize the division of *Pan* into two major groups is understandable since westerners saw few bonobos until the second half of the 20th century. In many ways, their behavior and anatomy was chimpanzee-like, such as in locomotion, diet and use of gestures. It took a careful examination of skulls from populations on either side of the Congo River to establish the biological differentiation of *Pan* into two species: bonobos to the south, chimpanzees to the north.

The tidy geographical separation of the two species makes their ecology a rich but still inadequately explored area for students of ape behavioral evolution. In particular, a question that continues to haunt our understanding is to what extent do the reported differences in behavior between bonobos and chimpanzees result from differences in their evolutionary ecology.

In theory the biological distinctions between the two species might be meaningless outcomes of geographical separation. Such a process is known in various primates where closely related species or subspecies straddle a river, as bonobos and chimpanzees do. Being separated merely by an erratic line of water, such sister taxa typically live in essentially the same type of habitat as each other, and as expected, therefore, behave in very similar ways. Yet they can be taxonomically distinct thanks to differences in traits such as coat-color or vocalizations. Several species of *Callitrichidae* and *Cercopithecus* illustrate this system, which is attributable to drift rather than to selection for adaptive traits (Ayres and Clutton-Brock 1992).

By analogy, therefore, an obvious possibility for the influence of ecology on bonobos is that their few anatomical differences from chimpanzees might be the result of drift rather than adaptation to a specific environment. If so, species' differences in behavior should be biologically unimportant. Stanford (1998) implicitly advocated a version of this hypothesis in a provocative paper suggesting that the two species differed less in behavior than generally thought. However, even in

---

[1] *Department of Anthropology, Harvard University, Peabody Museum 50B, 11 Divinity Avenue, Cambridge, Massachusetts, 02138 United States*

T. Furuichi and J. Thompson (eds.), *The Bonobos: Behavior, Ecology, and Conservation*    97
© Springer 2008

1998 there was substantial evidence from the wild that bonobos had different types of social relationships from chimpanzees. The evidence was particularly strong from Wamba, the longest-running and most influential source of data on wild bonobos. The Wamba studies revealed that bonobos differed from chimpanzees in patterns of aggression, sexual behavior, cooperation and tool-use. They even differed in grouping. Wamba bonobos traveled in larger and more stable parties than chimpanzees, with females being especially gregarious compared to chimpanzees (Kano 1992).

The significance of the Wamba differences from chimpanzees has in some cases been supported by other research in the wild. For example, female bonobos have also been shown to be relatively gregarious in Lomako. In other cases, however, the degree to which the Wamba data are representative of bonobos in general has been questioned. For example, at Lomako, bonobos tended to travel in smaller, less stable and more chimpanzee-like parties than those at Wamba (White 1988). This raised the possibility that the Wamba study might have represented an unusual site, such as a population adapted to an environment particularly rich in large herbs. Alternatively, perhaps the Wamba researchers' technique of provisioning bonobos with sugar-cane could have biased their results. It is critical that such questions be settled so that the significance of the Wamba research data can be properly evaluated. Several of the chapters in this section address these kinds of problems while also reviewing the state of knowledge of bonobo ecology and behavior.

In Chapter 5, Hashimoto et al. assemble the most complete picture to date of the pattern of intergroup transfer. Their study includes both observations before 1996 (when research was temporarily closed due to insecurity), and an exceptional set of subsequent events following the killing of many bonobos in neighboring groups. By combining genetic analyses with long-term observation, Hashimoto et al. confirmed that at Wamba, males tended to live in the groups where they were born, while nulliparous females transferred to breed elsewhere. Remarkably, however, after two or possibly three unit-groups (also known as communities) were eliminated by hunters, there were some cases of immigration into the study groups both by mothers and males. Hashimoto et al. suggest that such cases reflect a higher level of tolerance in bonobos than found in chimpanzees. Their chapter thus confirms the essentially chimpanzee-like community structure of the Wamba bonobos, while adding dramatic new evidence of social flexibility in response to novel circumstances.

In Chapter 6, Mulavwa et al. revisit the question of the impact of provisioning. To check whether the large subgroups previously reported at Wamba could have been affected by the fact that the bonobos were being provisioned at the time, Mulavwa et al. present new data from a period with no provisioning. They found essentially the same results as before. Once again, female bonobos emerge as more gregarious than chimpanzees, regardless of seasonal changes in fruit abundance. Their chapter thus provides important validation of the Wamba patterns of grouping. Both Chapters 5 and 6 in different ways support and enrich the notion that bonobos are behaviorally distinct from chimpanzees.

In Chapter 7, Furuichi et al. use Wamba data to try to solve a problem arising from the species differences in gregariousness. Previous hypotheses have suggested that scramble competition (the rate at which individuals experience reduced access to food merely because other individuals in the same group are feeding also) is less intense for bonobos than for chimpanzees. In the most direct test yet, Furuichi et al. support this by finding, in contrast to chimpanzees, no evidence that bonobos suffered reduced feeding success by being in large parties. Why this should be, however, remains a delightful puzzle. Furuichi et al. note that the distribution of plant resources looks very similar between chimpanzee and bonobo habitats. The answer that I find appealing is that the absence of gorillas in the geographical range occupied by bonobos is responsible for increasing the availability of edible herbs and might thereby sufficiently reduce scramble competition for bonobos as to cause many of the remarkable behavioral differences from chimpanzees (Wrangham and Peterson 1996, cf. Yamakoshi 2004). But the data are not yet available to decide among this and other ideas. As Chapters 5–7 show, bonobo social ecology is still at an exciting exploratory phase.

The same is true of the bonobos' population ecology. It has long been reported that population densities vary widely, but the habitat preferences and even the geographical range of bonobos are known only poorly. For those reasons, the last three chapters in this section present an important advance in our knowledge. Chapters 8, 9 and 10 all report on efforts to understand where bonobos live. All rely on the same technique: inferring densities from the distribution of bonobo nests.

In Chapter 8, Mohneke and Fruth describe their studies on nest density around Lui Kotal, in the southwestern sector of Salonga National Park. In this relatively small area (around 100–200 sq km), they were able to document in detail the effect of variability in nest-decay rates on the resulting estimate of bonobo density. Mohneke and Fruth produce valuable cautionary advice about the choice of techniques for nest surveys.

In Chapter 9, Reinartz et al. use such techniques to assess how bonobo density is related to habitat type across a wider area of Salonga. They suggest that previous classifications of Congolese forests have not captured the most relevant traits for bonobos. Instead of conventional classes like 'primary' and 'secondary,' Reinartz et al. used a two-way system to describe both trees and terrestrial herbs. They conclude that bonobos are indeed distributed patchily, partly because of their preference for mature dry forests with high edible herb density. Reinartz et al.'s chapter supports a traditional idea that bonobos are especially reliant on the presence of Marantaceae herbs, but also shows that the relationship is complex. Their chapter will be of particular interest to conservationists because they suggest that the optimal bonobo habitats can be recognized from satellite imagery. Their proposal could importantly accelerate the identification of key bonobo areas.

This would be immensely valuable given how little is still known. The recent state of ignorance is dramatically illustrated by Hart et al.'s statement in Chapter 10 that even in the 1980s, it was uncertain whether Salonga National Park contained any bonobos at all. Yet Hart et al. now estimate that this massive park, which was established in 1970 and covers 36,000 sq km, contains at least 8,000, and possibly as many as 28,500, bonobos – almost certainly the most important contiguous population of this threatened species.

Hart et al.'s chapter represents an astonishing achievement. Their survey teams walked 1869 km, and covered all major areas using a simple systematic procedure for recording nest densities. They also tested their methods by studying nest densities in detail in three smaller, but still relatively enormous, areas 2000–3000 sq km). Like the chapters by Mohneke and Fruth and Reinartz et al., Hart et al.'s results include large potential sources of error in their estimates. But by explicitly discussing this problem, all three chapters help advance the science of such estimation. Happily, the wide statistical range does not threaten Hart et al.'s major conclusions. They found that bonobo density is higher in the upland areas towards the east, lower in the black-water forests further west, and is often high near human occupation. Their findings will greatly aid the design of conservation strategies and give hope that such efforts will be worthwhile.

The big picture that emerges from Chapters 8, 9 and 10 is distinctly encouraging. The three chapters complement each other well by suggesting that different teams working in different areas of Salonga produce broadly comparable results. Thus, Mohneke and Fruth estimated a population density in the Lui Kotal area of 0.52–1.06 per sq km; in the rich Etate area Reinartz et al. found densities of 1–2 adults per sq km; and in the park as a whole, Hart et al. estimated overall densities between 0.23 and 0.80 per sq km. While these chapters present much material relevant to further refining comparisons among areas and research groups, a picture is now emerging of the habitat factors influencing bonobo densities across the landscape, as helpful for conservationists as for evolutionary ecology. For example, bonobos do not appear, as was once speculated, to be primarily swamp adapted. They are a species of mature forest that thrives with a dense herb layer.

This important conclusion does nothing to help solve the problem of why bonobo and chimpanzee sociality are in many ways different, given that chimpanzees also thrive in mature upland forests. But it reminds us of the essential challenge of explaining why, if bonobos are behaviorally distinct from chimpanzees, they have evolved to be different. The rapid growth of research revealed by Chapters 5–10 establishes a new level of confidence in our understanding of bonobo ecology.

# References

Ayres JM, Clutton-Brock TH (1992) River boundaries and species range size in Amazonian primates. Am Nat 140: 531–537

Kano T (1992) The Last Ape: Pygmy Chimpanzee Behavior and Ecology. Stanford Univ Press, Stanford

Stanford CB (1998) Social behavior of chimpanzees and bonobos: Empirical evidence and shifting assumptions. Current Anthropol 39: 399–420

White FJ (1988) Party composition and dynamics in *Pan paniscus*. Int J Primatol 9: 179–193

Wrangham RW, Peterson D (1996) Demonic Males: Apes and the Origins of Human Violence. Houghton Mifflin, Boston

Yamakoshi G (2004) Food seasonality and socioecology in *Pan*: are West African chimpanzees another bonobo? African Study Monogr 23: 43–60

# Avant-propos à la Section d'Etudes Ecologiques

Richard Wrangham[1]

Les bonobos et les chimpanzés se ressemblent tellement que lorsque dans les années 1920 Robert Yerkes étudia un spécimen juvénile de chacune des espèces, il est notoirement connu qu'il ignora totalement qu'ils appartenaient à des genres différents. L'incapacité initiale de la science à reconnaître la division de *Pan* en deux groupes majeurs s'explique du fait que les occidentaux ont vu peu de bonobos avant la deuxième moitié du 20ème siècle. Sous de nombreux aspects leur comportement et leur anatomie étaient semblables à ceux des chimpanzés, tout comme leur façon de se déplacer, leur régime alimentaire et leur gestuelle. Il a fallu un examen minutieux de crânes de populations issues des deux rives du fleuve Congo pour établir une différentiation biologique du genre *Pan* en deux espèces: les bonobos au sud et les chimpanzés au nord.

La séparation géographique nette des deux espèces fait de leur écologie un terrain riche mais encore insuffisamment exploré pour des étudiants s'intéressant à l'évolution comportementale des grands singes. Une question en particulier qui continue à hanter notre compréhension est de savoir jusqu'à quel point les différences rapportées entre le comportement des bonobos et celui des chimpanzés résultent des différences dans l'évolution de leur écologie.

En théorie, les distinctions biologiques entre les deux espèces pourraient être des conséquences insignifiantes de la séparation géographique. Un tel processus est connu chez d'autres primates où des espèces ou sous-espèces apparentées de façon proche se répartissent de part et d'autre d'une rivière, tout comme le font les bonobos et les chimpanzés. Etant simplement séparé par un cours d'eau, de tels genres apparentés vivent typiquement dans le même genre d'habitat les uns et les autres et, comme on s'y attendait, se comportent de ce fait de façon très similaire. Malgré cela, certains traits tels que la couleur du pelage ou les vocalisations peuvent les rendre taxonomiquement différents. Plusieurs espèces de Callitrichidae et de *Cercopithecus* illustrent ce système imputable à une dérive plutôt qu'à une sélection de traits adaptifs (Ayres and Clutton-Brock 1992). Par analogie, en ce qui concerne l'influence de l'écologie sur les bonobos, il est très possible que leur peu de différences anatomiques avec les chimpanzés soit le résultat d'une dérive plutôt que

---

[1]*Department of Anthropology, Harvard University, Peabody Museum 50B, 11 Divinity Avenue, Cambridge, Massachusetts, 02138 United States*

d'une adaptation à un environnement spécifique. Si tel est le cas, les différences comportementales entre les espèces devraient être biologiquement sans importance. Stanford (1998) a implicitement évoqué une version de cette hypothèse dans une publication provocatrice suggérant que, du point de vue comportemental, les deux espèces différaient moins qu'on le pensait. Néanmoins, il y avait encore en 1998 des preuves concrètes qu'à l'état sauvage, les bonobos avaient des types de relations sociales différentes de celles des chimpanzés. L'évidence était particulièrement claire à Wamba, la source de données la plus ancienne et la plus influente sur les bonobos sauvages. Les études menées à Wamba ont révélé que les bonobos différaient des chimpanzés dans les tendances agressives, le comportement sexuel, la coopération et l'utilisation d'outils. Ils différaient même dans leur façon de former des groupes. Les bonobos de Wamba se déplaçaient en groupes plus importants et plus stables que les chimpanzés, avec des femelles particulièrement grégaires comparées aux chimpanzés (Kano 1992).

L'importance des différences par rapport au chimpanzés relevées à Wamba a, dans certains cas, été confirmée par d'autres recherches sur le terrain. A Lomako, par exemple, les femelles bonobos ont également montré un comportement relativement grégaire Avec d'autres cas cependant, on s'est demandé jusqu'à quel point les données de Wamba étaient représentatives des bonobos en général. A Lomako, par exemple, les bonobos avaient tendance à se déplacer en groupes plus petits et moins stables (ressemblant donc plus aux groupes de chimpanzés) que ceux de Wamba, (White 1988). La possibilité que l'étude de Wamba ait représenté un site inhabituel, tel qu'une population adaptée à un environnement particulièrement riche en hautes herbes, a donc été émise. D'autre part, il est possible que l'approvisionnement des bonobos en cannes à sucre par les chercheurs à Wamba ait biaisé leurs résultats. Il est essentiel de répondre à ces questions afin que les données provenant de Wamba puissent être évaluées correctement. Plusieurs des chapitres de cette section abordent ce genre de problèmes tout en révisant le niveau de connaissance sur l'écologie et le comportement du bonobo.

Dans le chapitre 5, Hashimoto *et al.* nous présente l'image la plus complète des schémas de transfert entre groupes. Leur étude inclut les observations antérieures à 1996 (lorsque la recherche fut temporairement arrêtée pour cause d'insécurité) ainsi qu'une série exceptionnelle d'événements ultérieurs qui ont suivi le massacre de nombreux bonobos dans des groupes voisins. En combinant les analyses génétiques à des observations à long terme, Hashimoto *et al.* ont confirmé qu'à Wamba, les mâles ont tendance à vivre au sein du groupe dans lequel ils sont nés, alors que les femelles nullipares effectuaient des transferts pour aller se reproduire ailleurs. Cependant d'une façon remarquable, après que deux, ou peut être trois, groupes (aussi connus sous le nom de communautés) aient été décimés par les chasseurs, des cas d'immigration de mères ainsi que de mâles ont été observés dans les groupes étudiés. Hashimoto *et al.* ont suggéré que de pareils cas reflétaient un plus haut degré de tolérance chez les bonobos que chez les chimpanzés. Leur chapitre confirme donc une structure communautaire des bonobos de Wamba ressemblant à celle des chimpanzés, tout en ajoutant des nouvelles preuves dramatiques de flexibilité sociale en réponse à de nouvelles circonstances.

Dans le chapitre 6, Mulavwa *et al.* posent un nouveau regard sur l'impact de l'approvisionnement. Afin de vérifier si les grands sous-groupes signalés auparavant à Wamba auraient pu être affectés par le fait que les bonobos étaient approvisionnés à l'époque, Mulava *et al.* présentent de nouvelles données d'une période sans approvisionnement. Ils ont trouvé, pour l'essentiel, les mêmes résultats qu'auparavant. Une fois encore, les bonobos femelles apparaissent plus grégaires que les femelles chimpanzés, et cela peu importe les variations dans l'abondance en fruits en fonction des saisons. Leur chapitre apporte donc une confirmation importante sur la constitution des groupes à Wamba. Les chapitres 5 et 6, chacun à sa façon, soutiennent et enrichissent la notion que les bonobos sont distincts des chimpanzés du point de vue comportemental.

Dans le chapitre 7, Furuichi *et al.* utilisent les données de Wamba pour essayer de résoudre un problème issu de la différence de grégarité entre les espèces. Des hypothèses précédentes ont suggéré que la compétition sous forme de ruée (l'importance de la réduction de l'accès à la nourriture simplement parce que les autres individus du même groupe se nourrissent aussi) est moins intense chez les bonobos que chez les chimpanzés. Pourtant dans le test le plus direct, Furuichi *et al* soutiennent cette idée après avoir trouvé que, contrairement aux chimpanzés, il n'existe pas de preuves que les bonobos souffrent d'accès à la nourriture réduits du fait qu'ils vivent en larges groupes. La raison de cette observation reste une énigme intéressante. Furuishi *et al.* font remarquer que la distribution de ressources végétales est très similaire entre les habitats des chimpanzés et ceux des bonobos. J'aime l'explication selon laquelle l'absence de gorilles sur l'étendue géographique occupée par les bonobos serait responsable de la disponibilité plus élevée d'herbes comestibles et, de ce fait, réduirait suffisamment la compétition-ruée chez les bonobos pour être à l'origine de beaucoup des nombreuses différences remarquables de comportement avec les chimpanzés (Wrangham et Peterson 1996, cf. Yamakoshi 2004). Cela dit, les données n'étant pas encore disponibles, on ne peut décider entre cette idée ou d'autres. Comme le montrent les chapitres 5 à 7, l'écologie sociale du bonobo se trouve encore dans une excitante phase d'exploration.

Il en est de même en ce qui concerne leur écologie de population. Il a été dit depuis longtemps que les densités de population variaient grandement, mais nous ne possédons qu'une connaissance très pauvre des préférences au niveau de l'habitat ainsi que de l'aire de distribution géographique des bonobos. Pour ces raisons, les trois derniers chapitres de cette section représentent une avancée importante pour notre savoir. Les chapitres 8, 9 et 10 présentent tous trois les efforts fournis afin de comprendre où vivent les bonobos. Tous reposent sur la même technique: inférer des densités à partir de la distribution des nids de bonobos.

Dans le chapitre 8, Mohneke et Fruth décrivent leurs études sur la densité de nids autour de Lui Kotal, dans le secteur sud-ouest du Parc National de la Salonga. Sur cette surface relativement petite (environ 100–200 km²), ils ont réussi à documenter en détail l'effet de la vitesse de destruction des nids sur l'estimation résultante de densité des bonobos. Mohneke et Fruth donnent de précieux conseils de prudence sur le choix des techniques utilisées pour l'étude des nids.

Dans le Chapitre 9, Reinartz *et al.* utilisent de telles techniques pour démontrer comment la densité de bonobos est liée au type d'habitat sur une plus grande surface du Salonga. Ils suggèrent que les classifications précédentes des forêts congolaises n'ont pas retenu les caractéristiques les plus importantes pour les bonobos. Plutôt que les classifications conventionnelles « primaires » et « secondaires », Reinartz *et al.* utilisaient un système à double sens pour décrire aussi bien les arbres que les herbes terrestres. Ils ont conclu que les bonobos sont effectivement distribués de façon éparse, en partie du fait de leur préférence pour des forêts sèches matures à forte densité en herbes hautement comestibles. Le chapitre de Reinartz *et al.* soutient une idée traditionnelle selon laquelle les bonobos sont particulièrement dépendants de la présence d'herbes de la famille des Marantaceae mais montre également que la relation est complexe. Leur chapitre sera particulièrement intéressant pour les conservationnistes car ils suggèrent que les habitats optimaux pour les bonobos peuvent être reconnus à partir de l'imagerie provenant des satellites. Leur proposition pourrait accélérer de façon notoire l'identification des régions clés pour les bonobos.

Ceci serait immensément profitable vu le peu que l'on connaît. L'actuel niveau de notre ignorance est illustré de façon dramatique dans l'affirmation de Hart *et al.* dans le Chapitre 10 selon laquelle même dans les années 1980 on ne savait pas si le Parc National de la Salonga contenait un seul bonobo. Pourtant Hart *et al.* estiment aujourd'hui que ce parc gigantesque qui fut créé en 1970 et qui couvre 36.000 km$^2$ abrite au moins 8.000 et peut-être jusqu'à 28.500 bonobos, ce qui représente presque certainement la plus importante population contiguë de cette espèce menacée.

Le chapitre de Hart *et al.* représente un exploit étonnant. Leurs équipes ont parcouru 1869 km et ont couvert toutes les aires principales en utilisant une procédure systématique simple pour enregistrer les densités de nids. Ils ont également testé leurs méthodes en étudiant en détail les densités de nids dans trois plus petites zones (mais encore relativement énormes avec leurs 2 à 3000 km$^2$). Tout comme les chapitres écrits par Mohneke et Fruth ou par Reinharz *et al.* les résultats de Hart *et al.* contiennent de grandes sources d'erreurs potentielles dans leurs estimations. Mais en discutant explicitement de ce problème, ces trois chapitres contribuent à faire avancer la science de ces estimations. Heureusement, la large marge statistique ne menace pas les principales conclusions de Hart *et al.* Ils ont trouvé que la densité des bonobos est plus grande dans les zones d'altitude vers l'est et plus basse dans les forêts à eau noire à l'extrême ouest et qu'elle est souvent élevée près des zones d'occupation humaine. Leurs découvertes aidera grandement à établir des stratégies de conservation et donnent l'espoir que de tels efforts en vaudront la peine.

La grande image qui émerge des Chapitres 8.9 et 10 est positivement encourageante. Les trois chapitres se complètent bien en suggérant que les différentes équipes travaillant dans des zones différentes de la Salonga produisent des résultats largement comparables. Ainsi, Mohneke et Fruth ont estimé la densité de population dans la zone de Lui Kotal de 0.52 à 1.06 individus par km$^2$; dans la riche zone de Etate, Reinartz *et al.* ont trouvé des densités de 1 à 2 adultes au km$^2$ et dans le parc dans sa globalité, Hart *et al.* ont estimé des densités moyennes se situant entre

0.23 et 0.80 par km$^2$. Bien que ces chapitres présentent beaucoup de matière qui doit encore être affinée par des comparaisons entre les aires et les groupes de recherche, une image se dessine des facteurs de l'habitat qui influencent la densité des bonobos dans le paysage, aussi utile pour les conservationnistes que pour l'écologie de l'évolution. Par exemple, les bonobos ne semblent pas, comme on l'imaginait avant, être adaptés à l'origine aux marais. C'est une espèce de la forêt mature qui se développe avec un dense couvert herbacé.

Cette importante conclusion n'aide pas à résoudre le problème de savoir pourquoi la société des bonobos et celle des chimpanzés sont différentes sous bien des aspects, sachant que les chimpanzés prospèrent également dans les forêts matures d'altitude. Mais cela nous rappelle le défi essentiel consistant à expliquer pourquoi, si les bonobos sont différents des chimpanzés dans leur comportement, ils ont évolué afin de devenir différents. La rapide croissance de la recherche révélée dans les Chapitres 5 à 10 établit un nouveau niveau de confiance dans notre compréhension de l'écologie des bonobos.

# Longitudinal Structure of a Unit-group of Bonobos: Male Philopatry and Possible Fusion of Unit-groups

Chie Hashimoto[1], Yasuko Tashiro[2], Emi Hibino[1], Mbangi Mulavwa[3], Kumugo Yangozene[3], Takeshi Furuichi[1], Gen'ichi Idani[2], and Osamu Takenaka[1]

## Introduction

Bonobos and chimpanzees have a male-philopatric social structure (Nishida 1979, Itani 1985, Goodall 1986, Wrangham 1986, Pusey and Packer 1987, Kano 1992, Wallis 1997, Reynolds 2005, Furuichi 2006). Demographic data from long-term research sites show that all males remain in their natal groups throughout life, while most females leave their natal groups and join neighboring groups. Itani (1977, 1985) argued that female or male philopatry is a rigid, species-specific social structure. However, some researchers have reported cases in which male chimpanzees or bonobos joined non-natal groups (Nishida and Hiraiwa-Hasegawa 1985, Sugiyama 1999, 2004, Hohmann 2001). Thus, it is not clear how consistent male philopatry is in chimpanzees and bonobos, and under which circumstances male transfer occurs in these species.

We have been conducting research on wild bonobos at Wamba since 1974 and have studied various aspects of bonobo ecology and behavior. Since the original identification of all members of the main bonobo study group in 1976, all natal females have disappeared before maturity, and no males have immigrated into the study group (Furuichi 1989); thus, we had been confident that male philopatry was a rigid social structure for wild bonobos.

Our long-term research was interrupted by political disorder in 1991. When we resumed research in 1994, we found that some individuals of our study group had disappeared (Furuichi et al. 1998). In 1996, our research was again interrupted by civil war. When we visited Wamba to observe security conditions in 2002 during a ceasefire, we confirmed that the main study group, E1, had survived the war (Furuichi and Mwanza 2003).

---

[1] Primate Research Institute, Kyoto University, Inuyama, Aichi, 484–8506 Japan

[2] Great Ape Research Institute, Hayashibara Biochemical Laboratories, Japan

[3] Research Center for Ecology and Forestry, Ministry of Scientific Research and Technology, D.R. Congo

T. Furuichi and J. Thompson (eds.), *The Bonobos: Behavior, Ecology, and Conservation*
© Springer 2008

Although some adult individuals of E1 disappeared during the wars, the total number of bonobos in E1 did not decrease (20 individuals in 1996; 25 individuals in 2004–2005). To our surprise, the number of adult males in E1 was larger than expected, even if we assumed that all of the immature males present before the wars had survived. This data meant that at least some adult males had joined E1 by individual transfer or by fusion of groups.

We re-examined the tendencies of male philopatry and female transfer in wild bonobos using data from E1 spanning 30 years. Furthermore, via direct observation and DNA analysis, we explored the nature of the increase in number of males that occurred in E1 during the wars.

# Methodology

## Study Site and Group

Our study site is at Wamba, in the northern sector of the Luo Reserve, Democratic Republic of Congo (Kano 1992, Hashimoto and Furuichi 2001, Idani et al. 2008). The Wamba research camp is located at 0°11'08" N, 22°37'58" E (WGS1984). The northern sector contains five settlements, where vegetation is comprised of primary forest, old secondary forest, young secondary forest, swamp forest, and agricultural fields (Kano 1992, Idani et al. 1994, Hashimoto et al. 1998). Bonobos use all five types of vegetation. Until 1996, when our research was interrupted, six unit-groups (E1, E2, Bokela, Plantation, Kofola, and Sema) had been using the northern sector as their entire home range or as a part of their home range (Kuroda 1979, Idani 1990, Kano 1992). The history of the Luo Reserve is described by Kano et al. (1996), Hashimoto & Furuichi (2001), Furuichi & Mwanza (2003), and Idani et al. (2008).

In 1976, researchers identified all members of the main study group, E. Two subgroups of E were present from the beginning of the study, and they split into independent groups (E1 and E2) in about 1983 (Kuroda 1979, Kano 1982, Furuichi, 1989, Kano 1992, Furuichi et al. 1998). The main focus of our study is E1, which had been artificially provisioned for a part of each year until 1996 (Kuroda 1979, Hashimoto et al. 1998), after which the study was interrupted by civil wars. We resumed study of the E1 in 2003 with no provisioning.

## Observations

From 1974 to 1996, we observed E and then E1. After security conditions improved, our local assistants resumed observation of E1 in 2002. We resumed observation in August 2003, and thereafter two or more Japanese or Congolese researchers observed E1 continuously until January 2006. Throughout this period, we followed

E1 from sleeping site to sleeping site for 6 days a week whenever possible. We recorded the names of bonobos within visual range. See Mulavwa et al. (2008) for the number of observation days and mean hours of observation per day.

Though we did not employ artificial provisioning after resumption of the study in 2002, bonobos of E1 were well habituated by local assistants by August 2003. In March to April 2004, two of us observed E1 for about one month and identified all of its members. We continued to identify members of E1 until January 2006.

## DNA Analysis

In 1990 and 1994, we collected wadge samples (the fibrous residue of sugarcane spat out by bonobos after chewing) from all members of E1 except for some of the infants. We determined the sequence of the D-loop region of the mitochondrial DNA (Hashimoto et al. 1996). After resumption of the study in 2003, we collected fecal samples from most adult members in March to April 2004; we collected supplementary samples in January 2005.

We collected fresh feces samples from each individual during observation. We placed them in plastic tubes with silica gel and kept them at ambient temperature in the field. We used the QIAamp DNA Stool Mini Kit (QIAGEN) to extract genomic DNA from the samples. We amplified a segment of the D-loop region of mitochondrial DNA consisting of 416-base pairs via the polymerase chain reaction (PCR) with the primers 5'-TAAAC TATTC TCTGT TCTTT CA-3' and 5'-CGGGA TATTG ATTTC ACGGA GG-3'. We conducted PCR via the Expand High Fidelity PCR System (Roche) in a GeneAmp PCR System 9700 (ABI) as follows: 4 min at 94 °C, 35 cycles of 30 sec at 94 °C, 1 min at 47 °C, 2 min at 72 °C; and 10 min at 72 °C; hold at 4°C. We checked the product by gel electrophoresis via CYBR Green (TAKARA). We collected the target band and conducted a second PCR with the same set of primers, according to the same protocol with the exception that we used 40 cycles rather than 35. We labeled the products via the BigDye Terminator v1.1 Cycle Sequencing Kit (ABI) and sequenced them with an ABI PRISM 310 Genetic Analyzer (ABI). We aligned sequences via Genetyx ver 5.0.

## Results

### *Male Philopatry and Female Transfer*

We list the males that we confirmed as members of E or E1 by 1996 in Table 5.1. We excluded members of E that belonged to E2 after group fission, because only individuals that belonged to E1 are included in this study. Ten males were present in E when all of the members were identified in 1976, and 11 males were born into

**Table 5.1** Life history of males of E1 up to 1996

| Name | Birth year | Immigration Year | Age | Disappearance Year | Age | Supposed cause |
|------|------------|------------------|-----|--------------------|-----|----------------|
| Kake | Initial member in 1976 | | | 1989 | 39–44 | Death by old age |
| Kuro | Initial member in 1976 | | | 1991–92 | 36–42 | Death by old age |
| Hata | Initial member in 1976 | | | 1991 | 31–36 | Death by old age |
| Ika | Initial member in 1976 | | | – | | |
| Ibo | Initial member in 1976 | | | 1905/6/9 | 25 | ? |
| Mon | Initial member in 1976 | | | – | | |
| Tawashi | Initial member in 1976 | | | – | | |
| Goro | Initial member in 1976 | | | 1905/6/6 | | Death by poaching |
| Mitsuo | Initial member in 1976 | | | 1991–92 | 16–17 | ? |
| Ten | Initial member in 1976 | | | – | | |
| Haluo | 1977 | | | 1989 | 12 | ? |
| Senta | 1980 | | | 1992–94 | 12–14 | ? |
| Haku | 1982 | | | 2002 | | Death by poaching |
| Matsu | 1984 | | | 1988–89 | 4–5 | Death in immature age |
| Hayato | 1986 | | | – | | |
| Mao | 1986 | | | – | | |
| Shijimi | 1988 | | | – | | |
| Kikuo | 1988 | | | – | | |
| Bio | 1990 | | | – | | |
| Haze | 1990 | | | – | | |
| Maro | 1990 | | | 1995–1996 | 5–6 | Death in immature age |

E or E1 after 1976. No males immigrated from other groups to become permanent members of either E or E1.

Ten of the 21 males had died or disappeared by 1996. Six males (*Kake, Kuro, Hata, Goro, Matsu,* and *Maro*) are known to or are presumed to have died. We were unable to determine whether the other 4 males (*Ibo, Mitsuo, Haluo,* and *Senta*), whose ages ranged from 12 to 26 years, had died or emigrated. Since they were not observed in groups neighboring E1, and no case of immigration of adult males is confirmed for E1 or neighboring groups, they probably died.

We list the females that were confirmed as members of E or E1 by 1996 in Table 5.2. Eight females were present in E when we identified all members in 1976, and 15 females were born into E or E1 after 1976. Five additional females immigrated into E or E1 and became permanent members. Some other females visited E1 for a short period and then left; they are not included in Table 5.2. Nineteen of the 28 females had died or disappeared by 1996. Seven females (*Kame, Sen, Mitsu, Kameko, Naomi, Nako,* and *Midori*) are known to or are presumed to have died.

We could not determine whether the other 12 females had died or emigrated. Their ages ranged from 2 to 36 years, with a peak range of 6 to 10 years. Because at 6 to 10 years of age they were in good health and exhibited a tendency to stay on the periphery of the group when last observed in E1, they probably emigrated. The ages

**Table 5.2** Life history of females of E1 up to 1996

| Name | Birth year | Immigration | | Disappearance | | |
|------|-----------|------|-----|------|-----|---------------|
| | | Year | Age | Year | Age | Supposed cause |
| Kame | Initial member in 1976 | | | 1990 | 40–45 | Death by old age |
| Sen | Initial member in 1976 | | | 1992–94 | 42–47 | Death by old age |
| Mitsu | Initial member in 1976 | | | 1992–94 | 37–42 | Death by old age |
| Halu | Initial member in 1976 | | | | | |
| Shiro | Initial member in 1976 | | | 1991–92 | 34–35 | ? |
| Bihi | | 1978 | 14 | | | |
| Mayu | Initial member in 1976 | | | 1995–96 | 29–30 | ? |
| Nao | | 1983 | 12 | | | |
| Miso | | 1984 | 10 | | | |
| Kiku | | 1984 | 10 | | | |
| Shin | | 1992–96 | 10–14 | | | |
| Iku | Initial member in 1976 | | | 1980–81 | 9–10 | ? |
| Junko | Initial member in 1976 | | | 1980–81 | 9–10 | ? |
| Shiko | 1978 | | | 1987 | 9 | ? |
| Kameko | 1980 | | | 1981 | 1 | Death in immature age |
| Biko | 1981 | | | 1989 | 8 | ? |
| Mako | 1981 | | | 1988 | 7 | ? |
| Balu | 1982 | | | 1988 | 6 | ? |
| Toshi | 1984 | | | 1992–94 | 8–10 | ? |
| Naomi | 1985 | | | 1985 | 0 | Death in immature age |
| Bibi | 1986 | | | 1992–94 | 6–8 | ? |
| Miki | 1986 | | | 1992–94 | 6–8 | ? |
| Nasa | 1987 | | | 1995–96 | 8–9 | ? |
| Miho | 1990 | | | 1992 | 2 | Death in immature age |
| Nako | 1993 | | | 1995 | 2 | Death in immature age |
| Midori | 1993 | | | | | |
| Kino | 1994 | | | | | |
| Bina | 1996 | | | | | |

of disappearance from and immigration into E1 differed (6–10 and 10–14 years, respectively), suggesting that, after leaving their home group, females visit several groups before finally settling in a new group. In fact, some young females of ca. 8–10 years of age visited E1 for a brief period and then left. Because we have observed no case of female transfer in this age class, it is assumed that the 2 females that disappeared in prime adulthood (*Shiro* and *Mayu*) probably died. Some females with dependent offspring visited E1, but they eventually returned to their original groups.

A comparison by sex of the ages of appearance and disappearance from E1 is in Fig. 5.1. The appearance of males exclusively occurred only at birth. The appearance of females, except for births, exclusively occurred at the age of 10–15 years. In contrast, disappearance was extremely frequent among females 5–10 years old. This comparison supports the supposition that only females transfer between unit-groups in wild bonobos, whereas males remain in their natal groups.

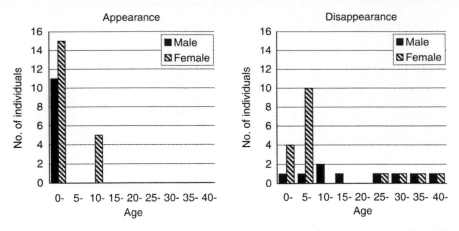

**Fig. 5.1** A comparison by sex of the ages of appearance in and disappearance from E and E1. Appearances in the age class 0–5 are births.

## Disappearance of Neighboring Groups and Possible Fusion of Unit-Groups

During our absence, the number of E1 group members decreased from 20 in 1996 to 17 in 2004 (Fig. 5.2). Our local assistants reported that soldiers killed a young adult male of E1, *Haku*, and that other cases of hunting by soldiers or local people might have occurred. Moreover, many adult individuals had new injuries, such as loss of digits, which appeared to be caused by wire snares set for bush pig or antelope. Some bonobos, especially young individuals, may have been killed by such injuries.

Bonobo deaths from hunting may have occurred less frequently in E1 than in other groups because E1 occupied the interior of the reserve, whereas the other groups occupied the periphery. Deaths from human activities likely caused the disappearance of some peripheral groups during our absence.

In 1991, prior to which year the political situation had been stable, six groups claimed home ranges in the northern sector of the Luo Reserve (Fig. 5.3). From direct observation of them, we estimated that the total number of bonobos was about 250. However, the number of bonobos decreased by half between 1991 and 1996, probably as a result of increased poaching during our absence between 1991 and 1994. In 2005, we observed only 3 of the 6 groups, E1, E2, and Plantation. Two other groups, Kofola and Bokela, probably disappeared due to poaching, given that we walked in their home ranges many times and found no trace of them. The presence of another group, Sema, is unclear because we visited their home range infrequently (Fig. 5.3).

During our absence from 1996 to 2003, E1 greatly extended their home range to the east and northeast, areas previously used by Bokela and Kofola (Fig. 5.4). In 2004, 6 individuals appeared to join E1 when E1 visited the area. We first observed *Nord*, *Yuki*, *Jacky*, and their infants in April 2004 in the eastern part of the E1 home

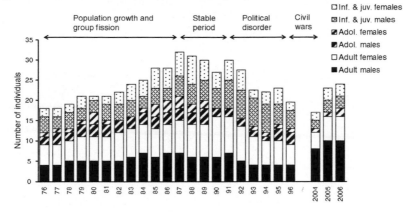

**Fig. 5.2** Demographic changes of E1. E1 split from the E group in or around 1983. Bars before 1982 indicate individuals in the southern subgroup of E1 that eventually became the independent E1 group.

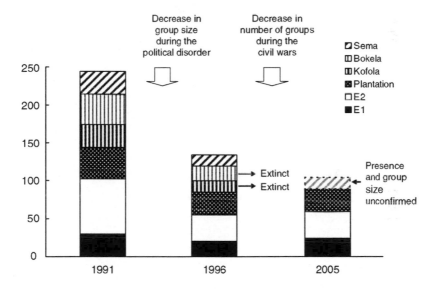

**Fig. 5.3** Changes in the estimated number of bonobos in the northern sector of the Luo Reserve.

range. They were all shy toward human observers, and they did not follow E1 members when they left the area for the western part of their home range. After September 2004, they became regular members of E1, and they ranged together even in the western area. We first observed *Dai* in E1 in September 2004 when E1 visited the eastern area. He was shy and avoided human observers. From November 2004, he became a more regular member of E1 and became accustomed to human observers.

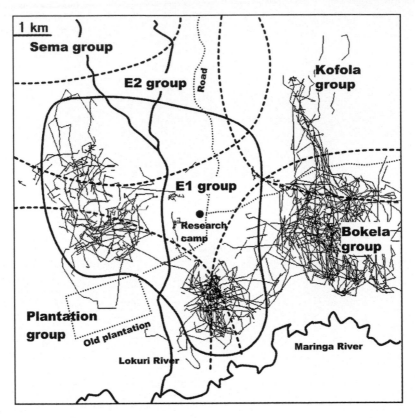

**Fig. 5.4** Ranging routes of E1 and vegetation of the ranging area. Each jagged line denotes a ranging route from one tracking session. The vegetation was mapped from Landsat data recorded on January 14, 1991 (Hashimoto et al., 1998).

## Re-Identification of E1 Members

Before our research was interrupted in 1996, we had identified 20 individuals (4 adult males, 6 adult females, and 10 immature individuals) in E1 (Table 5.3). After research resumed in 2003, we identified 24 individuals (10 adult males, 7 adult females, and 7 immature individuals). In 2004, we identified 2 of the adult males, *Ten* and *Tawashi*, by external characteristics, and we assumed the other 8 adult males (*Gausche, Mori, Noire, Nobita, Loboko, Jeudi, Nord,* and *Dai*) to be individuals that had been immature males with no conspicuous traits in 1996. Of the 7 immature males in 1996, one (*Haku*) was killed during the civil war. Therefore, in 2003, 6 males (*Hayato, Mao, Shijimi, Kikuo, Bio,* and *Haze*) may have still been present in E1; they were candidates for the newly named 8 adult males.

To confirm their identity, we analyzed mitochondrial DNA sequences from fecal samples. We matched 4 of the newly named 8 adult males to 4 immature males of 1996. The sequence of *Noire* matched that of *Mayu*, who is the mother of *Mao*.

**Table 5.3** Members of the E1 group in 1996 and in 2004–2006

| Sex | Members in 1996 | | Members in 2004–2006 | | Possible identity suggested by DNA sequence and other traits |
|---|---|---|---|---|---|
| | Name | Age in years in Jan 1996 | Name | Age in years in Jan 2005 | |
| Male | Ika | 34–36* | | | |
| | Mon | 29* | | | |
| | Tawashi | 22* | Tawashi | 29* | |
| | Ten | 26* | Ten | 35* | |
| | Haku | 13 | | | |
| | Hayato | 9 | (Hayato) | (18) | |
| | Mao | 9 | (Mao) | (18) | |
| | Shijimi | 7 | (Shijimi) | (16) | |
| | Kikuo | 7 | (Kikuo) | (16) | |
| | Bio | 5 | (Bio) | (14) | |
| | Haze | 5 | (Haze) | (14) | |
| | | | Gauche | 15–19* | |
| | | | Nord | 20* | |
| | | | Dai | 30* | |
| | | | Mori | 9–14* | Mori or Jeudi is probably identical to Bio |
| | | | Noire | 15–19* | Probably identical to Mao |
| | | | Nobita | 15–19* | Probably identical to Kikuo |
| | | | Loboko | 9–14* | |
| | | | Jeudi | 9–14* | Mori or Jeudi is probably identical to Bio |
| | | | Jiro | 3–4* | |
| | | | Kitaro | 0 | |
| | | | Shiba | 0 | |
| | | | Hokuto | Died at 2 years old in 2004 | |
| Female | Halu | 39* | | | |
| | Bihi | 32* | | | |
| | Nao | 25* | Nao | 34* | |
| | Miso | 22* | | | |
| | Kiku | 22* | Kiku | 31* | |
| | Shin | 14* | (Shin) | (23*) | |
| | Midori | 2 | | | |
| | Kino | 1 | | | |
| | Bina | 0 | | | |
| | | | Hoshi | 20–24* | Remains a possibility that Hoshi is Shin |
| | | | Sala | 13–14* | |
| | | | Yuki | 20–24* | |
| | | | Jacky | 15–19* | |
| | | | Moseka | 10* | |
| | | | Kirara | 7* | |
| | | | Nana | 4* | |
| | | | Yukiko | 2–3* | |
| | | | Nachi | 0 | |

* Age estimated. Names and ages in parentheses for members in 2004–2006 shows the individuals who might be existing but identified with different names.

The sequence of *Nobita* matched that of *Kiku*, who is the mother of *Kikuo*. No individuals present in 1996 had mothers whose sequence matched those of *Mayu* or *Kiku*, except for *Mao* and *Kikuo*. Therefore, *Noire* and *Mao* are likely the same individual, as are *Nobita* and *Kikuo*. The sequences of *Mori* and *Jeudi* are identical to that of *Bihi*, the mother of *Bio*. However, *Bihi* had only one son, *Bio*, in 1996, and no adult female in 1996 had a sequence that matched that of *Bihi*. Therefore, either *Mori* or *Jeudi* is likely to be the same individual as *Bio*. The other male probably joined E1 during or after the war.

We did not analyze the DNA of 4 adult males (*Gausche, Loboko, Nord,* and *Dai*) because of a lack of samples. If they were previously immature members of the 1996 E1 group, 3 of them might be *Hayato, Shijimi,* or *Haze,* and the fourth might be an immigrant into E1. Therefore, at least two males (either *Jeudi* or *Mori* and either *Gausche, Loboko, Nord,* or *Dai*) seem to have entered E1 during or after the war.

# Discussion

Like chimpanzees, bonobos have a male-philopatric social structure. Males remain in their natal unit-group, whereas females leave their natal unit-group before sexual maturity and transfer between groups (Wrangham 1986, Kano 1992, Furuichi, 2006). In our study group, E and E1, no cases of male transfer were observed between 1976 and 1996, though there were many cases of female emigration and immigration during the same period (Furuichi 1989, Kano 1992, this study). However, after an interruption of the study from 1996 to 2002, at least two males joined E1 from other groups.

In early studies of primate social systems, Itani (1977, 1985) argued that each primate species has a specific basic social structure that is strongly affected by its phylogenetic position, and male or female philopatry is a basis for primate social structures. In general, this claim is still valid, but there are some exceptional cases of transfer by males or females. For example, some researchers reported temporary visits of out-group males in male-philopatric unit-groups of bonobos and chimpanzees. Hohmann (2001) reported that 2 strange adult males visited and stayed in his bonobo study group at Lomako for 12 months, and that one of them developed friendly social relationships with resident males. In addition, one juvenile male chimpanzee at Mahale encountered members of another group when his mother temporarily joined the M group (Nishida and Hiraiwa-Hasegawa 1985). At Bossou, 2 strange adult male chimpanzees joined a semi-isolated group of chimpanzees and stayed there for several days, and another adult male joined and stayed for several months (Sugiyama 1999). Moreover, most adolescent and young adult males disappeared from Bossou, and at least some of them must have emigrated (Sugiyama, 2004).

Three cases of female transfer have been reported for Japanese monkeys, which have a matrilineal social structure. Takahata et al. (1994) reported that 2 females transferred to an adjacent troop when their troop rapidly decreased in size, leaving them as the last 2 surviving members. Sugiura et al. (2002) also reported 2 cases of transfer of a female when she became the last member of a declining troop.

Gibbons have monogamous social groups in which both males and females leave their natal groups and form new ones without joining other groups. However, a young adult male and a young adult female joined non-natal monogamous groups and settled there after their forested habitat was fragmented by forest fire (Oka & Takenaka 2001).

Since most of the cases described above occurred under unusual circumstances, immigration of the strange males to E1 in our study might have also occurred under these circumstances. When we resumed research on E1, we found that their home range had expanded into the eastern area previously used by the Kofola and Bokela groups, probably because the disappearance of these groups left their home ranges vacant. When we first observed *Nord*, *Yuki* and her infant, and *Jacky* and her infant in E1, the group was ranging in the eastern area. Even after they joined E1, for several months they remained in the eastern area when the main members of E1 went back to the west. We also first observed *Dai* when E1 was ranging in the eastern area. Not only immigration of males, but also that of females with infants was unusual. We observed no permanent immigration of adult females with infants in the first 20 years of study of E and E1. Thus, this case might be better understood as aggregation of declining groups, rather than a strict intergroup transfer of adult males. Exactly what caused the immigration of *Jeudy* or *Mori* in our absence during the war is unclear. However, it is possible that something similarly unusual happened when the local population of bonobos was severely impacted by human activities.

Although some cases of male transfer have been reported in chimpanzees, there is no record of permanent immigration of adult males. This fact may reflect the intolerant relationships between males from different groups of chimpanzees (Nishida 1985, Goodall 1986, Wrangham and Peterson 1996, Reynolds 2005). Contrarily, bonobos sometimes display affinitive relationships between different groups (Idani 1990). Different groups of bonobos sometimes forage together for as long as a week, and members of these groups exhibit affiliative social interactions. Though further observation of the new immigrants is needed, the high tolerance between different groups of bonobos might have enabled the permanent aggregation of fragmented groups, as observed in the above-mentioned cases of Japanese macaques.

**Acknowledgments**   We thank Drs. Takayoshi Kano and Toshisada Nishida for their continued support of the study at Wamba and for research guidance. We also thank Dr. Tetsuro Matsuzawa and Ms. Sally Coxe for aiding resumption of the study; Drs. Shin Nakamura and Akiko Takenaka for their support of our laboratory work; Dr. Mwanza and members of the Research Center for Ecology and Forestry (CREF) of the Democratic Republic of Congo for their support of our field work; and Dr. Shigeo Uehara and members of the Primate Research Institute, Kyoto University, for valuable discussion and advice. We are grateful to Mr. Nkoi Batolumbo and other local staff and villagers for their support during and after the war. This study was supported by the National Geographic Fund for Research and Exploration (#7511-03 to Furuichi), the JSPS core-to-core program HOPE (#15001 to Matsuzawa), JSPS Grant-in-Aid for Scientific Research (#17570193 to Hashimoto, #12575017 and 17255005 to Furuichi), and Japan Ministry of Environment Global Environment Research Fund (#F-061).

# References

Furuichi T (1989) Social interactions and the life history of female *Pan paniscus* in Wamba, Zaire. Int J Primatol 10: 173–197

Furuichi T (2006) Evolution of the social structure of hominoids: reconsideration of food distribution and the estrus sex ratio In: Ishida H, Tuttle T, Pickford M, Nakatsukasa M, Ogihara N (eds) Human origins and environmental backgrounds. Springer, New York, pp 235–248

Furuichi T, Mwanza N (2003) Resumption of bonobo studies at Wamba, the Luo reserve for scientific research. Pan Africa News 10: 31–32

Furuichi T, Idani G, Ihobe H, Kuroda S, Kitamura K, Mori A, Enomoto T, Okayasu N, Hashimoto C, Kano T (1998) Population dynamics of wild bonobos (*Pan paniscus*) at Wamba. Int J Primatol 19: 1029–1043

Goodall J (1986) The Chimpanzees of Gombe: patterns of behavior. Harvard University Press, Cambridge

Hashimoto C, Furuichi T (2001) Current situation of bonobos in the Luo reserve, equateur, Democratic Republic of Congo. In: Galdikas BMF, Briggs NE, Sheeran LK, Shapiro GL, Goodall J (eds) All apes great and small, vol 1: african apes. Kluwer Academic/Plenum, pp 83–93

Hashimoto C, Furuichi T, Takenaka O (1996) Matrilineal kin relationships and social behavior of wild bonobos (*Pan paniscus*): sequencing the D-loop region of mitochondrial DNA. Primates 37: 305–318

Hashimoto C, Tashiro Y, Kimura D, Enomoto T, Ingmanson E J, Idani G, Furuichi T (1998) Habitat Use and ranging of wild bonobos (*Pan paniscus*) at Wamba. Int J Primatol 19: 1045–1060

Hohmann G (2001) Association and social interactions between strangers and residents in bonobos (*Pan paniscus*). Primates 42: 91–99

Idani G (1990) Relations between unit-groups of bonobos at Wamba, Zaire: encounters and temporary fusions. African Study Monographs 11: 153–186

Idani G, Kuroda S, Kano T, Asato R (1994) Flora and vegetation of Wamba forest, central Zaire with reference to bonobo (*Pan paniscus*) foods. Tropics 3: 309–332

Idani G, Mwanza N, Ihobe H, Hashimoto C, Tashiro Y, Furuichi T (2008) Changes in the status of bonobos, their habitat, and the situation of humans at Wamba, in the Luo scientific reserve, democratic republic of the Congo. In: Furuichi T, Thompson J (eds) The bonobos: behavior, ecology, and conservation. Springer, New York, pp 291–302

Itani J (1977) Evolution of primate social structure. J Hum Evol 6: 235–243

Itani J (1985) The evolution of primate social structures. Man 20: 593–611

Kano T (1982) The social group of pygmy chimpanzees (*Pan paniscus*) of Wamba. Primates 23: 171–188

Kano T (1992) The last ape: pygmy chimpanzee behavior and ecology. Stanford Univ Press, Stanford

Kano T, Lingomo B, Idani G, Hashimoto C (1996) The challenge of Wamba. In: Cavalieri P (ed) The great ape project, Eecta & Animali 96/8, Milano, pp 68–74

Kuroda S (1979) Grouping of the pygmy chimpanzees. Primates 20: 161–183

Mulavwa M, Furuichi T, Yangozene K, Yamba-Yamba M, Motema-Salo B, Idani G, Ihobe H, Hashimoto C, Tashiro Y, Mwanza N (2008) Seasonal changes in fruit production and party size of bonobos at Wamba, in the Luo scientific reserve, democratic republic of the Congo. In: Furuichi T, Thompson J (eds) The bonobos: behavior, ecology, and conservation. Springer, New York, pp 121–134

Nishida T (1979) The social structure of chimpanzees of the Mahale Mountaïns. In: Hamburg DA, McCown ER (eds) The great apes. The Benjamin/Cummings Publishing Company, Menlo Park, pp73–121

Nishida T (1985) Group extinction and female transfer in wild chimpanzees in the Mahale National Park, Tanzania. Z Tierpsychol 67: 284–301

Nishida T, Hiraiwa-Hasegawa M (1985) Responses to a stranger mother-son pair in the wild chimpanzee: a case report. Primates 26: 1–13

Oka T, Takenaka O (2001) Wild gibbons' parentage tested by non-invasive DNA sampling and PCR-amplified polymorphic microsatellites. Primates 42: 67–73

Pusey AE, Packer C (1987) Dispersal and philopatry. In: Smuts BB, Cheney DL, Seyfarth RM, Wrangham RW, Struhsaker TT (eds) Primate Societies, The University of Chicago Press, Chicago and London, pp 250–266

Reynolds V (2005) The Chimpanzees of the Budongo Forest: ecology, behaviour, and conservation. Oxford Univ Press

Sugiura H, Agetsuma N, and Suzuki S (2002) Troop extinction and female fusion in wild Japanese macaques in Yakushima. Int J Primatol 23: 69–84

Sugiyama Y (1999) Socioecological factors of male chimpanzee migration at Bossou, Guinea. Primates 40: 61–68

Sugiyama Y (2004) Demographic parameters and life history of chimpanzees at Bossou, Guinea. Amer J Phys Anthropol 124: 154–165

Takahata Y, Suzuki S, Okayasu N, and Hill D (1994) Troop extinction and fusion in wild Japanese macaques in Yakushima Island, Japan. Amer J Primatol 33: 317–322

Wallis J (1997) A survey of reproductive parameters in the free-ranging chimpanzees of Gombe National Park. J Reproduction Fertility 109: 297–307

Wrangham R W (1986) Ecology and social relationships in two species of chimpanzee. In: Rubenstein DL, Wrangham RW (eds) Ecological aspect of social evolution. Princeton Univ Press, Princeton, pp 352–378

Wrangham RW, Peterson D (1996) Demonic males: apes and the origins of human violence. Houghton Miffin, Boston

# Seasonal Changes in Fruit Production and Party Size of Bonobos at Wamba

Mbangi Mulavwa[1], Takeshi Furuichi[2], Kumugo Yangozene[1], Mikwaya
Yamba-Yamba[1], Balemba Motema-Salo[1], Gen'ichi Idani[3], Hiroshi Ihobe[4],
Chie Hashimoto[2], Yasuko Tashiro[3], and Ndunda Mwanza[1]

## Introduction

Because chimpanzees (*Pan troglodytes*) have a unique fission-fusion social structure, many researchers have investigated the nature of foraging parties. They have reported that chimpanzees form foraging parties whose size and sex composition change flexibly, and that the sizes of parties may vary according to fluctuations in fruit abundance, the number of estrous females, or both (Wrangham 1977, Ghiglieri 1984, Isabyre-Basuta 1988, Stanford et al. 1994, Boesch 1996, Matsumoto-Oda et al. 1998, Newton-Fisher et al. 2000, Boesch and Boesch-Achermann 2000, Hashimoto et al. 2001). Researchers have also reported that females tend to join mixed-sex parties less frequently than males do, and that this likely occurs because ranging in large mixed-sex parties may not be beneficial to the feeding activities of females (Wrangham 1979, 2000, Janson and Goldsmith 1995, Williams et al. 2002, Reynolds 2005).

The closest relative of chimpanzees, bonobos (*P. paniscus*), have a similar fission-fusion social structure. However, the size and composition of the parties are different from those of chimpanzees. Previous studies of bonobos at Wamba and Lomako in the Democratic Republic of the Congo showed that bonobos form larger and more stable parties (Kuroda 1979, Kano 1982, 1992, Furuichi 1987, White 1988), and that the size of parties may be influenced by fruit abundance to a lesser extent. This is probably because either the seasonal changes in fruit abundance are small or because some foods are available year-round (Kano 1982, Kano and Mulavwa 1984, White 1998, Malenky and Stiles 1991, Malenky and Wrangham 1994, Chapman et al. 1994). Furthermore, bonobo females tend to join mixed-sex parties

[1] *Research Center for Ecology and Forestry, Ministry of Scientific Research and Technology, Mabali, Equateur, D.R. Congo*

[2] *Primate Research Institute, Kyoto University, Japan*

[3] *Great Ape Research Institute, Hayashibara Biochemical Laboratories, Japan*

[4] *School of Human Sciences, Sugiyama Jogakuen University, Japan*

T. Furuichi and J. Thompson (eds.), *The Bonobos: Behavior, Ecology, and Conservation*
© Springer 2008

more frequently than males do (Kano 1982, Furuichi 1987, White 1988). Because chimpanzees and bonobos form male-philopatric groups, it is not surprising that in chimpanzees, kin-related males aggregate more than unrelated females. Even in Taï, Côte d'Ivoire, where the sex difference in sociality among chimpanzees is less than at other sites, the composition of mixed parties is consistently male-biased (Boesch 1996). However, with bonobos, unrelated females aggregate more than related males. Although the higher sociality of female bonobos may be attributable to their high social status, which reduces the cost of contest competition, or to the higher density of food resources in bonobo habitat, which reduces the cost of scramble competition for slower-moving females, the hypotheses have not been examined quantitatively.

Thus, the differences in party size, party composition, and especially the grouping pattern of females, are key issues in understanding the ecological adaptations of *Pan* species. However, many constraints are imposed in examining proposed hypotheses to explain the differences between chimpanzees and bonobos. First, unlike chimpanzees, very limited information is available on the relationship between fruit abundance and party size in bonobos. Kuroda (1979) suggested the possibility of a relationship between seasonal fruit abundance and party size, but his study did not present quantitative data on fruit abundance. Although White (1998) and Hohmann and Fruth (2002) at Lomako presented quantitative analyses on the relationship between fruit abundance and party size, quantiative studies on more populations of bonobos are needed to reveal the relationship between the ranging pattern and seasonal changes in fruit abundance. Second, although Chapman et al. (1994) compared party size and fruit abundance between chimpanzees and bonobos, they indicated that their results were provisional, because party sizes may vary due to differences in observation methods. In fact, various methods and definitions have been proposed and used for the study of chimpanzee party size, and the different methods and definitions tend to yield very different results on party size (Hashimoto et al. 2001, Reynolds 2005).

Our study had three purposes. The first was to examine the relationship between fruit abundance and party size of bonobos. This is the first study at Wamba that provides quantitative data on fruit production and party size for a period of more than a year. As mentioned below, bonobos of our study group had not been artificially provisioned for 7 years when we started observation for this study. Therefore, our study provides valuable information on the ecology of bonobos under natural conditions.

The second purpose was to provide data for comparative studies between chimpanzees and bonobos by using the same definition of party size and fruit production. We employed methodologies that had been developed for studies on chimpanzees in the Kalinzu Forest, Uganda (Furuichi et al. 2001, Hashimoto et al. 2001), which would allow for accurate interspecific comparison of the relationship between party size and fruit production.

The third purpose was to reevaluate the results of studies on bonobos at Wamba. Since 1976 when all members of group E, which split into the E1 and E2 groups before 1983, were identified, they were provisioned with artificial food until 1996 when the study was interrupted by civil wars (Furuichi 1989, Kano 1992). Because the bonobos had been given only a small amount of artificial food for a limited time, researchers working at Wamba assumed that the tendencies observed in E1, such as large and stable party size and gregariousness of females, reflected the nature of wild bonobos in an unbiased way. However, there is reason to challenge that these tendencies may have appeared due to the influence of the artificial provisioning. By comparing current and past tendencies of grouping patterns, we may be able to evaluate the extent to which artificial provisioning influenced past studies at Wamba.

## Methods

### Study Group

We observed the E1 group (unit-group or community, van Elsacker et al. 1995) of wild bonobos at Wamba in the northern section of the Luo Scientific Reserve, D.R. Congo. The history of E1 and the details of the study site are described by Furuichi (1989), Kano (1992), Hashimoto et al. (1998, 2008), and Idani et al. (2008). In January 2005, E1 included 10 adult males, 6 adult females, 1 adolescent female, 2 juvenile females, 3 infant males, and 1 infant female. Hashimoto et al. (2008) have described the more recent changes in membership of the group.

We observed E1 from September 2003 to December 2005. We attempted to locate parties of E1 and to follow them from sleeping site to sleeping site 6 days per week, for a total of 711 days. During this time, we directly observed bonobos for 484 days, or 68% of the total working days. The total time of direct observation, excluding time spent tracing bonobos by footprints or vocalizations, was 2,216 hours. On average, bonobos were within sight of observers for 4.6 hours per day on days we conducted direct observations.

### Monitoring Fruit Abundance

To monitor fruit abundance, we used five line transects and reconnaissance paths, the total length of which is 22,550 m (Fig. 6.1). We used some reconnaissance paths for our survey because they were very narrow trails that ran without avoiding particular types of vegetation and did not seem to affect the growth and fruiting of trees along them. Although the fruit trails were set arbitrarily, they covered various vegetation types in the home range of E1 with minimal bias (Fig. 6.1). We recorded daily rainfall at the research camp, which was situated in the center of the home range.

We walked each trail twice a month. We recorded the number of clusters of fallen fruit that were found within 1 m on each side of the trail, the number of fruits in each cluster, species of fruit, and whether they were ripe or unripe, following the methods used in a study of wild chimpanzees in the Kalinzu Forest, Uganda (Furuichi et al. 2001). As recommended by Furuichi et al. (2001), we evaluated fruit abundance based on the number of clusters of ripe fallen fruits per km of the trail. Although a list of scientific names of 510 plant species found in the study site was available (Idani et al. 1994), a considerable number of unidentified species

**Fig. 6.1** Vegetation in the ranging area of E1 and trails for monitoring fruit abundance. The vegetation image was made from Landsat data recorded on 14 January 1991 (Hashimoto et al. 1999). Pale areas include agricultural fields and young secondary forest. Medium-colored areas include old secondary forest and primary forest. Dark areas around the river represent swamp forest.

were recorded during our study. Therefore, we used identifications based on vernacular names to represent the number of fruit species.

## Party Size and Composition

To obtain data comparable to those for chimpanzees, we employed the definition of the 1-hour party size proposed by Hashimoto et al. (2001) for evaluating party sizes of chimpanzees. While following a party, we recorded the number of bonobos within each 1-hour segment. We recorded the names of bonobos in sight at the beginning of each hour and continued recording bonobos that appeared in sight until the end of the hour. Thus, the 1-hour party represents the minimum number of bonobos that were present in the party during each 1-hour observation. We also recorded the number of minutes for which bonobos were in sight of observers in each 1-hour segment. Through this information, we were able to attain reliable data on the 1-hour party size.

For the comparison between chimpanzees and bonobos, we used data for chimpanzees of the M group in the Kalinzu Forest Reserve, Uganda that we collected during the 1997-1998 study period, using a similar 1-hour party method (Hashimoto et al. 2001). Because members of M had not yet been fully identified at that time, for the calculation of relative party size, we used the group size and composition recorded in 2005: 19 adult males, 22 adult females, 4 adolescent males, 9 adolescent females, 6 juvenile and infant males, 9 juvenile and infant females, and 1 infant of unidentified sex. We assumed that this 2005 group composition was not largely different from that during 1997–1998 because we did not observe any dramatic change in the membership.

# Results

## Seasonal Changes in Fruit Abundance

The monthly rainfall and fruit abundance are illustrated in Fig. 6.2. The average annual rainfall was 2,843 mm in 2004 and 2,922 mm in 2005. Although no clear rainy or dry season occurred, more rain in the area occurs around October/November, and less around January/February.

The fruit abundance showed fairly irregular changes. There is no significant correlation between the abundance of ripe fruit and rainfall, or between the abundance of ripe food fruit and rainfall (Figs. 6-3a,b). Food fruits include the species that bonobos ate during the current or past study periods. There is correlation between the abundance of ripe fruit and that of ripe food fruit (Fig. 6-3c). The number of species of ripe food fruits on the trails also showed irregular changes that paralleled

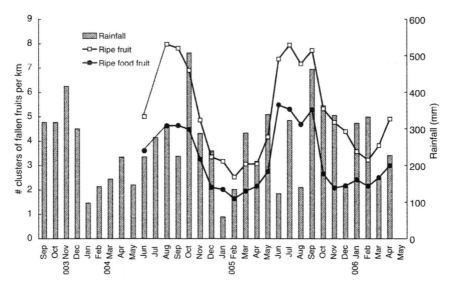

**Fig. 6.2** Monthly rainfall and fruit abundance. The data for fruit abundance were available only for June 2004 and from August 2004 to December 2005.

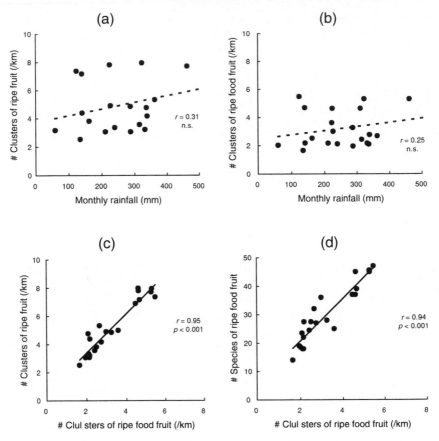

**Fig. 6.3** Correlations between the abundance of ripe fruit and rainfall: (a) the abundance of ripe food fruit and rainfall, (b) the abundance of ripe food fruit and the abundance of ripe fruit, (c) and the number of species of ripe food fruit and the abundance of ripe food fruit (d). Each dot represents 1 month.

the changes in fruit abundance (Fig. 6-3d). These results suggest that rainfall cannot be used as an indicator of fruit abundance as expressed by the number of clusters or species of ripe fruits or ripe food fruits, and that the actual abundance of fruits must be monitored during the same study period to observe the influence of food abundance on ranging patterns or feeding behavior of bonobos.

## Size and Composition of the 1-hour Party

Figure 6.4 shows the relationship between the number of bonobos that were observed in each 1-hour segment and the number of minutes for which bonobos were in sight of observers in the same 1-hour segment. The number of bonobos increased

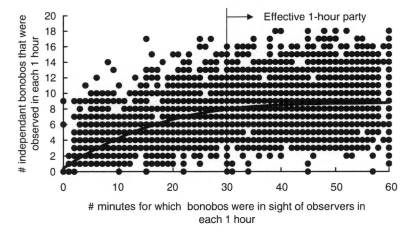

**Fig. 6.4** Relationship between the number of bonobos that we observed in each 1-hour segment and the number of minutes during which bonobos were in sight of observers in the same 1-hour segment. Each dot represents the data for one 1-hour observation.

with the amount of observation time until 30 minutes and then became saturated. Thus, it took about 30 minutes of observation until we confirmed most of the bonobos in a party. We therefore excluded data for 1-hour parties with less than 30 minutes of observation, which may have underestimated the party size. This result also suggests that scanning observations may largely underestimate party size. We termed the 1-hour party size with observation of 30 min or more the effective 1-hour party size. We obtained at least one effective 1-hour party size on each of 465 days; the mean number of effective 1-hour party size per day was 5.1.

We calculated the 1-hour party size for each observation day, daily 1-hour party size, by averaging the effective 1-hour party sizes for that day (Table 6.1). The mean number of bonobos in daily 1-hour parties was 11.2. The mean number of independent bonobos in daily 1-hour parties was 8.7, which included 4.1 adult males and 3.2 adult females. Thus, as reported in previous studies at Wamba, parties of bonobos in E1 consistently included similar numbers of males and females (Kano 1983, Furuichi 1987).

We also obtained the party size of chimpanzees of the M group in the Kalinzu Forest, Uganda, using the same 1-hour party method. The mean of daily 1-hour party size was 5.9 independent individuals, which included $3.5 \pm 1.7$ (S.D.) adult males and $1.2 \pm 0.8$ adult females (N = 53 days).

Figure 6.5 is a comparison of the 1-hour party size between bonobos of E1 and chimpanzees of M. As stated above, the 1-hour party of bonobos includes similar numbers of males and females. However, because E1 contained fewer female members during the observation period, the relative party size, which is the percentage of individuals in the party to the number of all individuals in the unit-group (Boesch

**Table 6.1**  Mean number of E1 bonobos in the daily
1-hour party

|                                        | Mean | S.D. |
| -------------------------------------- | ---- | ---- |
| Adult male                             | 4.1  | 1.6  |
| Adult female                           | 3.2  | 1.1  |
| Adolescent male                        | 0.2  | 0.4  |
| Adolescent female                      | 0.2  | 0.3  |
| Juvenile                               | 0.8  | 0.6  |
| Unidentified                           | 0.2  | 0.4  |
| *Subtotal / Independent individuals*   | 8.7  | 2.8  |
| Infant                                 | 2.4  | 1.0  |
| *Total / All individuals*              | 11.2 | 3.6  |

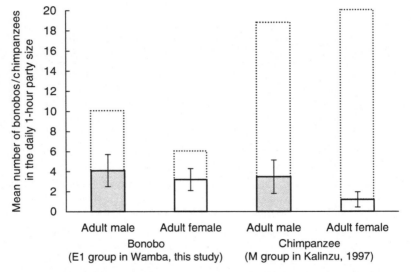

**Fig. 6.5** Comparison of party size and composition between chimpanzees and bonobos. Bars show the mean number of adult males or females in the 1-hour party, and error bars indicate the standard deviation. Bars with dotted lines show the number of members of each sex in the studied unit-groups.

1996), is higher for females. The relative party size was 41% for adult males, 53% for adult females, and 51% for all independent individuals. In chimpanzees, the 1-hour party included a similar number of adult individuals as did the bonobos. However, because the number of unit-group members was much larger for chimpanzees than for bonobos, the relative party size was much lower than in bonobos. In particular, female chimpanzees show a much lower tendency to join parties. The relative party size for M was 20% adult males, 6% adult females, and 13% all independent individuals.

## Relationship of Party Size to Fruit Abundance

Figure 6.6 shows the daily 1-hour party size of all independent bonobos, adult males, and adult females throughout the study period. Although the sexual composition of the party was quite stable, the party size seemed to show seasonal fluctuations. Therefore, we calculated monthly means of the daily 1-hour party size and compared them with the abundance of ripe food fruits in each month as expressed by the number of clusters of fallen fruit on the trails. As shown in Fig. 6.7a, the number of all independent individuals is significantly correlated with food fruit abundance. The number of adult males and the number of adult females are also significantly correlated with the abundance of food fruit. In addition, we analyzed the correlation of the party size with the number of species of food fruit that were found on the fruit trails each month. The number of all independent individuals, the number of adult males, and the number of adult females are all significantly correlated with the number of species of food fruit (Fig. 6.7b).

## Discussion

We found that E1 formed a large stable party that included both males and females. The mean 1-hour party size was 11.2 for all individuals and 8.7 for independent individuals. Although the methods of estimating party size differ between studies, E1

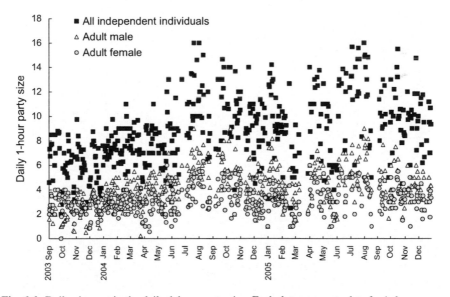

**Fig. 6.6** Daily changes in the daily 1-hour party size. Each dot represents data for 1 day.

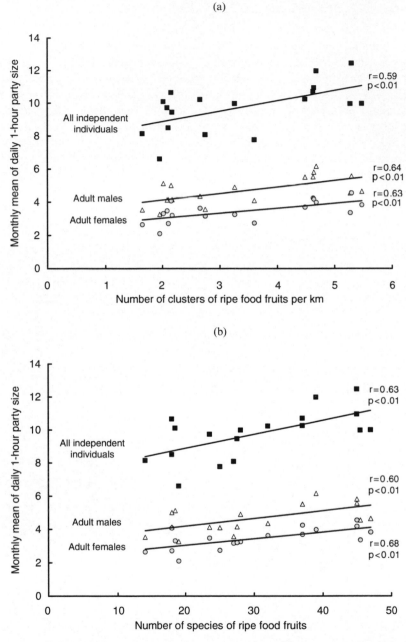

**Fig. 6.7** Relationship of monthly average of the daily 1-hour party size to (a) the abundance of ripe food fruit in a month and (b) to the number of species of ripe food fruit in a month. Each dot represents data for 1 month.

mean party size is within the range of those for bonobos in previous studies. It is smaller than the 16.9 (Kuroda 1979) and 18.9 (Kano 1982) previously reported for E from which E1 split, and it is larger than the 8.5 mean party size for bonobos at Yalosidi, D.R. Congo (Kano 1983). It is also larger than the 7.9 (Badrian and Badrian 1984), 5.4 (independent individuals, White 1988), 5.4 (independent individuals, Malenkey and Stiles 1991), 6.4 (independent individuals, Chapman et al. 1994), and 4.9 (independent individuals, Hohmann and Fruth 2002) mean party sizes of Lomako bonobos.

The party size of bonobos in our study is larger than that of the chimpanzees of M group in the Kalinzu Forest, and of chimpanzees in most other populations (Chapman et al. 1994, Reynolds 2005). Boesch (1996) suggested that mean party size should be expressed as a percentage of the total community (or unit-group) size, and showed that the relative mean party sizes of chimpanzees over the study periods were between 9 and 21% of the unit-group size. The relative mean party size of chimpanzees of the M group in Kalinzu (13%) falls within this range, and Hohmann and Fruth (2002) reported that the known relative party size for chimpanzees falls between 9 and 30%. However, the relative party size of Lomako bonobos is in the higher part of the range (27%, Hohmann and Fruth 2002), and that of E1 (51%) is far beyond it. Thus, the party size of E1 bonobos seems to be larger than that of chimpanzee populations both in absolute number and in percentage of the unit-group size.

For bonobos of E1, the relative party size was larger for adult females than adult males, indicating that individual females joined mixed parties more frequently than did individual males. This result corresponds with past reports for E1 (Kano 1982, 1992, Furuichi 1987), and reports for Lomako bonobos (White 1988). This is a marked difference from chimpanzees of M group in Kalinzu, in which the relative party size was much smaller for females than for males. The lower gregariousness of females is a common feature among chimpanzees (Wrangham 1979, 2000, Janson and Goldsmith 1995, Williams et al. 2002, Reynolds 2005). Thus, the differences in grouping patterns between chimpanzees and bonobos are not only reflected in the party size, but also in the reversed sociality of males and females. Many researchers have argued about these differences with reference to the high social status of females, prolonged estrus of females, and moderate contest competition for food resources (Furuichi 1989, Wrangham 1986, Kano 1992, Chapman et al. 1994, Furuichi and Hashimoto 2002), suggesting that more studies are needed to resolve the issue.

The relationship between seasonal changes in fruit abundance and party size has also been debated by many researchers. A general consensus exists that party size may increase during periods of fruit abundance (Janson and Goldsmith 1995). Some studies have shown that chimpanzees may form larger parties during seasons with higher fruit abundance (Wrangham 1977, Ghiglieri 1984, Isaberya-Basuta 1988, Boesch 1996, Matsumoto-Oda et al. 1998). However, there are contradictory findings for different populations of chimpanzees. Isaberya-Basuta (1988) reported that the positive relationship between fruit abundance and party size did not hold when multiple sources of important foods were available. Stanford et al. (1994)

reported that the party size of chimpanzees in Gombe, Tanzania was largest in the dry season when the food supply was restricted. Boesch (1996) suggested that marked differences occurred among years in the relationships between fruit abundance and party size. Newton-Fisher et al. (2000) and Hashimoto at al. (2001) reported that in chimpanzees of Budongo and Kalinzu, Uganda, fruit abundance did not have a significant influence on the party size, while the presence or number of estrous females did. Basabose (2004) also reported no positive relationship between fruit abundance and party size in Kahuzi-Biega, D.R. Congo.

Our results on bonobos indicated a significant correlation between food fruit abundance or number of species and party size in terms of the number of adult males, adult females, and independent individuals, though the increase in party size with the increase in fruit abundance was very small. These results are compatible with earlier reports on bonobos. Kuroda (1979) reported that party size of Wamba bonobos became larger when preferred fruits were abundant. For Lomako bonobos, White (1998) showed that the number of independent individuals and adult males in the party significantly increased with fruit abundance, while there was no significant correlation for adult females. Hohmann and Fruth (2002) reported that for Lomako bonobos, the number of independent individuals, adult males, or adult females in the party had no significant correlation with fruit abundance, but the number of adult females significantly correlated with the number of fruit species consumed by bonobos in each month.

Although it may be premature to conclude a general tendency for bonobos based on such a small number of reports, it seems that the variation in the correlation of party size and fruit abundance is smaller for bonobos than for chimpanzees. In chimpanzees, the presence or absence of estrous females dramatically influences the number of adult males that join a party (Matumoto-Oda et al. 1998, Newton-Fisher et al. 2000, Hashimoto et al. 2001). This factor may mask the influence of fruit abundance on party size in some studies. In bonobos, however, some estrous females usually are present because many females exhibit pseudo-estrus even during the infertile periods (Wrangham 1986, Furuichi 1987, Kano 1992, Furuichi and Hashimoto 2002). Furuichi and Hashimoto (2002) reported that in a group of Wamba bonobos, a fairly constant number of estrous females were present throughout the study period (3.1 ± 1.1 [S.D.], range 1–5, N = 57 days in 1985/1986; 3.1 ± 1.9, range 1–7, N = 48 days in 1987/1988; 4.1 ± 1.5, range 2–8, N = 43 days in 1990/1991). Although Hohmann and Fruth (2002) reported that parties in which mating occurred were larger than average, the difference seemed to be very small. The usual presence of estrous females and the moderate fluctuation in their numbers in the unit-group or community of bonobos may cause the correlation between fruit abundance and party size to appear as it does.

We conducted our study on the bonobos of E1 under completely natural conditions. Because artificial provisioning had been terminated 7 years before the study, we assumed that its influence on the ecology or behavior of the bonobos was negligible. The results of this study, including the formation of large mixed parties, higher sociality of females, and positive correlation of party size with seasonal fruit abundance, closely resembled the results of studies that had been conducted while the bonobos were artificially provisioned. This is probably because bonobos received only a small amount of food during a limited period of time when they were artificially provisioned (Furuichi 1989, Kano 1992).

**Acknowledgments**  We thank Dr. Takayoshi Kano and Dr. Toshisada Nishida for their continued support of the study at Wamba, and Dr. Tetsuro Matsuzawa and other researchers of Kyoto University for their support for the research project and valuable discussions on the results of this study. We also thank the staff of the Research Center for Ecology and Forestry (C.R.E.F.) of the D.R. Congo for supporting the field studies. This study was supported by the National Geographic Fund for Research and Exploration (#7511-03 to Furuichi), the Japan Society for the Promotion of Science (JSPS) core-to-core program HOPE (#15001 to Matsuzawa), the JSPS Grant-in-aid for Scientific Research (#12575017, # 17255005 to Furuichi), and the Cooperation Research Program of Primate Research Institute, Kyoto University.

# References

Badrian AJ, Badrian NL (1984) Group composition and social structure of *Pan paniscus* in the Lomako Forest. In: Susman RL (ed) The pygmy chimpanzee: evolutionary biology and behaviour. Plenum Press, New York, pp 173–181

Basabose AK (2004) Fruit availability and chimpanzee party size at Kahuzi montane forest, Democratic Republic of Congo. Primates 45: 211–219

Boesch C (1996) Social grouping in Taï chimpanzees. In: McGrew WC, Marchant LF, Nishida T (eds) Great ape societies. Cambridge Univ Press, Cambridge, pp 101–113

Boesch C, Boesch-Achermann H (2000) The Chimpanzees of the Taï Forest: behavioural ecology and evolution. Oxford Univ Press, Oxford

Chapman CA, White FJ, Wrangham RW (1994) Party size in chimpanzees and bonobos: a reevaluation of theory based on two similarly forested sites. In: McGrew WC, de Waal FBM, Heltne PG (eds) Chimpanzee cultures. Harvard University Press, Cambridge, Massachusetts, London, pp 41–57

van Elsacker L, Vervaecke H, Verhegen RF (1995) A review of terminology on aggregation patterns Move to before Furuichi T (1987) in bonobos (*Pan paniscus*). Int J Primatol 16: 37–52

Furuichi T (1987) Sexual swelling, receptivity and grouping of wild pygmy chimpanzee females at Wamba, Zaïre. Primates 28: 309–318

Furuichi T (1989) Social interactions and the life history of female *Pan paniscus* in Wamba, Zaire. Int J Primatol 10: 173–197

Furuichi T, Hashimoto C (2002) Why female bonobos have a lower copulation rate during estrus than chimpanzees. In: Boesch C, Hohmann B, Marquardt L (eds) Behavioral diversity of chimpanzees and bonobos. Cambridge Univ Press, New York, pp 156–167

Furuichi T, Hashimoto C, Tashiro Y (2001) Fruit availability and habitat use by chimpanzees in the Kalinzu Forest, Uganda: examination of fallback foods. Int J Primatol 22: 929–945

Ghiglieri MP (1984) The Chimpanzees of Kibale Forest. Columbia Univ Press, New York

Hashimoto C, Tashiro Y, Kimura D, Enomoto T, Ingmanson EJ, Idani G, Furuichi T (1998) Habitat use and ranging of wild bonobos (*Pan paniscus*) at Wamba. Int J Primatol 19: 1045–1060

Hashimoto C, Furuichi T, Tashiro Y, Kimura D (1999) Vegetation of the Kalinzu Forest, Uganda: ordination of forest types using principal component analysis. African Study Monogr 20: 229–239

Hashimoto C, Furuichi T, Tashiro Y (2001) What factors affect the size of chimpanzee parties in the Kalinzu Forest, Uganda?: examination of fruit abundance and number of estrous females. Int J Primatol 22: 947–959

Hashimoto C, Yasuko T, Hibino E, Mulavwa M, Yangozene K, Furuichi T, Idani G, Takenaka O (2008) Longitudinal structure of a unit-group of bonobos: male philopatry and possible fusion of unit-groups. In: Furuichi T, Thompson J (eds) The bonobos: behavior, ecology, and conservation. Springer, New York, pp 107–119

Hohmann G, Fruth B (2002) Dynamics of social organization of bonobos (*Pan paniscus*) In: Boesch C, Hohmann G, Marquardt L (eds) Behavioral diversity of chimpanzees and bonobos. Cambridge Univ Press, New York, pp 138–150

Idani G, Kuroda S, Kano T, Asato R (1994) Flora and vegetation of Wamba forest, central Zaire with reference to bonobo (*Pan paniscus*) foods. Tropics 3: 309–332

Idani G, Mwanza N, Ihobe H, Hashimoto C, Tashiro Y, Furuichi T (2008) Changes in the status of bonobos, their habitat, and the situation of humans at Wamba, in the Luo Scientific Reserve, Democratic Republic of the Congo. In: Furuichi T, Thompson J (eds) The bonobos: behavior, ecology, and conservation. Springer, New York, pp 291–302

Isabirye-Basuta G (1988) Food competition among individuals in a free-ranging chimpanzee community in Kibale Forest, Uganda. Behaviour 105: 135–147

Janson CH, Goldsmith ML (1995) Predicting group size in primates: foraging costs and predation risks. Behav Ecol 6: 326–336

Kano T (1982) The social group of pygmy chimpanzees (*Pan paniscus*) of Wamba. Primates 23: 171–188

Kano T (1983) An ecological study of the pygmy chimpanzees (*Pan paniscus*) of Yalosidi, Republic of Zaire. Int J Primatol 4: 1–31

Kano T (1992) The last ape: pygmy chimpanzee behavior and ecology. Stanford Univ Press, Stanford

Kano T, Mulavwa M (1984) Feeding ecology of the pygmy chimpanzees (*Pan paniscus*) of Wamba. In: Susman RL (ed) The pygmy chimpanzee. Plenum, New York, pp 233–274

Kuroda S (1979) Grouping of the pygmy chimpanzees. Primates 20: 161–183

Malenky RK, Stiles EW (1991) Distribution of terrestrial herbaceous vegetation and its consumption by *Pan paniscus* in the Lomako Forest, Zaire. Am J Primatol 23: 153–169

Malenky RK, Wrangham RW (1994) A quantitative comparison of terrestrial herbaceous food consumption by *Pan* paniscus in the Lomako Forest, Zaire and *Pan troglodytes* in the Kibale Forest, Uganda. Am J Primatol 32: 1–12

Matsumoto-Oda A, Hosaka K, Huffman MA, Kawanaka K (1998) Factors affecting party size in chimpanzees of the Mahale Mountains. Int J Primatol 19: 999–1011

Newton-Fisher NE, Reynolds V, Plumptre AJ (2000) Food supply and chimpanzee (*Pan troglodytes schweinfurthii*) party size in the Budongo Forest Reserve, Uganda. Int J Primatol 21: 613–628

Reynolds V (2005) The Chimpanzees of the Budongo Forest: ecology, behaviour, and conservation. Oxford Univ Press

Stanford CB, Wallis J, Mpongo E, Goodall J (1994) Hunting decisions in wild chimpanzees. Behaviour 131: 1–18

White FJ (1988) Party composition and dynamics in *Pan paniscus*. Int J Primatol 9: 179–193

White FJ (1998) Seasonality and socioecology: the importance of variation in fruit abundance to bonobo sociality. Int J Primatol 19: 1013–1027

Williams JM, Hsien-Yang L, Pusey AE (2002) Costs and benefits of grouping for female chimpanzees at Gombe. In: Boesch C, Hohmann G, Marquardt L (eds) Behavioral diversity of chimpanzees and bonobos. Cambridge Univ Press, New York, pp 192–203

Wrangham RW (1977) Feeding behaviour of chimpanzees in Gombe National Park, Tanzania. In: Clutton-Brock TH (ed) Primate ecology: studies of feeding and ranging behaviour in lemurs, monkeys and apes. Academic Press, London, New York, San Francisco, pp 503–538

Wrangham RW (1979) Sex differences in chimpanzee dispersion. In: Hamburg DA, McCown ER (eds) The great apes. Benjamin/Cummings Publishing, Menlo Park, California, pp 481–489

Wrangham RW (1986) Ecology and social relationships in two species of chimpanzee. In: Rubenstein DL, Wrangham RW (eds) Ecological aspect of social evolution. Princeton Univ Press, Princeton, pp 352–378

Wrangham RW (2000) Why are male chimpanzees more gregarious than mothers?: a scramble competition hypothesis. In: Kappeler PM (ed) Primate males: causes and consequences of variation in group composition. Cambridge Univ Press, Cambridge, pp 248–258

# Relationships among Fruit Abundance, Ranging Rate, and Party Size and Composition of Bonobos at Wamba

Takeshi Furuichi[1], Mbangi Mulavwa[2], Kumugo Yangozene[2], Mikwaya Yamba-Yamba[2], Balemba Motema-Salo[2], Gen'ichi Idani[3], Hiroshi Ihobe[4], Chie Hashimoto[1], Yasuko Tashiro[3], and Ndunda Mwanza[2]

## Introduction

As close relatives, chimpanzees (*Pan troglodytes*) and bonobos (*Pan paniscus*) share some important characteristics in their social structure. Both species form male philopatric unit-groups, with males remain in their natal group and females transfer between groups before they reach sexual maturity (Nishida 1979, Kano 1982, 1992, Goodall 1986, Wrangham 1986, 1987, Pusey and Packer 1987, Nishida et al. 1990, Furuichi 1989, 2006, Wallis 1997, Reynolds 2005). In addition, unit-groups of both species split into foraging parties of flexible size and composition (Kuroda 1979, Nishida 1979, Wrangham 1979, Goodall 1986, Kano 1982, 1992). However, the species show marked differences in their association patterns. In chimpanzees, males tend to join larger parties more frequently than females do, while females tend to range alone or in smaller parties (Nishida 1979, Wrangham 1979, 2000, Goodall 1986, Janson and Goldsmith 1995, Boesch 1996, Reynolds 2005, Thompson and Wrangham 2005). In contrast, female bonobos tend to join parties more frequently than males do (Kano 1982, 1992, Furuichi 1987, 1989, White 1988, Mulavwa et al. 2008). Many researchers have debated why unrelated females aggregate more than related males do, and they have proposed several hypotheses on this matter, presented below (White and Wrangham 1988, Kano 1992, Wrangham 2000, Furuichi 2006).

To clarify the nature of aggregation in female bonobos, we must first examine why female chimpanzees tend to aggregate less than males do. The two hypotheses are that the benefit of grouping is greater for males, and that the cost of grouping is greater for females.

[1] *Primate Research Institute, Kyoto University, Inuyama, Aichi, 484-8506 Japan*

[2] *Research Center for Ecology and Forestry, Ministry of Scientific Research and Technology, D.R. Congo*

[3] *Great Ape Research Institute, Hayashibara Biochemical Laboratories, Japan*

[4] *School of Human Sciences, Sugiyama Jogakuen University, Japan*

Male chimpanzees may want to aggregate because alliance formation with some males through daily association may be aimed towards competition with the other males within the group (Goodall 1986, de Waal 1991, Nishida and Hosaka 1996, Watts 1998, Reynolds 2005). Furthermore, close association among males in a group helps overcome severe competition between different groups (Nishida 1985, Goodall 1986, Wrangham and Peterson 1996). Furuichi and Hashimoto (2002) proposed a hypothesis that the severe intragroup and intergroup sexual competition due to the high estrous sex ratio (or operational sex ratio, given by the number of adult males per female in estrus) may have enforced the association and alliance among kin-related males that led to the evolution of male philopatry in *Pan* species.

However, female chimpanzees may suffer from several kinds of grouping costs, including harassment by males such as coercion and infanticide, contest competition for food resources with dominant males, and scramble competition for food resources (Wrangham 2000, 2002). Females may be able to avoid harassment by ranging apart from males, but at the same time, females may need to form close associations with males to prevent harassment by them (Goodall 1986, Hamai et al. 1992, Arcadi and Wrangham 1999, Wrangham 2002). As for the contest competition, aggressive interactions with males or displacement over food do not seem to be frequent enough to keep females out of the party, though there is insufficient quantitative data to evaluate this hypothesis (Goodall 1986, Wrangham 2000). Thus, harassment and contest competition may not be significant factors in the low gregariousness of female chimpanzees.

Scramble competition may be more persuasive in preventing female chimpanzees from joining large parties (Wrangham 1979, 2000, Isabirye-Basuta 1988, Janson and Goldsmith 1995, van Schaik 1999, Boesch and Boesch-Achermann 2000). If many individuals range together in a party, they would consume a food patch more quickly and would need to travel between patches more frequently. Such ranging may incur a higher cost for females. Females, especially those carrying infants, may move more slowly than males and spend more time and energy traveling between patches. Moreover, when they arrive at a new patch, the food may have already been consumed or the preferred positions for feeding may be occupied by fast-arriving males. Thus, the cost for ranging in a large party may be highest for females with dependent infants, second highest for cycling females, and lowest for males; consequently, females may be less gregarious. However, if scramble competition explains the lower gregariousness of female chimpanzees, it remains unclear why female bonobos tend to aggregate more than males do. Does the hypothesis hold for chimpanzees but not for bonobos?

The scramble competition hypothesis provides some important predictions that can be tested in chimpanzees and bonobos. If the cost of foraging affects the gregariousness of females, seasonal changes in the feeding environment, such as food abundance or distribution, may influence the party size. If the higher ranging cost restrains females from joining parties, they may be reluctant to join larger parties that are likely to move faster or cover a larger distance in a day. To test these predictions, we first examined the relationships among fruit abundance, party size, and daily mean ranging rate. We then examined whether females' attendance in parties is affected by changes in the party size and ranging rate. Finally, we discuss whether the difference in the ranging cost may explain the difference between chimpanzees and bonobos in the female ranging pattern.

## Methodology

### Study Site and Group

We observed a unit-group, or community (van Elsacker 1995, Reynolds 2005), of wild bonobos, called E1, at Wamba (0°11′8″N, 22°37′58″E), in the northern section of the Luo Scientific Reserve, Democratic Republic of the Congo. The ranging area of E1 was comprised of primary forest, old secondary forest, young secondary forest, swamp forest, and agricultural fields. We recorded the daily rainfall at the Wamba research camp throughout the study period. Annual rainfall was 2843 mm in 2004 and 2922 mm in 2005.

E1 split from E around 1983 (Furuichi 1987, Kano 1992). Researchers had studied E since 1974, and all members were identified in 1976. E, and E1 after the split, had been provisioned with sugar cane for some part of each year until 1996, when the study was interrupted by consecutive civil wars. We resumed observations of E1 in 2002, without artificial provisioning.

Our current study is based on observations of E1 from September 2003 to December 2005. In January 2005, E1 comprised 23 members: 10 adult males, 6 adult females, 1 adolescent female, 2 juvenile females, 3 infant males, and 1 infant female. Further details on the location of the study site, vegetation, history of the study group, and current membership are described by Furuichi (1987), Kano (1992), Idani et al. (1994, 2008), and Hashimoto et al. (1998, 2008).

### Observations

We tried to locate parties of E1 on 6 days of each week. We made direct observations of bonobos on 484 days, which represented 68% of the working days during the study period. We followed the parties from sleeping site to sleeping site when possible. On average, we observed bonobos directly for 4.6 h per day on days when we conducted direct observations.

We started a tracking session when we found bonobos at a sleeping site in the morning or while they were ranging in the forest. While tracking a party, we recorded their position at 30-min intervals via a global positioning system (GPS) receiver. When we lost sight of the bonobos, we followed their tracks or vocalizations. If we could find them again within 1 h, we assumed that we had been successfully tracking them on their ranging route. If >1 h passed, we assumed that we had lost them, and began a new tracking session when we found them again. We calculated the daily mean ranging rate by dividing the total ranging distance by the total time spent for all tracking sessions of the day. Due to some mechanical problems with the GPS, data for ranging rate were only available for September 2003-January 2004, March-November 2004, and October-December 2005.

While following a party, we recorded members using a 1-h party method (Hashimoto et al. 2001). We recorded all bonobos in sight at the beginning of each 1-h segment and continued recording bonobos that appeared in sight until the end of that hour. We calculated the party size and composition for each day by averaging the data recorded for each observation hour on that day. Data on party size and composition are available for the whole study period.

To census fruit abundance, we established five trails totaling 22,550 m, which covered the home range of E1, and we walked the trails twice per month. When we found a cluster of fallen fruit within 1 m on either side of the trail, we recorded the name of the species and whether the fruits were ripe or unripe. We estimated fruit abundance from the number of clusters of ripe food fruit per km of trails (Furuichi et al. 2001). The food fruit included only the species that we observed bonobos eating in the current or past study periods. Data for fruit abundance are available for June 2004 and from August 2004 to December 2005. Further details on the observations of the 1-h party size and fruit abundance are described by Furuichi et al. (2001), Hashimoto et al. (2001), and Mulavwa et al. (2008).

## Results

### *Relationship of Party Size to Fruit Abundance*

Figure 7.1 shows the results of Mulavwa et al. (2008) on the relationship of the party size to the abundance of ripe food fruit. The number of all independent individuals, the number of adult males, and the number of adult females significantly correlate with fruit abundance. However, the increase in the number of males or females with fruit abundance is fairly limited. The difference in the expected number of individuals between the highest fruiting season and the lowest fruiting season is 1.5 for adult males (36.7% of the expected number in the lowest fruiting season) and 1.1 for adult females (37.6% of the expected number in the lowest fruiting season).

### *Ranging Rate*

We overlaid the ranging routes that we recorded during the entire observation period on a satellite image of the ranging area (Fig. 7.2). In the satellite image, the pale-colored areas are agricultural fields and young secondary forests; the

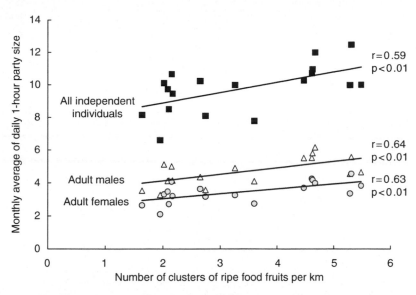

**Fig. 7.1** Relationship of the monthly average of the daily 1-h party size to the abundance of ripe food fruit in a month. Each dot represents data for 1 month (n = 18 months for which data on fruit abundance were available; see Mulavwa et al. (2008).

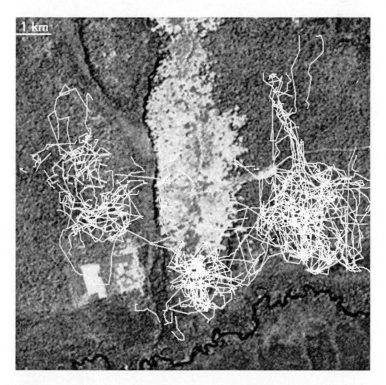

**Fig. 7.2** Ranging routes of E1 and vegetation in the ranging area. Each line fragment shows the ranging route of a tracking session. The vegetation image was made from Landsat data recorded on 14 January 1991 (Hashimoto et al. 1998).

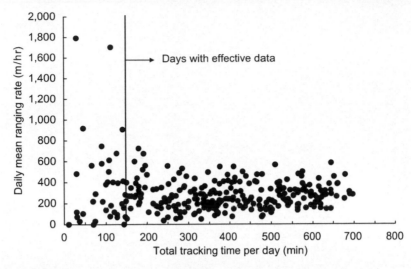

**Fig. 7.3** Relationship of daily mean ranging rate to the total tracking time per day. Each dot represents one observation day (N=288). Days with <150 min of tracking time showed larger variance in the mean ranging rate. The number of days on which total tracking time was 150 min or more was 250.

medium-colored areas are old secondary forest and primary forest; and the dark-colored areas around the river are swamp forests.

E1 ranged in a U-shaped area around the village of Wamba. Although they mainly used the primary forest and old secondary forest, they sometimes used the young secondary forest and swamp forest (see also Hashimoto et al. 1998). An analysis of the ranging area according to a timeline revealed that E1 usually remained in one area, viz., to the west, south, or south-east of the village, for several weeks. After that, probably when the available fruit resource in an area was depleted, they moved to another location and stayed there for several weeks.

Figure 7.3 shows the relationship of the daily mean ranging rate and the total tracking time for a day. The days with a short total tracking time showed a large variation in the mean ranging rate because the shorter tracking sessions may have accidentally coincided with times of rapid movement or resting. Therefore, we omitted days with <150 min of total tracking time from the analyses.

The number of observation days with ≥150 min of tracking time is 250. The mean tracking time per day is 413 ± 144 (S.D.) min, and the average of the daily mean ranging rate is 274 ± 124 (S.D.) m/hr. Bonobos ranged with a slower mean rate on days when they stayed in one area, and ranged with a faster mean rate on days when they shifted areas, which may explain the large standard deviation. On average, bonobos started traveling at 07:26 h and began making nests at 16:57 h. If we assume that the daily ranging time is 9 h 31 min, the estimated daily travel distance is 2608 m. Because the daily ranging distance is represented by the daily mean ranging rate, we use the daily mean ranging rate as a parameter of ranging.

## Relationship of Ranging Rate to Party Size and Fruit Abundance

Figure 7.4 shows the relationship between daily mean ranging rate and the daily mean 1-h party size, as expressed by the number of independent individuals. Data for both ranging rate and party size are available for 234 days. There is significant correlation between the two variables (Pearson's $r = 0.23$, $p < 0.001$, $N = 234$). However, the party size explained only a small proportion of variance in the ranging rate, and the expected daily mean rate increased by only 10.6 m/hr with the addition of each individual.

If the ranging rate increases with the party size, and the party size increases with fruit abundance, then the ranging rate may increase with fruit abundance. To examine this hypothesis, we compared the monthly mean ranging rate with the fruit abundance for that month. Data for both fruit abundance and ranging rate are available for 8 months. Although the ranging rate tended to increase with fruit abundance, the correlation is not significant (Pearson's $r = 0.67$, $p = 0.07$, $N = 8$; Fig. 7.5a). For supplementary information, we also tested the relationship between ranging rate and rainfall, but again there is no significant correlation (Pearson's $r = 0.20$, $p = 0.43$, $N = 17$; Fig. 7.5b). Thus, the ranging rate remained at a fairly constant level regardless of the seasonal changes in fruit abundance and rainfall.

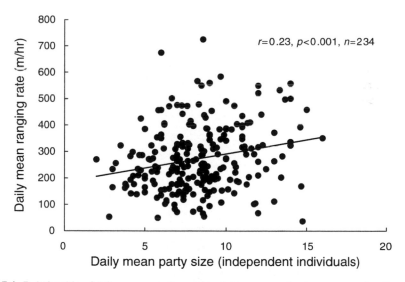

**Fig. 7.4** Relationship of daily mean ranging rate to daily mean party size, expressed as the mean number of independent individuals in a 1-h party. Each dot represents data for 1 day. N = 234 days, on which effective 1-h party size and effective ranging speed both were recorded; see Mulavwa et al. (2008) for details on the effective 1-h party size.

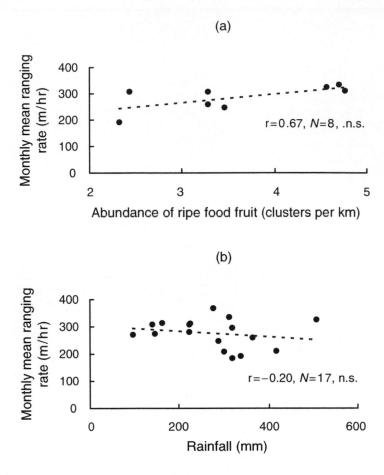

**Fig. 7.5** Relationship of monthly mean ranging rate with the abundance of ripe food fruit (a) and rainfall (b). Each dot represents the data for 1 month. Monthly mean ranging rate was calculated by averaging the daily mean ranging rate for each month.

## *Attendance of Females to Parties with Different Size and Ranging Rate*

To explore whether females change their attendance to parties according to the size and ranging rate, we examined the relationship of the attendance ratio of males and females with these two factors. We calculated the attendance ratio as the proportion of the number of males or females present in the party to the total number of males or females in the unit-group, which is equivalent to the relative party size proposed by Boesch (1996). Thompson and Wrangham (2005) proposed the comparison between the attendance ratio and party size as a method to reduce bias in the comparison of attendance between sexes and between study sites.

(a)

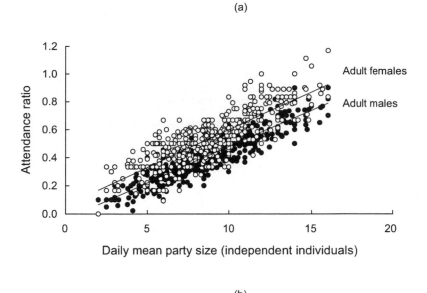

(b)

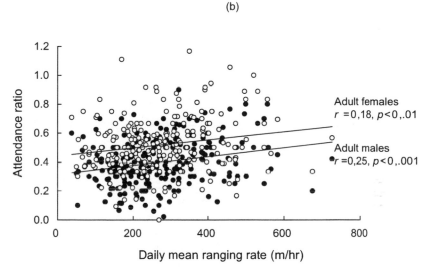

**Fig. 7.6** Relationship of the attendance ratio of males and females to party size (a) and ranging rate (b). The attendance ratio of males or females was calculated by dividing the number of males or females in the party by the total number of males or females in the unit-group. One dot represents data for 1 day. Analysis for (a) was based on 465 days, on which at least one effective 1-h party size was recorded (see Mulavwa et al. 2008), and analysis for (b) was based on 234 days, on which both the effective 1-h party size and the effective ranging rate were recorded (see Mulavwa et al, 2008 for details on the effective 1-h party size). The relative number of females exceeded 1 on some days, probably because of temporary visits by strange females.

   The attendance ratio of both sexes increased with party size with a similar slope, and the attendance ratio of females was always higher than that of males (Fig. 7.6a). This means that females were more willing to join a party than were males, irrespective of the changes in party size.

   There is similar tendency for the relationship between the attendance ratio and the ranging rate (Fig. 7.6b). The attendance ratio of both males and females increased with the ranging rate with a similar slope, and the attendance ratio of females was always higher than that of males. This means that females were more willing to join a party than were males, irrespective of the changes in ranging rate.

## Discussion

Figure 7.7 is a summary of our results. The monthly average of the party size of E1 significantly increased with the abundance of ripe food fruit. The daily mean ranging rate, and hence the daily ranging distance, significantly increased with that of party size. The monthly average of the daily mean ranging rate also increased with the abundance of ripe food fruit, but the correlation was not significant. Many studies on chimpanzees have suggested positive relationships among fruit abundance, party size, and daily mean ranging rate or daily ranging distance (e.g., Chapman et al. 1994, Janson and Goldsmith 1995, Wrangham 2000, Williams et al. 2002), with which the results of our study are generally compatible.

   Party size may increase with fruit abundance because favorite fruits may attract many chimpanzees, and fruiting trees may serve as large food patches in which a large number of chimpanzees feed together. Although there is some variability among studies, our results generally agree with those of previous studies of wild bonobos (Kuroda 1979, White 1998, Hohmann and Fruth 2002, Mulavwa et al. 2008). The ranging rate or distance may increase with party size, because parties including many individuals may consume fruits in a patch more quickly and, therefore, may shift between food patches more frequently. On the other hand, the correlation of ranging rate/distance with fruit abundance is unclear, because the ranging rate/distance may be influenced not only by abundance of fruit, but also by its distribution. If fruit abundance increases with the fruiting of favorite species that are distributed evenly or randomly, then ranging rate or distance will increase as the apes seek out their preferred fruits. In contrast, if fruit abundance increases with an increase in the fruiting of large trees or trees that show a clumped distribution, then ranging rate/distance may decrease.

   Preceding studies of chimpanzees have suggested that the cost of ranging in large parties would be greater for females, which may explain the tendency of female chimpanzees to range alone or in smaller parties (Wrangham 2000, Williams et al. 2002, Pontzer and Wrangham 2006). Females may suffer from harassment and contest competition when they feed with dominant males in large mixed parties. Females may also suffer more from scramble competition than males would. Females, particularly those with dependent offspring, may move more slowly and

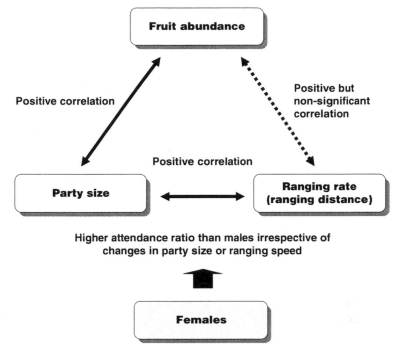

**Fig. 7.7** Relationships among fruit abundance, party size, and ranging rate, and attendance of females to parties with different size and ranging rate/distance.

therefore need to spend more time traveling between food patches, and may arrive at a new patch later than fast-moving males. However, in bonobos, females are more likely than males to join mixed parties (Kano 1982, 1992, Furuichi 1987, 1989, White 1988, White and Wrangham 1988, Mulavwa et al. 2008). Our study showed that females did not avoid joining larger or fast-moving parties. Why is there such a difference between chimpanzees and bonobos? Is the theory developed to explain female dispersion of chimpanzees not a valid explanation for variations found in *Pan* species? Or are there bonobo-specific factors that affect the ranging pattern of females?

Earlier studies suggested that larger fruit patches, a higher density of fruit patches, the existence of feed-as-you-go foods such as terrestrial herbaceous vegetation, and less marked seasonality of food abundance and distribution in the bonobo habitat may prevent an increase in ranging rate and allow the formation of large stable parties (White and Wrangham 1988, Malenky and Stiles 1991, Kano 1992, Chapman et al. 1994, Malenkey and Wrangham 1994, White 1998, Wrangham 2000, Furuichi and Hashimoto 2002). Our results seem to agree with this hypothesis. Although party size of bonobos changes with the seasonal change in fruit abundance, the expected party size in the highest fruiting season increased only 37% relative to that of the lowest fruiting season. Similarly, although the ranging rate changed with the change in party size, it increased by only 10.6 m/hr with the addition of each new individual. Thus in bonobos, the change in party size or ranging rate may be too small to suppress female attendance in mixed-sex parties.

However, there is a problem in the hypothesis that attributes the difference in female ranging pattern between chimpanzees and bonobos to the differences in food availability and scramble competition. The abundance, distribution, and seasonality of food resources in bonobo habitats may not be different enough from those of chimpanzee habitats to explain the differences in female ranging patterns. For example, Chapman et al. (1994) reported that the density and diameter at breast height (DBH) of food trees are not different between Lomako, D.R. Congo, for bonobos and Kibale, Uganda, for chimpanzees. Chimpanzees inhabit a wide variety of habitats that vary from rain forest in Taï, Côte d'Ivoire (Boesch and Boesch-Achermann 2000), to semi-arid areas in Senegal and Tanzania (Moore 1996, Pruetz et al. 2002), and the habitats of bonobos may be included in the range of this variation.

Another factor that may explain the difference in female ranging patterns between chimpanzees and bonobos is the high social status of female bonobos. Many studies of captive bonobos indicate that female bonobos have higher social status than males (Parish 1996, Paoli and Palagi 2008, Stevens et al. 2008). Studies of wild bonobos at Wamba indicated that males and females had almost equal status, and that females had priority for feeding (Kano 1992, Furuichi 1997). The high social status substantially reduces the cost of contest competition for females that join mixed-sex parties.

Moreover, the high social status may reduce the cost of scramble competition for females. Ranging of mixed parties of E1 is usually controlled by females. Because they ranged in one area for up to several weeks, the daily ranging distance is not very long. Even when males want to travel a longer distance to shift the foraging area, they often give up the attempts if females do not want to do so, and return to the females by evening. If females can control the ranging pattern, then the ranging rate may not exceed the acceptable level for females. Therefore, females can attend mixed-sex parties without incurring the additional cost of scramble competition that comes from their lower mobility.

If the preceding hypotheses do not sufficiently explain the differences in female ranging patterns between the two species, we may need to further investigate the influence of the high social status of female bonobos on the cost of both contest and scramble competition. For that purpose, we may need to carry out ecological studies at more sites of chimpanzees and bonobos, using common methodologies for quantitative comparisons.

**Acknowledgements** Dr. Takayoshi Kano and Dr. Toshisada Nishida have provided continuous support and advice for the study at Wamba. Dr. Tetsuro Matsuzawa offered invaluable support for the resumption of the study after the civil wars. Discussions with Dr. Juichi Yamagiwa, Dr. Shigeru Suzuki, Dr. Yuji Takenoshita, and researchers in the Laboratory of Human Evolution and Primate Research Institute, Kyoto University, were most helpful. Staff of the Research Center for Ecology and Forestry (CREF) and the Luo Scientific Reserve in the D.R. Congo helped with various aspects of our study. We sincerely thank them all. This study was financially supported by the National Geographic Fund for Research and Exploration (#7511-03 to Furuichi), the Japan

Society for the Promotion of Science (JSPS) core-to-core program HOPE (#15001 to Matsuzawa), and JSPS Grants-in-aid for Scientific Research (#12575017, #17255005 to Furuichi).

# References

Arcadi AC, Wrangham RW (1999) Infanticide in chimpanzees: review of cases and a new within-group observation from the Kanyawara study group in Kibale National Park. Primates 40: 337–351

Boesch C (1996) Social grouping in Taï chimpanzees In: McGrew WC, Marchant LF, Nishida T (eds) Great ape societies. Cambridge Univ Press, Cambridge, pp 101–113

Boesch C, Boesch-Achermann H (2000) The Chimpanzees of the Taï Forest: behavioural ecology and evolution. Oxford Univ Press, Oxford

Chapman CA, White FJ, Wrangham RW (1994) Party size in chimpanzees and bonobos: a reevaluation of theory based on two similarly forested sites. In: McGrew WC, de Waal FBM, Heltne PG (eds) Chimpanzee cultures. Harvard University Press, Cambridge, Massachusetts, London, pp 41–57

van Elsacker L (1995) A review of terminology on aggregation patterns in bonobos (*Pan paniscus*). Int J Primatol 16: 37–52

Furuichi T (1987) Sexual swelling, receptivity and grouping of wild pygmy chimpanzee females at Wamba, Zaire. Primates 28: 309–318

Furuichi T (1989) Social interactions and the life history of female *Pan paniscus* in Wamba, Zaire. Int J Primatol 10: 173–197

Furuichi T (1997) Agonistic interactions and matrifocal dominance rank of wild bonobos (*Pan paniscus*) at Wamba. Int J Primatol 18: 855–875

Furuichi T (2006) Evolution of the social structure of hominoids: reconsideration of food distribution and the estrus sex ratio. In: Ishida H, Tuttle R, Pickford M, Nakatsukasa M, Ogihara N (eds) Human origins and environmental backgrounds. Springer, New York, pp 235–248

Furuichi T, Hashimoto C (2002) Why female bonobos have a lower copulation rate during estrus than chimpanzees. In: Boesch C, Hohmann G, Marquardt L (eds) Behavioral diversity of chimpanzees and bonobos. Cambridge Univ Press, New York, pp 156–167

Furuichi T, Hashimoto C, Tashiro Y (2001) Fruit availability and habitat use by chimpanzees in the Kalinzu Forest, Uganda: examination of fallback foods. Int J Primatol 22: 929–945

Goodall J (1986) The chimpanzees of Gombe: patterns of behavior. Harvard University Press, Cambridge

Hamai M, Nishida T, Takasaki H, Turner LA (1992) New records of within-group infanticide and cannibalism in wild chimpanzees. Primates 33: 151–162

Hashimoto C, Tashiro Y, Kimura D, Enomoto T, Ingmanson EJ, Idani G, Furuichi T (1998) Habitat use and ranging of wild bonobos (*Pan paniscus*) at Wamba. Int J Primatol 19: 1045–1060

Hashimoto C, Furuichi T, Tashiro Y (2001) What factors affect the size of chimpanzee parties in the Kalinzu Forest, Uganda? Examination of fruit abundance and number of estrous females. Int J Primatol 22: 947–959

Hashimoto C, Yasuko T, Hibino E, Mulavwa M, Yangozene K, Furuichi T, Idani G, Takenaka O (2008) Longitudinal structure of a unit-group of bonobos: male philopatry and possible fusion of unit-groups. In: Furuichi T, Thompson J (eds) The bonobos: behavior, ecology, and conservation. Springer, New York, pp 107–119

Hohmann G, Fruth B (2002) Dynamics of social organization of bonobos (*Pan paniscus*). In: Boesch C, Hohmann G, Marquardt L (eds) Behavioral diversity of chimpanzees and bonobos. Cambridge Univ Press, New York, pp 138–150

Idani G, Kuroda S, Kano T, Asato R (1994) Flora and vegetation of Wamba forest, central Zaire with reference to bonobo (*Pan paniscus*) foods. Tropics 3: 309–332

Idani G, Mwanza N, Ihobe H, Hashimoto C, Tashiro Y, Furuichi T (2008) Changes in the status of bonobos, their habitat, and the situation of humans at Wamba, in the Luo Scientific Reserve, Democratic Republic of the Congo. In: Furuichi T, Thompson J (eds) The bonobos: behavior, ecology, and conservation. Springer, New York, pp 291–302

Isabirye-Basuta G (1988) Food competition among individuals in a free-ranging chimpanzee community in Kibale Forest, Uganda. Behaviour 105: 135–147

Janson CH, Goldsmith ML (1995) Predicting group size in primates: foraging costs and predation risks. Behav Ecol 6: 326–336

Kano T (1982) The social group of pygmy chimpanzees (Pan paniscus) of Wamba. Primates 23: 171–188

Kano T (1992) The last ape: pygmy chimpanzee behavior and ecology. Stanford Univ Press, Stanford

Kuroda S (1979) Grouping of the pygmy chimpanzees. Primates 20: 161–183

Malenky RK, Stiles EW (1991) Distribution of terrestrial herbaceous vegetation and its consumption by Pan paniscus in the Lomako Forest, Zaire. Am J Primatol 23: 153–169

Malenky RK, Wrangham RW (1994) A quantitative comparison of terrestrial herbaceous food consumption by Pan paniscus in the Lomako Forest, Zaire, and Pan troglodytes in the Kibale Forest, Uganda. Am J Primatol 32: 1–12

Moore J (1996) Savanna chimpanzees, referential models and the last common ancestor. In: McGrew WC, Marchant LF, Nishida T (eds) Great ape societies. Cambridge Univ Press, Cambridge, pp 275–292

Mulavwa M, Furuichi T, Yangozene K, Yamba-Yamba M, Motema-Salo B, Idani G, Ihobe H, Hashimoto C, Tashiro Y, Mwanza N (2008) Seasonal changes in fruit production and party size of bonobos at Wamba. In: Furuichi T, Thompson J (eds) The bonobos: behavior, ecology, and conservation. Springer, New York, pp 121–134

Nishida T (1979) The social structure of chimpanzees of the Mahale Mountaïns. In: Hamburg DA, McCown ER (eds) The great apes. The Benjamin/Cummings Publishing Company, Menlo Park, pp 73–121

Nishida T (1985) Group extinction and female transfer in wild chimpanzees in the Mahale National Park, Tanzania. Z Tierpsychol 67: 284–301

Nishida T, Hosaka K (1996) Coalition strategies among adult male chimpanzees of the Mahale Mountaïns, Tanzania. In: McGrew WC, Marchant LF, Nishida T (eds) Great ape societies. Cambridge Univ Press, pp 114–134

Nishida T, Takasaki H, Takahata Y (1990) Demography and reproductive profiles. In: Nishida T (ed) The Chimpanzees of the Mahale Mountaïns: sexual and life history strategies. Univ of Tokyo Press, Tokyo, pp 63–97

Paoli T, Palagi E (2008) What does agonistic dominance imply in bonobos? In: Furuichi T, Thompson J (eds) The bonobos: behavior, ecology, and conservation. Springer, New York, pp 39–54

Parish AR (1996) Female relationships in bonobos (Pan paniscus): evidence for bonding, cooperation, and female dominance in a male-philopatric species. Hum Nat 7:61–96

Pontzer H, Wrangham RW (2006) Ontogeny of ranging in wild chimpanzees. Int J Primatol 27: 295–309

Pruetz JD, Marchant LF, Arno J, McGrew WC (2002) Survey of savanna chimpanzees (Pan troglodytes verus) in southeastern Senegal. Amer J Primatol 58: 35–43

Pusey AE, Packer C (1987) Dispersal and philopatry. In: Smuts BB, Cheney DL, Seyfarth RM, Wrangham RW, Struhsaker TT (eds) Primate societies. The University of Chicago Press, Chicago and London, pp 250–266

Reynolds V (2005) The chimpanzees of the Budongo Forest: ecology, behaviour, and conservation. Oxford Univ Press

van Schaik CP (1999) The socioecology of fission-fusion sociality in orangutans. Primates 40: 69–86

Stevens J, Vervaecke H, van Elsacker L (2008) The bonobo's adaptive potential: social relations under captive conditions. In: Furuichi T, Thompson J (eds) The bonobos: behavior, ecology, and conservation. Springer, New York, pp 19–38

Thompson ME, Wrangham RW (2005) Comparison of sex differences in fission-fusion species: reducing bias by standardizing for party size. In: Newton-Fischer NE, Notman H, Paterson JD, Reynolds V (eds) Primates of western Uganda. Springer, pp 209–226

de Waal FBM (1991) Chimpanzee politics. Jonathan Cape, London

Wallis J (1997) A survey of reproductive parameters in the free-ranging chimpanzees of Gombe National Park. J Reproduction Fertility 109: 297–307

Watts DP (1998) Coalitionary mate guarding by male chimpanzees at Ngogo, Kibale National Park, Uganda. Behav Ecol Sociobiol 44: 43–55

White FJ (1988) Party composition and dynamics in *Pan paniscus*. Int J Primatol 9: 179–193

White FJ (1998) Seasonality and socioecology: the importance of variation in fruit abundance to bonobo sociality. Int J Primatol 19: 1013–1027

White FJ, Wrangham R W (1988) Feeding competition and patch size in the chimpanzee species *Pan paniscus* and *Pan troglodytes*. Behaviour 105:148–164

Williams JM, Liu HY, Pusey A (2002) Costs and benefits of grouping for female chimpanzees at Gombe. In: Boesch C, Hohmann G, Marquardt L (eds) Behavioral diversity of chimpanzees and bonobos. Cambridge Univ Press, New York, pp 192–203

Wrangham RW (1979) Sex differences in chimpanzee dispersion. In: Hamburg DA, McCown ER (eds) The great apes. Benjamin/Cummings Publishing, Menlo Park, California, pp 481–489

Wrangham RW (1986) Ecology and social relationships in two species of chimpanzee. In: Rubenstein DL, Wrangham RW (eds) Ecological aspect of social evolution. Princeton Univ Press, Princeton, pp 352–378

Wrangham RW (1987) Evolution of social structure. In: Smuts BB, Cheney DL, Seyfarth RM, Wrangham RW, Struhsaker TT (eds) Primate societies. The University of Chicago Press, Chicago and London, pp 282–296

Wrangham RW (2000) Why are male chimpanzees more gregarious than mothers? a scramble competition hypothesis. In: PM Kappeler (ed) Primate males: causes and consequences of variation in group composition. Cambridge Univ Press, Cambridge, pp 248–258

Wrangham RW (2002) The cost of sexual attraction: is there a trade-off in female *Pan* between sex appeal and received coercion? In: Boesch C, Hohmann G, Marquardt L (eds) Behavioral diversity of chimpanzees and bonobos. Cambridge Univ Press, New York, pp 204–215

Wrangham RW, Peterson D (1996) Demonic males: apes and the origins of human violence. Houghton Miffin, Boston

# Bonobo (*Pan paniscus*) Density Estimation in the SW-Salonga National Park, Democratic Republic of Congo: Common Methodology Revisited

Meike Mohneke[1] and Barbara Fruth[2]

## Introduction

Worldwide biodiversity has declined rapidly during the last decades due to pressures on environments from human population growth, resulting deforestation, habitat destruction, and bushmeat trade. Many nonhuman primates, including orangutans (*Pongo pygmaeus*), gorillas (*Gorilla gorilla*), chimpanzees (*Pan troglodytes*), and bonobos (*Pan paniscus*), are endangered or close to extinction (Tutin and Fernandez 1984, Sugiyama and Soumah 1988, Hoppe-Dominik 1991, Wilkie et al. 1992, Marchesi et al. 1995, Chapman et al. 1999, Wilkie and Godoy 2000, Moore 2001, Barnes 2002,Bennett et al. 2002, Draulans and Van Krunkelsven 2002, Ling et al. 2002, Muoria et al. 2003, Whitfield 2003). Bonobos are listed as highly vulnerable in the IUCN/SSC Action Plan for African Primate Conservation (Oates 1986), and as endangered in the IUCN Red List of Threatened Species (IUCN 2004). They are endemic to the Congo basin in the Democratic Republic of Congo. The species is officially protected by Congolese and international laws and is listed in Appendix 1 of CITES (CITES 2005) and on Class A of the African Convention on the Conservation of Nature and Natural Resources (African Union 1968).

Although highly ranked for conservation action, knowledge about the bonobos' actual distribution and status is largely based on estimates that lack solid investigations on the ground. Estimates of the potential area of bonobo distribution range from 343,000 km$^2$ based on observational data (Butynski 2001) to 840,400 km$^2$, with the Lualaba River as the eastern and the Sankuru River as the southern boundary (Thompson-Handler et al. 1995). Population size estimates covering large distribution areas range from 0.12 individuals per km$^2$ (Kortlandt 1995) to 0.4 individuals per km$^2$ (Thompson-Handler et al. 1995, Kano 1984). For a more realistic assessment of the

[1] *Rheinische Friedrich Wilhelms University, Department of Zoology, 53115 Bonn, Germany*

[2] *Max Planck Institute for Evolutionary Anthropology, Department of Primatology, Germany*

T. Furuichi and J. Thompson (eds.), *The Bonobos: Behavior, Ecology, and Conservation*
© Springer 2008

specific status of bonobos in the wild with respect to effective conservation policies, a detailed census over a wide range, taking into account factors influencing the
bonobo distribution pattern, such as habitat and seasonality, is urgently needed
(Susman et al. 1981, Kano 1984).

First attempts have been made for smaller areas of investigation. Hashimoto and
Furuichi (2001) investigated Luo Scientific Reserve, an area of 481 km², in 1996,
where the smaller northern part represents the long term research site of Kyoto
University, Wamba, while the larger southern part was added in 1992. They covered
over 100km of census routes, and found a density of 0.28 – 0.54 individuals per km².
Between 1996 and 1998, Eriksson (1999) conducted the first systematic census of
bonobos outside a study area between the Lomako and Yekokora Rivers covering an
area of about 1200km². He found a density of 1.3 – 1.4 weaned individuals per km².
From October 2000 to May 2002, Reinartz et al. (2006) investigated 48 line
transects totalling 67.8 km across 9 sites in both sectors of Salonga National Park.
They found densities ranging between 0 and 2.8 individuals per km². Between 2003
and 2006, a project with major focus at the Monitoring of Illegal Killing of
Elephants (MIKE) in collaboration with the Wildlife Conservation Society (WCS)
investigated bonobo abundance for the entire Salonga National Park. Grossmann et al.
(2008) used standing crop nest counts along 260km of line transects. They found
0.29 – 0.90 nest-building bonobos per km². Unfortunately, methodological issues
leading to an extremely low nest encounter rate make their significance disputable
(MIKE 2004).

Here, we present a small scale intensive rather than large scale extensive survey
to reveal the status of bonobos within and outside the study site Lui Kotale, in the
southwest corner of the Salonga National Park (Fig. 8.1), established in 2002 under
the auspice of the Max Planck Society (Hohmann and Fruth 2003). We conducted
the survey from August 2003 to February 2004, applying several methods for
comparison. For the implementation of the survey, we applied the standard
line-transect sampling method, which is the common method to estimate primate
densities. Based on the cryptic behavior and resulting low detectability of great
apes, researchers considered nest instead of ape counts as the best method for
density estimations. Pioneers such as Ghiglieri (1984) and Tutin and Fernandez
(1984) used the traditional standing crop nest count method. To achieve higher
accuracy, we took into account life-span of nests in the focal area as well as nest
construction rates per individual. We then transformed the method into variable
width transect sampling (Buckland et al. 1993). Because the method has been applied
in a number of surveys, it provides directly comparable population estimates
(Sabater Pi and Vea 1990, Hoppe-Dominik 1991, Hashimoto 1995, Marchesi et al.
1995, Bermejo 1999, Eriksson 1999, Blom et al. 2001, Van Krunkelsven 2001,
Hashimoto and Furuichi 2001, Poulsen and Clark 2004). The method was further
modified by Plumptre and Reynolds (1996) to the marked nest count. It provides a
more precise alternative in that it avoids calculating nest decay rate by repeating
transect walks. We applied both methods in order to achieve comparability and to
test for precision.

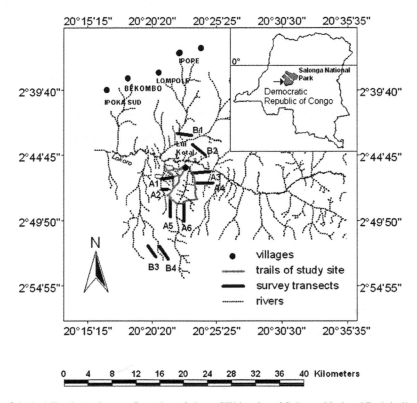

**Fig. 8.1** Lui Kotale study area. Location of site at SW border of Salonga National Park indicated by arrow within DRCongo overview. Villages closest to study site and the research camp are indicated by dot and name. Trail system and standardized sample transects are indicated by grey and black lines respectively.

# Methods

## Study Site

The Lui Kotale study site is situated at the fringe of Salonga National Park (1°00′–3°20′S, 20°–22°30′E) in the Democratic Republic of Congo (Fig. 8.1). The Park was created in 1970, and comprising 36,000 km², it is Africa's largest area of rain forest protection. Since 1984, it has been on the list of world heritage sites representing the largest area for the protection of bonobos and other endemic species. The bonobo's presence has been known by missionaries since the early 70s (Eriksson pers. comm.) although it became scientifically confirmed much later (Meder et al. 1988, Van Krunkelsven et al. 2000, Van Kunkelsven 2001, Reinartz 2003). Recent data on the status of the bonobos within the Park's boundaries became available and showed that hunting represents the major threat for large mammals in general and bonobos in particular (Ilambu and Grossmann 2004, Hart et al. 2008, Eriksson 2007). The area of investigation is the southwestern border of the park with its camp site Lui Kotale (02° 45.610′S, 20° 22.723′E) south of the Lokoro River. The study area consists of 100% primary forest with an average maximal temperature of 27°C during day (6:00h-17:59h) and an average minimal temperature of 21°C at night (18:00-5:59h) for the years 2003/04. We conducted the survey between August 2003 and February 2004, encompassing both rainy (5 month) and dry seasons (2 month).

## Area Surveyed

We investigated bonobo distribution via 26.3 km of standardized line transects (Fig. 8.1). We established six transects (A1-6) within the study site extending beyond the surveyed area. A1 was 2 km long, A2 was 1 km long, and A3 to A6 were 3 km long each. For comparison, we cut 4 more transects outside the study area: B1 and B2 were about 3 km north of the Lokoro River, and B3 and B4 were about 5 km south. Each of them was 3 km long except for B4, which was 2.3 km long.

## Line Transect Sampling

Transects were established by one person following a predestined compass bearing. That individual was followed by a local field assistant, cutting a walk-able path through the dense forest via machete. Two more persons marked transects every 50 m via 50-m tape measure. We geo-referenced transects by taking GPS co-ordinates every 500 m (GARMIN 12XL).

Data collection followed Buckland et al. (1993, 2001). We used single nests as sampling units instead of nest groups, because bonobos sometimes return to the same nest site in short periods of time so that the nest aggregations can then be easily mistaken as one nest group. Allocation of nests to one nest group is also difficult when the nests are further apart and/or show different decay stages as sampling units.

We conducted nest counts with the help of one or two local field assistants. We checked transects for nests by walking with a speed of ca. 1 km/h. When we detected a nest, we measured the perpendicular distance from transect to the nest.

## Habitat Types

We recorded habitat types every 50 m: (1) heterogeneous primary forest on terra firme; (2) homogenous primary forest dominated by *Monopetalanthus sp.*; (3) homogenous primary forest dominated by *Gilbertiodendron dewevrei;* (4) heterogeneous forest temporarily inundated; or (5) heterogeneous forest permanently inundated.

## Standing Crop Nest Count

The standing crop nest count method is based on the nests discovered during the first walk along transects. We sampled all 10 transects (total length = 26.3km). We pooled within and outside of study site transects in order to obtain a sufficient sample size for a precise density estimate.

Our method considers the total number of detected nests ($N$), the nest decay rate as represented by mean survival time of nests ($dr$), nest production rate ($pr$), total length of transects ($L$), and the effective strip width given by distance ($2w$) in order to estimate the population density of weaned bonobos ($D$):

$$D = N / dr * pr * L * 2w$$

As production rate, we used 1.37 nests/individual/day (see below).

## Marked Nest Count

The marked nest count method is based on the nests discovered during consecutive walks along transects, i.e. after those discovered during the first walk. We ran within study site transects (A1-6) twice per month between October 2003 and February 2004, resulting in 11 consecutive walks (total 165 km). During each walk, we counted and marked only new, unmarked nests. The method considers the total number of detected nests ($N$), the total number of days elapsed between the first and the last census walk ($t$), the nest production rate ($pr$), the total transect length ($L$), and the effective strip width given by distance ($2w$) in order to estimate the population density of weaned bonobos ($D$):

$$D = N / t * pr * L * 2w$$

## Nest Decay Rate

Direct observation of bonobo nest construction at night or indirect evidence in the form of nests made up of fresh green leaves only, and urine and feces beneath them, revealed about two fresh nest groups per week, wherein the exact date of construction is known. We monitored decay states of all nests within the groups once a week until they were decomposed.

We used the days elapsing from construction date until complete decomposition to estimate the decay rate. We submitted data to a Kaplan-Meier Estimation (survival analyses in SPSS 13.0 advanced models).

## Nest Production Rate

Every evening, each mature bonobo builds a nest for the night. Day nest construction, however, is less predictable. At Lomako, Fruth (1995) observed 192 day nests made by mature individuals. In contrast to night nests, frequency of day nest construction differed between sexes: While 1.86 nests per 10 observation hours were built by females, only 0.39 nests were built by males. Considering the ratio of males to females in 485 day travel parties, each party had 3.5 females and 1.6 males on average ($SD_{females}$ = 2.123; $SD_{males}$ = 1.295; $M_{females}$ = 3; $M_{males}$ = 1). Thus, females outnumbered males in day-travel parties by 2.2:1.0 on average. Thus, one female built 0.53 and one male built 0.24 nests per 10 observation hours. Since a nest-to-nest observation day consisted of 9.7 hours on average, a rate of 0.37 nests/mature individual was added to the habitual night nest, resulting in a nest production rate of 1.37 nests/mature individual.

## Choice of Detection Function

In order to estimate effective strip width for density calculation, we analyzed data in DISTANCE 4.1 Release 2 (Thomas et al. 2003). To improve model fitting, we grouped the perpendicular distances measured between each nest and transect line into 6 equal intervals ranging from 0 to 40 (Buckland et al. 1993, 2001). We selected a detection probability model based on the Akaike's Information Criterion (AIC) (Burnham and Anderson 1998, Buckland et al. 2001). To enlarge the sample size for fitting the detection function, we performed a post-stratification by pooling detected nests of standing crop and marked nest count.

# Results

## Nest Decay Rate

We surveyed 24 fresh nest groups ranging from 2 to 21 nests/group (nest group size: x = 10.5 nests/group; SD = 5.5; M = 10) over a period of 28 weeks. Of a total of 218 nests, 173 (79.4%) were decomposed by the end of the study. The rest (45 nests) was still visible when Mohneke left and was monitored by other staff until complete decomposition in May 2004. Following the Kaplan-Maier Estimation procedure, nest decay rate represented by mean survival time of bonobo nests is 75.5 days (SD = 3.6; 95 % confidence interval = 68.4–82.5) with a median of 62 (SD = 7.1; 95% confidence interval = 48.2–75.8).

## Habitat Types

Table 8.1 shows the habitats' distribution along transects. Bonobos' choice for a specific habitat type for nest building is significantly different from availability (Chi-square-test: chi$^2$ = 80.94, df = 4, p < 0.001). The majority (97%) of n = 261 nests were in mixed primary forest. The number is significantly higher than expected by chance, regarding the availability of mixed primary forest (binomial test: z = 8.92, p < 0.0001).

In contrast, swamp forests revealed significantly lower numbers of nests than expected (binomial test: z = −7.83, p < 0.0001). With respect to the availability of homogeneous forest habitat, the number of recorded nests is also significantly lower than expected by chance (binomial test: z = −3.69, p < 0.001).

Outside the study area, all recorded nests were along the southern transects (B3–4). The distribution of the recorded nests suggests a gradient from north to south, with most nests (86 %) concentrated in the southern area. Our findings are supported by dry heterogeneous primary forest increasing and swamp forest decreasing from north to south (Fig. 8-2).

Proportion of habitat types for transects in the North (B1-2), Middle (A1-6) and South (B3-4) of investigated area. (d-het-pf) Heterogeneous primary forest on terra firme soil; (w-temp-het) heterogeneous primary forest temporarily inundated; (w-perm-het) heterogeneous primary forest permanently inundated; (d-hom-Mono) homogeneous primary forest dominated by *Monopetalanthus ssp.*; (d-hom-Gilb) homogeneous primary forest dominated by *Gilbertiodendron dewevrei*.

## Densities

*Standing Crop Nest Count:* A total of 48 nests along 26.3 km of line transects served as the analyses of the standing crop count method. This represents a nest

**Table 8.1** Availability and choice of habitat types for nest construction

| | Habitat characteristics | | | Habitat | | |
| | | | | availability | choice | |
| Class | Forest | Composition | Dominated by | % Transects | # of nests | % of nests |
|---|---|---|---|---|---|---|
| 1 | dry | heterogenous | | 73.0% | 255 | 97.4% |
| 2 | | homogeneous | Monopetalanthus | 3.3% | – | – |
| 3 | | | Gilbertiodendron | 3.1% | 1 | 0.4% |
| 4 | wet (temporarily) | heterogenous | | 3.6% | – | – |
| 5 | wet (permanently) | heterogenous | | 17.0% | 5 | 2.2% |

Availability of habitat types (class 1–5) along transects (in %) and habitat choice represented by the absolute number (#) or relative proportion (%) of nests found in each habitat type. Habitat types are classified as follows: 1. heterogeneous primary forest on *terra firme* soil; 2. homogeneous primary forest dominated by *Monopetalanthus spp.*; 3. homogeneous primary forest dominated by *Gilbertiodendron dewevrei;* 4. heterogeneous primary forest temporarily inundated; 5. heterogeneous primary forest permanently inundated.

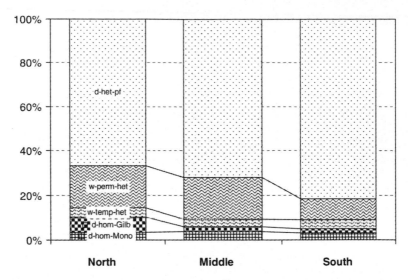

**Fig. 8.2** Habitat representation from North to Southx. For abbreviations see legend of Table 8.1.

encounter rate of 1.83 nests per km. The hazard-rate key function was selected by the AIC best fitting model. Regarding the estimate without applying a post-stratification, the density is 0.73 weaned individuals/km² (95% CI 0.29–1.85). After the post-stratification, the estimated density represents 0.49 weaned individuals/km² (95% CI 0.20–1.21). The encounter rate from within-site transects is much lower than from transects outside the study site (1.6 nests/km for A1-6 versus 2.12 nests/km for B 1–4). Due to low sample size of the individual transect groups from inside and outside the study area, we abandoned the plan to estimate densities separately for the respective areas.

*Marked Nest Count:* During the marked nest count survey, we detected a total of 105 nests during 11 repeated transect walks (15 km each) representing a total length of 165 km over a period of 142 days. The half-normal key function was selected by the AIC best fitting model. The estimated density based on it is 0.92 weaned individuals/km² (95% CI 0.36–2.36) and 1.01 weaned individuals/km² (95% CI 0.39–2.61) after post-stratification.

## Influence of Decay Rate

Table 8.2 is a list of surveys from the last 20 years that made density estimates of chimpanzees (*Pan troglodytes*) or bonobos (*Pan paniscus*) by applying standard line transect methods. As indicated in the table, the majority of the surveys used nest decay rate estimates made at study sites other than their own.

Table 8.2 Overview of studies based on standing crop nest counts presenting density estimations

| Species | Country | Study Area | Decay Rate Area Specific (yes/no) | Decay Rate Taken From | Reference |
|---|---|---|---|---|---|
| **Pan paniscus** | DR Congo | Lokofe-Lilungu Region | No | Gabon (Tutin & Fernandez 1984) | Sabater Pi & Vea (1990) |
| | | Equateur Province | Yes | DRC (Fruth pers. comm.) | Eriksson (1999) |
| | | Luo Reserve | No | Uganda (Ghiglieri 1984) | Hashimoto & Furuichi (2001) |
| | | Salonga National Park | No | DRC (Fruth pers. comm.) | Van Krunkelsven (2001) |
| | | | No | DRC (Fruth pers. comm.) | Reinartz et al. (2006) |
| | | | No | DRC (Fruth pers. comm.) | Grossmann et al. (2008) |
| | | | Yes | | This study |
| **Pan troglodytes** | Ivory Coast | Nation wide | No | Gabon (Tutin and Fernandez 1984) | Hoppe-Dominik (1991) |
| | | Nation wide | (yes) | Ivory Coast (Fruth 1990) | Marchesi et al. (1995) |
| | Cameroon | Campo Ma'an Area | Yes | | Matthews and Matthews (2004) |
| | Gabon | Nation wide | (yes) | Gabon (Tutin and Fernandez 1984) | Tutin and Fernandez (1984) |
| | | Petit Loango Reserve | No | Gabon (Tutin and Fernandez 1984) | Furuichi et al. (1997) |
| | Republic of Congo | Odzala National Park | No | Gabon (Tutin and Fernandez 1984) | Bermejo (1999) |
| | | Lac Tele Community Reserve | No | Gabon (Tutin and Fernandez 1984) | Poulsen and Clark (2004) |
| | Uganda | Kibale Forest | Yes | | Ghiglieri (1984) |
| | | Kalinzu Forest | No | Uganda (Ghiglieri 1984) | Hashimoto (1995) |
| | | Budongo Forest | Yes | | Plumptre and Reynolds (1996) |
| | | Nation wide | (yes) | Plumptre and Reynolds (1996) | Plumptre and Cox (2006) |

Table indicates species under focus, country and area of investigations, as well as reference to origin of applied decay rate. Yes: decay rate has been established and used for site under investigation; Yes: decay rate has been established for one site but applied nation wide; No: decay rate was taken from other study.

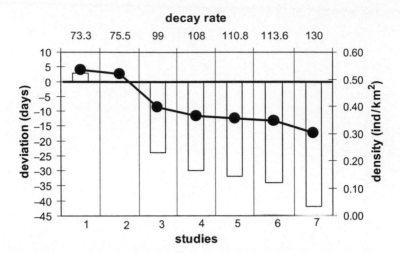

**Fig. 8.3** Influence of decay rate on density estimation. Deviation (left axes; bars) shows differ-ence in days between decay rates from other original studies (1; 3–7) to decay rate of present study (2, bold reference line = 75.5 days = 0). Density (right axis; black dots) shows average number of weaned individuals per km² calculated from nest density of this study with decay rates taken from other studies (1;3–7): (1) Marchesi et al. (1995), (2) Mohneke & Fruth (this study), (3) Eriksson (1999), (4) Fruth (unpublished data), (5) Ghiglieri (1984), (6) Tutin and Fernandez (1984), (7) Matthews and Matthews (2004).

Figure 8.3 shows the potential variation in density estimations calculated for Lui Kotale when applying nest decay rates from other studies. The relative deviation from the density calculated for our study increases with increasing difference of estimated decay rates to our observed decay rate.

# Discussions

## *Densities*

Summarizing the results obtained from both methods, the estimated density of weaned individuals in Lui Kotale is ca. 0.73 individuals/km², ranging between 0.49 (standing crop) and 0.92 (marked nest count), as calculated after post-stratification. Although standing crop nest count gives a density estimate equal to half of the estimate of the marked nest count, their confidence intervals overlap. The large difference in densities estimated by the standing crop nest count before and after post-stratification indicates the importance of a large sample size for best fitting detection function and precise estimates.

We consider the marked nest count as more reliable, because along the 26.3 km surveyed, the number of sightings required for a reliable analysis could not be achieved in a one-time walk. The 48 detected nests represent <50% of the minimum sample size (100 sightings) suggested by Plumptre (2000) and ≥80% of the minimum suggested by Buckland (1993).

In addition, the decay rate that we observed might not exactly suit this census, because it started not long before the standing crop survey. Thus, most of the 48 nests were likely built during the dry season, while decay observation considered more nests built during the rainy season. This bias is probably responsible for the large difference between average densities estimated by standing crop and marked nest count. The finding of Plumptre and Reynolds (1996) that the mean nest survival time is shorter for the dry season, suggests that our estimated mean survival time should have been <75.5 days. This would lead to a higher density of weaned bonobos per km².

Due to time constraints, we walked transects only once for the standing crop count before starting the marked nest count. All nests found during that walk were marked in order to exclude them from the subsequent walk where only newly produced nests were taken into account. Accuracy is said to be higher when transects are checked several times before starting the marked nest count survey, since each nest previously built artificially inflates density. Here, however, we found only three nests in advanced decay stages (3 or 4) during the first round of the marked nest count (= the second round after the standing crop count), making up for 2.8% of all recorded nests. This proportion is negligible and potential biases can be excluded.

Former bonobo density estimates and their confidence intervals range within and overlap with the confidence interval given by either method here. Sabater Pi and Vea (1990) estimated a density of 0.43 bonobos/km² for the Lokofe-Lilungu-Ikomaloki Region; Van Krunkelsven (2001) estimated 1.15 bonobos/km² for the Salonga National Park; and Eriksson (1999) estimated a density of 1.3 bonobos/km² for the Lomako Forest. Hashimoto & Furuichi (2001) estimated 0.49 bonobos/km² for the entire Luo Reserve. The MIKE project (2004) estimated a population between 1,000 and 10,000 individuals for the Salonga National Park. Reinartz et al. (2006) estimated 0.72 nest-builders/km² for the sites of investigation in Salonga National Park, while Grossmann et al. (2008) extrapolate the results achieved for their three inventory blocks to 0.41 bonobos/km² for the entire Salonga National Park. The differences in density estimations show a high level of variability that in some cases may indeed reflect density differences due to habitat quality or hunting impact. It may, however, also reflect differences in sampling methodology and thus bring uncertainty into appreciation of bonobo densities.

## Nest Decay Rate

Our mean survival time (= nest decay rate) of nests build by *Pan paniscus* is 75.5 days. For the Lomako study site, Eriksson (1999) calculated 99 days on average. We emphasize that decay rates strongly rely on the time and spatial scale in which data were collected. Factors such as sample size, rainfall, humidity, temperature, habitat type, tree species, and type of construction influence decay and explain variation in survival time of nests (Wrogemann 1992, Barnes 1993, Tutin et al. 1995, Plumptre and Reynolds 1996, Walsh and White 2005). Because decay rate strongly influences density estimation, longer decay rates that result in lower density estimations require caution when simply adopting rates calculated for a different season and/or for sites far from the region of investigation.

## Habitat Types

Nests are not distributed across habitat types as expected by availability. Heterogeneous primary forest on *terra firme* soil is the preferred habitat type for nest construction by *Pan paniscus* at Lui Kotale. Bonobos seem to avoid nest construction in swamp forests and homogenous habitats. The fact that we found no nest in the homogenous forest of *Monopetalanthus sp.* though the species is the fourth often selected for nest construction in heterogeneous associations, may demonstrate that proximity to food resources is more important than tree species' nest selection.

The increase of nest density from north to south is likely linked to change in habitat composition. Swamp forest increases are proportional towards the Lokoro River in the north and mixed primary forest increases are proportional towards the south.

## Comparison of Methods

The standing crop method has been a favoured approach because it requires only one walk after transects have been opened. This is of particular interest for large-scale surveys, as shown in Grossmann et al. (2008). Several factors, however, need particular consideration as they greatly influence precision: The length of transects has to be sufficient in order to detect enough nests for precise and unbiased estimates of encounter rate and detection probability. In addition, nest group size has to be taken into account when looking at a temporal scale, because nest group size varies across seasons and thus can lead to differences in detectability.

As outlined above, an additional problem in the application of this method is high variability in nest decay rates. Adopting decay rates from studies of other eco-regions/countries concerning nests of other ape-species often results in wrong density estimates as shown above. If we require, however, that each study should collect the area and season specific decay rates, the argument for economizing time looses its power. Furthermore, application of the extra correction factor implies an increase of variation resulting in a larger confidence interval. Therefore, decay rate has to be estimated with high precision, viz. with large samples. With high precision, the standing crop method can yield reliable estimates.

With the marked nest count method in which one visits transects regularly, the accumulation of nests is recorded, and correction for decay rate is unnecessary. Repeated counts of newly built nests clearly give a more precise density estimate when enough time is available (Hashimoto 1995, Plumptre and Reynolds 1996, Furuichi et al. 2001). However, application of the method on a larger scale will involve a high amount of time and effort. Consequently, there is a trade-off between losses of precision versus high investment. In terms of future large-scale surveys, the standing crop method is a useful tool that, carefully interpreted, helps us to learn more about the density and distribution of great apes. In addition, it allows comparison to those sites already surveyed in the past.

## Conservation implications

The study gives a first account of bonobo density in the south western corner of the Salonga National Park, covering an area of about $250 km^2$. With an average survey intensity of 1 km transect/$10 km^2$ of survey area compared to 1 km/$27 km^2$ (Grossmann et al. 2008), our study has an intensity that is almost 3 times above that of the large scale investigation. While the undeniable value of the large-scale extensive survey lies in the area covered and, in the case of Salonga National Park, the appreciation of bonobos' considerable presence within the Park's borders, the value of our small-scale intensive study lies apart from the appreciation of methodologies in the generation of a basis for the bio-monitoring of the given area. Increasing pressure by poachers in recent years has left a considerable impact on the presence and density of large mammals, particularly elephants, ungulates, and monkeys (Reinartz et al. 2008, Eriksson 2007, own unpubl. data). Bonobos, which are so far occasional or by-product prey, will become increasingly targeted with the decreasing density of conventional prey.

Despite the research site's location in the white sands of the Lokoro, with extended areas of swamp forest, the area of investigation appears to be a high density area that deserves monitoring of the bonobo population. Continuous investigations of the Max-Planck long-term investigations focus on the socio-ecology and demography of currently one community. Our study provides important groundwork for the monitoring and protection of the adjacent communities that are not under continuous observation. In addition, it may help to set large-scale results, such as those provided by Reinartz et al. (2006, 2008) and Grossmann et al. (2008), into relation and, thus, to contribute to reliable statements for the total number of bonobos inside Salonga National Park.

**Acknowledgements** We thank the "Institut Congolais pour la Conservation de la Nature" (ICCN) for granting permission to work in Salonga National Park, and Lompole village for granting permission to investigate the forest of their ancestors. Research was conducted under the auspices and with financial help of the Max-Planck Society and the German Ministry of Education and Research (BMBF, BIOLOG 01LC0022). The "Ministère de l'Intérieur" (DRC) issued the 'ordre de mission.' We thank Gottfried Hohmann for his constructive support during the start of the survey. Special thanks go to Falk Grossmann, Jamie Kemsey, Amy Cobden Veronica Vecellio, and Martin Surbeck for their supportive help in the field and to the local assistants, Mangos, Endou, Kabongo, Modizo, Lambert and Mara. Without their help, the survey would have been impossible. We wish to thank Peter Walsh, Hjalmar Kühl, and Falk Grossmann for providing constructive comments on the early manuscript. Jo Thompson and Takeshi Furuichi as well as two anonymous reviewers are thanked for detailed comments and careful editing on later stages of the manuscript.

# References

African Union (1968) African convention on the conservation of nature and natural resources. *http://www.africa-union.org/root/au/Documents/Treaties/Text/Convention_Nature%20&%20 Natural_Resources.pdf* , accessed 28 May 2007

Barnes RFW (1993) Indirect methods for counting elephant in the forest. Pachyderm, 16:24–30

Barnes RFW (2002) The bushmeat boom and bust in west and central Africa. Oryx 36:236–242

Bennett EL, Miner-Gulland EJ, Bakarr M, Eves HE, Robinson JG, Wilkie DS (2002) Hunting the world's wildlife to extinction. Oryx 36: 328–329

Bermejo M (1999) Status and conservation of primates in Odzala National Park, Republic of the Congo. Oryx 33:323–331

Blom A, Almasi A, Heitkönig IMA, Kpanou JB, Prins HHT (2001) A survey of the apes in the Dzanga-Ndoki National Park, Central African Republic: a comparison between the census and survey methods of estimating gorilla (*Gorilla gorilla gorilla*) and chimpanzee (*Pan troglodytes*) nest group density. Afr J Ecol 39:98–105

Buckland ST, Anderson DR, Burnham KP, Laake, JL (1993) Distance sampling: estimating abundance of biological populations. Chapman & Hall, London

Buckland ST, Anderson DR, Burnham KP, Laake JL, Borchers DL, Thomas L (2001) Introduction to distance sampling. Oxford University Press, Oxford

Burnham KP, Anderson DR (1998) Model selection and multimodel inference, 2nd ed. Springer-Verlag, New York

Butynski TM (2001) Africa's great apes. In: Beck BB, Stoinski TS, Hutchins M, Maple TL, Norton B, Rowan A, Stevens F, Arluke A (eds) Great apes and humans: the ethics of coexistence. Smithsonian Institution Press, Washington and London

Chapman CA, Balcomb SR, Gillespie TR, Skorupa JP and Struhsakers TT (1999) Long-term effects of logging on african primate communities: a 28-year comparison from Kibale National Park, Uganda. Conservation Biology 14:207–217

CITES (2005) Appendices I, II and III. *http://www.cites.org/eng/app/appendices.pdf, accessed 08* February 2006

Draulans D, Van Krunkelsven E (2002) The impact on war on forest areas in the Democratic Republic of Congo. Oryx 36:35–40

Eriksson J (1999) A survey of the forest and census of the bonobo (*Pan paniscus*) population between the Lomako and the Yekokora rivers in the Equateur province, D R Congo. MSc thesis, University of Uppsala, Uppsala, Sweden

Eriksson J (2007) Large scale bush-meat trade in Salonga National Park (SNP), south block. Unpublished report for UNESCO

Fruth B (1990) Nussknackplätze, Nester, und populationsdichte von Schimpansen: untersuchungen zu regionalen unterschieden im süd-westen der elfenbeinküste (rép. de côte d'Ivoire). Diplomarbeit (master-thesis): Ludwig-Maximilians-University Munich

Fruth B (1995) Nests and nest groups in wild bonobos (*Pan paniscus*) ecological and behavioral correlates. Shaker-Verlag, Aachen, Germany.

Furuichi T, Inagaki H, Angoue-Ovono S (1997) Population density of chimpanzees and gorillas in the Petit Loango Reserve, Gabon: employing a new method to distinguish between nests of the two species. Int J Primatol 18:1029–1046

Furuichi T, Hashimoto C, Tashiro Y (2001) Extended application of a marked -nest census method to examine seasonal changes in habitat use by chimpanzees. Int J Primatol 22:913–928

Ghiglieri MP (1984) The chimpanzees of Kibale forest: a field study of ecology and and social structure. Columbia Press, New York

Grossmann F, Hart J, Vosper A, Ilambu O (2008) Range occupation and population estimates of bonobo in the Salonga National Park: range occupation and population estimates of bonobo in the Salonga National Park: results of a large scale, multi phase inventory. In: Furuichi T, Thompson J (eds) The bonobos: behavior, ecology, and conservation. Springer, New York, pp 189–216

Hart J, Grossmann F, Vosper A, Ilanga J (2008) Human hunting and its impact on bonobo in the Salonga National Park, D.R. Congo. In: Furuichi T, Thompson J (eds) The bonobos: behavior, ecology, and conservation. Springer, New York, pp 245–271

Hashimoto C (1995) Population census of the chimpanzees in the Kalinzu Forest, Uganda: comparison between methods with nest counts. Primates 36:477–488

Hashimoto C, Furuichi T (2001) Current situation of bonobos in the Luo reserve, Equateur, democratic republic of Congo. In: Galdikas BMF, Briggs NE, Sheeran LK, Shapiro GL, Goodall J (eds) All apes great and small, vol 1: african apes. Kluwer, Academic/Plenum, New York, pp 83–93

Hohmann G, Fruth B (2003) Lui Kotale: a new site for field research on bonobos in the Salonga National Park. Pan Africa News 10:25–27

Hoppe-Dominik B (1991) Distribution and status of chimpanzees (Pan troglodytes verus) on the Ivory Coast. Primate Report 31:45–75

Ilambu O, Grossmann F (2004) Inventaires MIKE/WCS à la Salonga: Résultats préliminaires. Unpublished report to ICCN & WCS Kinshasa

IUCN (2004) IUCN Red List of Threatened Species. http://www.iucnredlist.org, accessed 08 February 2006

Kano T (1984) Distribution of pygmy chimpanzees (Pan paniscus) in the Central Zaire Basin. Folia Primatologica 43:36–52

Kortlandt A (1995) A survey of the geographical range, habitats and conservation of the pygmy chimpanzee (Pan paniscus): an ecological perspective. Primate Conservation 16:21–36

Ling S, Kümpel N, Albrechtsen L (2002) No new recipes for bushmeat. Oryx 36:330

Marchesi P, Marchesi N, Fruth B, Boesch C (1995) Census and distribution of chimpanzees in Côte D'Ivoire. Primates 36:591–607

Matthews A, Matthews A (2004) Survey of gorillas (Gorilla gorilla gorilla) and chimpanzees (Pan troglodytes troglodytes) in Southwestern Cameroon. Primates 45:15–24

Meder A, Burgel P, Boesch C (1988) Pan paniscus in Salonga National Park. Primate Conservation 9:110–111

MIKE (2004) Central african forests: final report on population surveys (2003-2004). Wildlife Conservation Society

Moore PD (2001) The rising cost of bushmeat. Nature 409:775–776

Muoria PK, Karere GM, Moinde NN, Suleman MA (2003) Primate census and habitat evaluation in the Tana delta region, Kenya. Afr J Ecol 41:157–163

Oates JF (1986) Action plan for African Primate Conservation: 1986–1990. IUCN/SSC Primate Specialist Group, Stoney Brook, New York

Plumptre AJ (2000) Monitoring mammal populations with line transect techniques in African forests. J Applied Ecol 37:356–368

Plumptre AJ, Cox D (2006) Counting primates for conservation: primate surveys in Uganda. Primates 47:65–73

Plumptre AJ, Reynolds V (1996) Censusing chimpanzees in the Budongo Forest, Uganda. Int J Primatol 17:85–99

Poulsen JR, Clark CJ (2004) Densities, distributions, and seasonal movements of gorillas and chimpanzees in swamp forest in Northern Congo. Int J Primatol 25:285–305

Reinartz GE (2003) Conserving *Pan paniscus* in the Salonga National Park, Democratic Republic of Congo. Pan Africa News 10:23–25

Reinartz GE, Bila Isia I. Ngamankosi M, Wema Wema L (2006) Effects of forest type and human presence on bonobo (*Pan paniscus*) density in the Salonga National Park. Int J Primatol 27: 603–634

Reinartz GE, Guislain P, Bolinga M, Isomana E, Bila Isia I, Bokomo N, Ngamankosi M, Wema Wema L (2008) Ecological factors influencing bonobo density and distribution in the Salonga National Park: applications for population assessment. In: Furuichi T, Thompson J (eds) The bonobos: behavior, ecology, and conservation. Springer, New York, pp 167–188

Sabater Pi J, Vea JJ (1990) Nest-building and population estimates of the bonobo from the Lokofe-Lilungu-Ikomaloki Region of Zaire. Primate Conservation 11:43–48

Sugiyama Y, Soumah AG (1988) Preliminary survey of the distribution and population of chimpanzees in the Republic of Guinea. Primates 29:569–574

Susman RL, Badrian N, Badrian A, Thompson-Handler N (1981) Pygmy chimpanzees in peril. Oryx 16:179–184

Thomas L, Laake JL, Strindberg KP, Marques FFC, Buckland ST, Borchers DL, Anderson DR, Burnham KP, Hedley SL, Pollard JH, Bishop JRB (2003) Distance 4.1, release 2. research unit for Wildlife Population Assessment, University of St. Andrews, UK. Available from:< http://www.ruwpa.stand.ac.uk./distance/

Thompson-Handler N, Malenky RK, Reinartz GEE (1995) Action plan for *pan paniscus*: report on free-ranging populations and proposals for their preservation. WC: Zoological Society of Milwaukee County, Milwaukee

Tutin CEG, Fernandez M (1984) Nationwide census of gorilla (*Gorilla g. gorilla*) and chimpanzee (*Pan t. troglodytes*) populations in Gabon. Amer J Primatol 6:313–336

Tutin CEG, Parnell RJ, White LJT, Fernandez M (1995) Nest building by lowland gorillas in the Lope Reserve, Gabon: environmental influences and implications for censusing. Int J Primatol, 16:53–76

Van Krunkelsven E, Inogwabini BI, Draulans D (2000) A survey of bonobos and other large mammals in the Salonga National Park, democratic republic of Congo. Oryx 34:180–187

Van Krunkelsven E (2001) Density estimation of bonobos (*Pan paniscus*) in Salonga National Park, Congo. Biological Conservation 99:387–391

Walsh PD, White LJT (2005) Evaluating the steady state assumption: simulations of gorilla nest decay. Ecological Applications 15:1342–1350

Whitfield J (2003) The law of the jungle. Nature 421:8–9

Wilkie DS, Sidle JG, Boundzanga GC (1992) Mechanized logging, market hunting and a bank loan in Congo. Conservation Biology 6:570–482

Wilkie DS, Godoy RA (2000) Economics of bushmeat. Science 287:973

Wrogemann D (1992) Wild chimpanzees in Lope, Gabon: census method and habitat use. PhD thesis, University of Bremen, Bremen, Germany

# Ecological Factors Influencing Bonobo Density and Distribution in the Salonga National Park: Applications for Population Assessment

Gay Edwards Reinartz[1], Patrick Guislain[1], T. D. Mboyo Bolinga[1], Edmond Isomana[1], Bila-Isia Inogwabini[2], Ndouzo Bokomo[1], Mafuta Ngamankosi[3], and Lisalama Wema Wema[3]

## Introduction

Bonobos (*Pan paniscus*) and mountain gorillas (*Gorilla beringei*) are Africa's two most endangered great apes (Butynski 2001). Conservation action plans for bonobos (Thompson-Handler et al. 1995, Coxe et al. 2000) emphasize the need for regional surveys in order to determine species distribution and abundance, to identify priority populations for protection, and to develop a range-wide conservation strategy (Susman 1995, Wolfheim 1983). Little is known about environmental factors limiting bonobo population distribution within their range. The distribution is thought to be patchy and discontinuous even in areas where seemingly suitable forests exist (Horn 1980, Kano 1984, Malenky et al. 1989, Thompson-Handler et al. 1995). Salonga National Park (SNP), the first and largest federally protected area for the bonobo, is a priority survey site. Created in 1970 as a reserve for bonobos and forest elephants (*Loxodonta africana cyclotis*), the SNP covers ca. 36,000 km$^2$, potentially harboring the largest area of undisturbed and legally protected bonobo habitat (D'Huart 1988, Thompson-Handler et al. 1995). In Salonga, several bonobo distribution hot-spots have been discovered by large-scale and site-based surveys (Blake 2005, Reinartz et al. 2006). Some authors consider them to be discrete populations (Inogwabini and Ilambu 2005). However, apart from natural barriers to dispersal, e.g., rivers, it is unknown whether true boundaries exist between proposed populations and what ecological parameters may determine their limits.

Early assessments of variables affecting the distribution of bonobos in the Central Congo Basin, such as those undertaken by Horn (1980) and Kuroda (1979, 1980) and later supplemented by Kano (1983, 1984), Badrian and Malenky (1984), Sabater Pi

[1] Zoological Society of Milwaukee, 10005 West Blue Mound Road, Milwaukee, Wisconsin, 53226 United States

[2] World Wide Fund for Nature, WWF-DRC Program, D.R. Congo

[3] Congolese Institute for Nature Conservation, D.R. Congo

and Vea (1990), White (1989, 1992), Malenky and Stiles (1991), Idani et al. (1994), Fruth (1995), Thompson (1997), Hashimoto et al. (1998), and Hashimoto and Furuichi (2001) have suggested the following ecological correlates: bonobos inhabit mature, semideciduous forests punctuated by areas of evergreen forests on *terra firma* soils [terminology from Evrard (1968) and interpreted by Kortlandt (1995)] (Horn 1980, Kano 1983, White 1989, Kortlandt 1995). In heterogeneous/disturbed environments, they occupy a wide ecological niche: bonobos frequently forage in younger secondary and swamp forests, i.e., inundated or seasonally inundated (Kano 1983, Sabater Pi and Vea 1990, Hashimoto et al. 1998), and occur in the forest-savannah mosaics in the southern portion of their range (Thompson 1997). Terrestrial herbaceous vegetation, particularly species of Marantaceae such as *Haumania liebrechtsiana*, is an important year-round food source preferentially consumed by bonobos (Kano 1983, Malenky and Stiles 1991) and is likely to influence species distribution, as is the availability of preferred fruiting and nesting tree species (Kuroda 1979, Kano 1983, Idani et al. 1994, Fruth 1995, Hashimoto et al. 1998). Moreover, the relative abundance of bonobos in areas accessed and hunted by humans is less than that in areas where no human hunting sign exists (Horn 1980, Kano 1983, 1984, Dupain et al. 2000, Dupain and Van Elsacker 2001).

Surveys are needed, not only to identify populations for protection, but to compare sites, assess habitat quality, and monitor populations. Historically, the characterization of bonobo habitat and density estimates have come from site-specific studies, e.g., Lac Tumba, Wamba, Lomako, Yalosidi, Lilungu, and Lukuru, and *post hoc* intersite comparisons: Kano and Mulavwa (1984), Badrian and Malenky (1984), White (1992). While they provide a general overview, their different methodologies limit interpretation of intersite comparisons at divergent locations. More recently, Mohneke and Fruth (2008) and Grossmann et al. (2008) use nonstratified surveys of different scales to estimate bonobo density and population size in the Salonga National Park. Because bonobo distribution is patchy and difficult to predict, large-scale, nonstratified surveys result in large variances in encounter rates and density estimates (Blake 2005, results Phase II of Grossmann et al. 2008), and they potentially miss core nesting areas. To refine precision of density estimates, Grossmann et al. (2008) used a multiphase, nonstratified survey approach: first to systematically locate general areas of bonobo concentrations and then to intensify transect sampling within these areas, finally extrapolating park-wide estimates of bonobo density. This approach yields improved regional estimates for areas with high bonobo density. However, little is known about habitat distribution, the overall variance in density estimates is high, and the approach is labor intensive. What is still required is a means of further stratifying surveys in order to allocate effort efficiently, gain precision of the estimate, and standardize intersite comparisons. Identifying ecological correlates of bonobo density provides a basis to stratify surveys by habitat type.

In this chapter, we identify forest types that have greater nest density and provide a model for survey stratification that reallocates greater survey effort to areas of preferred nesting habitat. In Phase 1 of the present study, we assess the effects of forest type and human activity on the relative densities of bonobos at multiple geographic sites throughout the SNP in order to quantify how these factors influence species distribution. Using the correlates from Phase 1 of our study, in Phase 2 we further explore how bonobo nesting habitat may be located *a priori* in order to stratify survey design and concentrate effort in areas that have a higher probability of bonobo nest occurrence. We choose areas on Landsat TM maps based on image texture and color changes and preliminarily test whether they contain a higher proportion of nest-forest types and higher bonobo density versus transects placed randomly with respect to habitat conditions.

## Methods

### *Study Area*

The SNP, located primarily within the Equateur Province, DRC, is Africa's largest tropical forest park (D'Huart 1988, Kempf and Wilson 1997). It is divided into northern and southern sectors of approximately equal size, separated by a swath of land roughly 45 km wide. Major rivers lead into and bound most of the Salonga (Fig. 9.1), which is a low plateau in the lower latitudes (350 m elevation) gradually

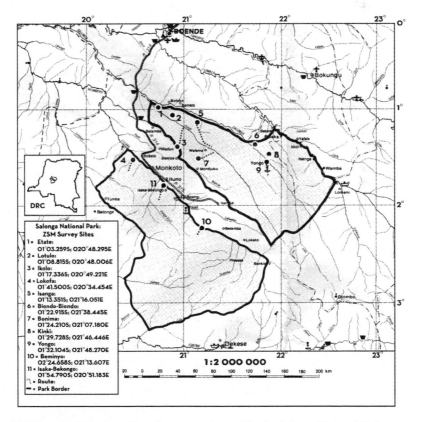

**Fig. 9.1** Name and location of bonobo survey sites within the Salonga National Park.

increasing in elevation southwards (up to 500m) (Evrard 1968, D'Huart 1988). Evrard (1968) characterized the predominant forest type on dry land as semideciduous (frequently dominated by *Scorodophloeus zenkeri*), occasionally interspersed with mono-dominant stands of the evergreen *Gilbertiodendron dewevrei* on lower slopes (Kortlandt 1995).

We conducted the first phase of our study from October 2000 to May 2002 in both sectors of the Salonga National Park (Fig. 9.1). We selected 11 survey areas from radar images and nongeoreferenced Landsat TM to locate major hydrological gradients. We further selected sites on the basis of accessibility and distribution along river routes and footpaths so that maximum distances could be traveled into the interior of the park and forest blocks. We conducted Phase 2 during the period from October 2004 to June 2006 at the Etate patrol post/ZSM research station, which is located S 01deg 3.255min, E 20deg 48.288min in the northern sector at the northwestern tip between the Salonga and Yenge Rivers (Fig. 9.1). We divided the study area into 3 sampling sectors based on their increasing distance from the patrol post: Etate 1, Etate 2, and Bofoku Mai (Fig. 9.2).

## Survey Methods (Phases 1 & 2)

### Reconnaissance and Transect Sampling

Reconnaissance walks (recces) covered various distances throughout a study area, using existing footpaths, old roads, animal tracks, or off-trail compass headings, using a hip-chain to measure distances. We noted signs of bonobos, other large mammals, and human presence, and we recorded the forest type for each sign.

We collected systematic data on bonobo nests and nest sites via variable-width line transect sampling (Buckland et al. 2001). For both reconnaissance walks and transects, we recorded each bonobo sign: direct sightings, food remains, tracks, dung, nests and nest sites. To gauge levels of human activities and hunting, we noted recent signs of human presence such as snares, traps, footpaths, machete cuts, campsites, shotgun cartridges, and direct sightings.

*Phase 1:* We placed transects in 9 study sites where recces indicated the presence of large mammals. We did not cut transects at either Nkinki or Bekongo because we found no signs of large mammals or bonobos on initial recces. Within the sites, we located the first transect randomly; subsequently, we systematically aligned a set of replicate transects (ranging from 500m to 2000m) with the first transect at 1 – 1.5 km intervals. We oriented transects parallel to major hydrological gradients and perpendicular to human trails. We sampled each transect only once, except for two transects at the Lokofa site, where we sampled 2 overlapping sets of transects in December 2000 and in May 2002.

For transect data, we analyzed differences in encounter rates (number of signs/ km) of human signs among sites with high and 0-low bonobo density using the nonparametric Kruskal-Wallis test (Sokal and Rohlf 1995, Remis 2000).

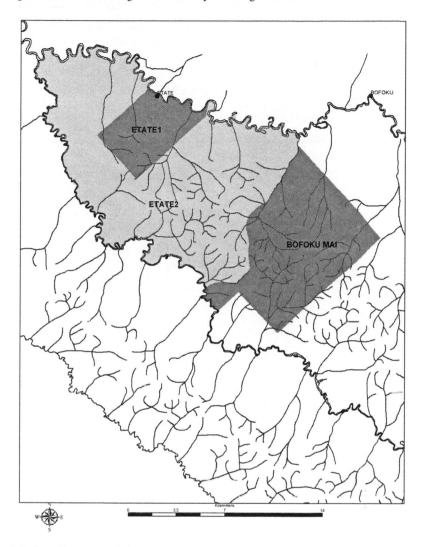

**Fig. 9.2** Sampling sectors in Phase 2 within the Etate study area.

*Phase 2:* We viewed Landsat-5 mosaic images (543-RGB, version 1, courtesy of Nadine Laporte, University of Maryland, and Zoological Society of Milwaukee) with ArcView 8.3 (Environmental Systems Research Institute, Inc. 1999–2002) to detect topography and changes in vegetation cover as denoted by textural and, to a lesser degree, by color variation. We mapped forest type waypoint data on to satellite images in order to preliminarily discern to what extent forest types may be associated with texture and color differences in the image. We identified blocks of *terra firma* on the satellite image which we predicted to contain mixed mature forests and selected these for ground truthing analysis. In order to locate preselected blocks in the field and place transects, we obtained geographic coordinates for points on the image using ArcView, and we located these points via GPS (Garmin 12XL).

We conducted reconnaissance sampling in 2 ways: by random recces wherein we followed paths of least resistance along compass headings without regard to forest types, and by intentional recces, wherein we followed compass headings in preselected forest blocks.

Within the Etate study area, we constructed random transects without regard to forest type and oriented them parallel to hydrological gradients and perpendicular to human trails. Likewise we constructed intentional transects in preselected forest blocks, i.e., transects placed intentionally within what we assumed *a priori* to be nest forest types. The Etate 1 sector contained 5 randomly located parallel transects, 1.5 – 2.5 km long, spaced ca. 1 km apart. The Etate 2 sector contained both random and intentional transects (Table 9.5), while Bofoku Mai had predominantly intentional transects. We sampled Etate 1 three times at 4 – 6 month intervals during the study period; in contrast, we sampled Etate 2 and Bofoku Mai only once. We estimated bonobo density for random and intentional transects separately via variable-width line transect sampling methods (Buckland et al. 2001). To assess human presence, we combined all data from transects and recces to calculate overall encounter rates of human signs within each sector.

**Density Estimates**

We used individual nest counts to estimate bonobo density (density of nest builders) following Buckland et al. (2001), assuming that (a) weaned bonobos make one nest per night per individual, and (b) nests last on average 99 (±5) days as calculated for the Lomako Forest, Equateur (Fruth pers. comm. in Van Krunkelsven 2001). Plumptre (2000) emphasized the inherent error related to converting nest density to individual animal density caused by the variability of nest decay rates. A decay rate of 99 days was the only estimate available to date for the mean life-span of bonobo nests. Acknowledging potential errors in decay rate, we converted the nest density data to nest-builder densities in order to make intersite comparisons; the same results are obtainable from nest density.

To estimate nest densities, we measured the perpendicular distance of individual nests (visible from the transects) to the transect line. A nest site is a group of nests of the same age (Hall et al. 1998) not separated by > 20 m (Fruth and Hohmann 1993). To estimate nest site density, we calculated the perpendicular distance from the geometric center of each nest site to the transect line.

The frequency distribution of the perpendicular distances was used by the computer program DISTANCE (version 4.1, Release 2: Thomas et al. 2003) to model the probability of detection for nest sites and nests and to calculate the effective strip width for the area sampled. The probability of detection changes with different forest types, so to obtain asymptotically unbiased estimates of bonobo density, we included forest type as a covariate in the model (Buckland et al. 2001, Marques and

Buckland 2003, Thomas et al. 2003). Following recommendations of Buckland et al. (2001), 5% of the observations detected at the greatest distance were truncated per stratum in order to improve the model fit. We also tested whether nest group size was independent of detection distance. To avoid pseudoreplication (Hurlbert 1984), we combined observations for transects resampled at Lokofa (Phase 1) and Etate 1 (Phase 2) and took their average density weighted by 1/number of sampling visits (Buckland et al. 2001, Buckland pers. comm.). In Phase 1, we analyzed differences in nest site and nest counts among survey sites via the nonparametric Kruskal-Wallis test (Sokal and Rohlf 1995, Remis 2000).

## Forest Description and Sampling

We designated forest classes in accordance with Evrard (1968) and modified from Kortlandt (1995), White and Abernethy (1997), and White and Edwards (2000). Because forest classification systems vary among authors describing bonobo and other great ape habitat (Kortlandt 1995), we adopted a forest description method that defines broad discrete categories of forest tree canopy, understory, and hydro-logical conditions encountered in Salonga. We confined our categories to forests occupying *terra firma* or seasonally inundated soils. We allocated no sampling effort completely to inundated forests due to the logistical difficulty of cutting transects and information that bonobos infrequently use inundated forests for nesting (Kano 1983, 1984, Kortlandt 1995).

The forest tree canopy classes (Reinartz et al. 2006) are: mixed mature (mm), old secondary (os), young secondary (ys), monodominant (md), and open (o) forests. The understory designations include woody (w), Marantaceae (M), liana (l), and open (o). Forest types consisted of overstory and understory combinations (e.g., mm/w) further defined by soil conditions and canopy conditions.

*Phase 1:* At points every 100 meters along transects and for each nest site and bonobo sign, we recorded the forest tree canopy class, understory, percent canopy cover, and soil type. For the combined area sampled and the area of sites, we calculated the proportion of forest types as the number of counts of a given forest type/the total number of forest counts.

*Phase 2:* Along transects and recces in the Etate study area, we measured the distance of each forest type and calculated the proportion of forest types as the total length of a given forest type (km)/the total effort (km). In order to determine the representative forest types in the study area, we calculated the proportions of forest types for only randomly placed transects and recces.

For Phase 1, a contingency table analysis (site x forest type G-test using counts of forest types observed at each site) tested whether forest types were equally represented among sites (Sokal and Rohlf 1995). For Phases 1 and 2, a single classification G-test for goodness of fit (based on assumptions extrinsic to the data: Sokal and Rohlf 1995) tested whether bonobo nests or nest sites or both were distributed uniformly within forest types (pooled across sites). In addition, in Phase 1, we pooled forest site data into categories according to whether the sites exhibited high or low-zero bonobo nest site density. A contingency table analysis (density category x forest type G-test) tested whether sites with high and low-zero bonobo density consisted of dissimilar forest types. All tests are two-tailed.

# Results

## *Phase 1*

We constructed 48 transects in 9 study sites throughout the sampling period. The sampling distance totaled 170.1 km and 67.8 km for reconnaissance walks and transects, respectively. Bonobo signs occurred in 8 of 11 locations, but the frequency of signs varied widely. We found no evidence of bonobos at Biondo Biondo (site 6), Lotulo (site 2), or Bekongo (site 11) on either the recces or transects (Fig. 9.1).

On transects placed in 9 survey sites, we observed a total of 49 nest sites composed of 270 nests. The Etate, Lokofa and Beminyo sites accounted for approximately 78% and 93% of the total number of nest sites and nests encountered, respectively (Table 9.1).

### Nest Site, Nest, and Bonobo Density

For nest site and nest density estimates calculated by DISTANCE, 94% – 98% of the total variance of the estimate was due to the variance in the encounter rate of nest sites and nests, respectively, thus demonstrating a nonuniform, patchy bonobo distribution among and within the locations sampled. Of transects sampled, ca. 48% had a 0 nest encounter rate; in contrast, one transect had an encounter rate of 92 nests/km (Lokofa site). For nest site and nest density, the hazard rate model best fit the detection function (Buckland et al. 2001, Thomas et al. 2003).

For the pooled sample of nest sites, the model estimated an overall density of 17.5 nest sites/km$^2$ (95% CI = 9.7–31.3). The 3 locations having the highest nest site density (Table 9.1) ranged from 18.7 to 55.1 nest sites/km$^2$, and zero to low density sites ranged from 0 to 14.4 nest sites/km$^2$. Nest density for the pooled sample is 71.5 nests/km$^2$ (95% CI = 34.9 – 146.5). Three locations with the highest bonobo nest occurrence ranged from 135 to 275 nests/km$^2$ while the remaining locations ranged from 0 to 18.5 nests/km$^2$. Due to small sample sizes and large variances in encounter rates within each site, the site density estimates are associated with large errors ($CV_{Global}$ = 29.7% for nest sites and 36.8% for nests). While these estimates are useful for comparative purposes, they reflect densities only for the areas sampled and not for the wider region. Despite the high density of bonobos observed at 3 sites, nest counts do not differ significantly among all sites because of the clumped distribution of nest sites and nests within survey locations (Kruskal-Wallis H = −2.56; $X^2$ = 13.362; d.f. = 8, P > 0.1).

**Table 9.1** Estimates of bonobo densities and bonobo and human sign encounter rates at survey sites in the Salonga National Park. Densities were calculated by DISTANCE 4.1, release 2 (Thomas et al., 2003). Encounter Rates = No. signs/total transect length

| Survey Sites | High Bonobo Density | | | | | | Zero-low Bonobo Density | | | | | |
| --- | --- | --- | --- | --- | --- | --- | --- | --- | --- | --- | --- | --- |
| | LOKOFA | ETATE | BEMINYO | ISANGA | YONGO | IKOLO | BONIMA | BIONDO | LOTULO | NKINKI | BEKONGO | GLOBAL |
| No. Transects | 7 | 5 | 4 | 5 | 8 | 6 | 5 | 1 | 5 | 0 | 0 | 46 |
| Transect Length (km) | 10.76 | 7.50 | 2.75 | 10.00 | 8.83 | 9.00 | 10.00 | 1.50 | 7.50 | – | – | 67.84 |
| No. of Nest Sites (observed) | 24 | 11 | 3 | 4 | 6 | 1 | 0 | 0 | 0 | – | – | 49 |
| No. of Nests (observed) | 171 | 58 | 22 | 5 | 11 | 3 | 0 | 0 | 0 | – | – | 270 |
| Nest Site Density (sites/km2) | 55.1 | 37.7 | 18.7 | 10.5 | 14.4 | 2.9 | 0 | 0 | 0 | – | – | 17.5 |
| 95% CI | – | – | – | – | – | – | – | – | – | – | – | (9.73–31.33) |
| Nest Density (nest/km2) | 275 | 160 | 135 | 10 | 18.5 | 5.9 | 0 | 0 | 0 | – | – | 71.5 |
| 95% CI | – | – | – | – | – | – | – | – | – | – | – | (34.93–146.51) |
| Mean Group Size [recent] | 8.0 | 6.3 | 11.0 | 1.3 | 2.3 | 3.0 | 0 | 0 | 0 | – | – | 6.5 |
| Nest Builder/km2 (nest counts)[1] | 2.78 | 1.61 | 1.36 | 0.10 | 0.19 | 0.06 | 0.0 | 0.0 | 0.0 | – | – | 0.72 |
| Overall Bonobo Sign Enc. Rate[2] | 0.58 | 1.41 | 2.8 | 0.43 | 2.83 | 0 | 0.13 | 0 | 0 | 0.26 | 0 | 0.55 |
| Human Sign Encounter Rates[3] | 0.65 | 0.53 | 0 | 0.10 | 0 | 9.33 | 1.00 | 3.40 | 0.53 | – | – | 1.70 |

[1] Nest density/99; Mean nest duration = 99 days.
[2] Calculated for recce distances, excluding number of nests.
[3] Calculated for transect distances.
5% Truncation of transect maximum width data, reducing the number of observations included in the analysis.

## Forest Sampling and Bonobo Density

Consistent with Evrard's (1968) description of the Tshuapa region, we most frequently encountered mixed mature forests during our study (72.3 %; Table 9.2). Old secondary forests were 21% of the forest types sampled; however, the proportion may be overestimated because we selected sites based on their accessibility, e.g., along river ports and trails connecting old village sites where old secondary forests predominated. Covering 65.18 km of transects (Table 9.2), we identified 11 forest types, or overstory/understory combinations. The most common types were mm/M (36.8%), mm/w (23.6), and os/M (10.5%).

On transects, we observed bonobo nests only in the mm/M, mm/w, and os/M forest types (Table 9.3) – called nest-forest types for convenience vs. other forest types. The majority of bonobo nest sites (64%) and nests (75%) occurred in the mm/M forest, most commonly on *terra firma* soils with 50–75% canopy cover (Table 9.3). Nest sites (n = 47) were not equally distributed among forest types but occurred in the most common forest types (mm/M, mm/w, and os/M) in numbers greater than expected (single classification goodness of fit based on hypothesis extrinsic to the sample data: $G = 37.980$, d.f. = 11, $P<0.001$, Table 9.3). However, after excluding the area of other forest types, there is no significant difference in the number of nest sites occurring in these 3 forest types ($G = 5.633$, d.f. = 2, $P>0.05$). Nests (n = 257) also were not distributed evenly among forest types ($G = 257.516$, d.f. = 11, $P<0.001$). In contrast to nest sites, the number of nests occurring in the mm/M forest is significantly higher than in all other forest types even after omitting the area of other forest types ($G = 80.577$, d.f. = 2, $P<0.001$), suggesting that the number of nest sites may not differ across forest types but that mean nest group size may be larger in the mm/M forest. However, the large variance in overall nest group size and small sample size precluded detection of a significant difference in group size among forest types (Kruskal-Wallis $H_{adj} = 3.507$, d.f. = 2, $0.05<P<0.1$).

The densities of nest sites and nests were higher in the mm/M (32.3 nest sites/km$^2$ and 151.3 nests/km$^2$, respectively) than in either the mm/w or os/M forests

**Table 9.2** Percentage of forest types encountered across all sampling locations

| | Tree Canopy Class | | | | |
|---|---|---|---|---|---|
| Understory | Mixed Mature (mm) % | Old Secondary (os) % | Young Secondary (ys) % | Mono-dominant (md) % | Open (o) % |
| Woody (w) | mm/w 23.6 | os/w 2.7 | – | md/w 0.4 | – |
| Marantaceae (M) | mm/M 36.8 | os/M 10.5 | ys/M 0.1 | – | o/M 2.4 |
| Liana (l) | mm/l 6.4 | os/l 7.8 | – | – | – |
| Open (o) | mm/o 4.4 | – | – | md/o 3.7 | – |
| TOTALS[1] | 72.3[1] | 21.0 | 0.1 | 4.1 | 2.4 |

[1] Out of the total forest sample, 1.1% was mixed mature forest with no understory designation. This is included in the total for mixed mature forest.

**Table 9.3** Proportion of forest types sampled and nest site, nest, and bonobo density within forest types

| Forest Types<br>Nest Sites | Proportion<br>of Forest[1] | Observed<br>No. Nest<br>Sites | Expected<br>No. Nest<br>Sites[2] | Area<br>Sampled[3]<br>(km²) | No. Nest Sites/<br>km² (SD) | | No.<br>Bonobos/<br>km² [4] |
|---|---|---|---|---|---|---|---|
| Mm/w | 0.236 | 9 | 11.1 | 0.635 | 14.2 | (11.69) | 0.8 |
| Mm/M | 0.368 | 32 | 17.3 | 0.991 | 32.3 | (19.41) | 2.4 |
| Os/m | 0.105 | 6 | 4.9 | 0.283 | 21.2 | (3.97) | 0.4 |
| "Other" | 0.291 | 0 | 13.7 | 0.784 | – | – | – |
| Combined | 1.000 | 47 | 47.0 | 2.692 | 17.5 | (13.73) | 1.2 |
| Nests | | | | | | | |
| Mm/w | 0.236 | 48 | 60.6 | 0.847 | 56.7 | (23.75) | 0.6 |
| Mm/M | 0.368 | 200 | 94.5 | 1.322 | 151.3 | (93.56) | 1.5 |
| Os/m | 0.105 | 9 | 27.0 | 0.378 | 23.8 | (4.92) | 0.2 |
| "Other" | 0.291 | 0 | 74.8 | 1.046 | – | – | – |
| Combined | 1.000 | 257 | 257.00 | 3.593 | 71.5 | (59.80) | 0.7 |

N = 704 or total number of forest type observations.
[1] Proportion of forest = No. of observations of each forest type /N.
[2] Expected no. nest sites = Total no. nest sites or nests observed × forest proportion.
[3] Area of forest type sampled = Total transect length (67.84km) × 2 × Effective strip width (0.01984 km for nest sites, 0.02648 km for nests) × Proportion of forest type.
[4] Bonobo density converted from nest site density was calculated using mean nest group size; Density nest builders = nest density/99 or correction for mean nest lifespan (Nest density = No. nest sites × mean group size).

(Table 9.3). Between the latter two forest types, nest site density was slightly greater in the os/M forest than in the mm/w forest (21.2 nest sites/km² vs. 14.2 nest sites/km²), but, in contrast, nest density was more than 2 times lower in the os/M forest than in the mm/w (23.8 nests/km² vs. 56.7 nests/km²). The os/M forest had the highest proportion of nest sites with single nests.

The proportions of forest types were not homogeneous across sample locations (test for homogeneity: $G_H = 51.179$, d.f. = 24, P<0.001) (Table 9.4). Nest site density increased significantly as the proportion of nest-forest types increased at a given location (Fig. 9.3, R = 0.7562, d.f. = 7, P<0.05). Lokofa, Beminyo, and Etate were composed of greater proportions (≥80.0%) of the nest-forest types and had the highest bonobo density. In locations composed of <70% of the nest-forest types, where transects often ran through a mosaic of different forest types, nest site density within the mm/M, mm/w, and os/M forest types was lower than the density at locations where the same forest types existed in greater proportion. Mean nest group size also increased significantly with increased proportion of nest-forest types (R = 0.9598, d.f. = 3, P<0.01). While forest composition differed between locations with high vs. 0-low bonobo density ($G_H = 88.018$, d.f. = 3, P << 0.001), locations with high bonobo density had a significantly higher proportion of mm/M forests ($G_H = 22.270$, d.f. = 1, P< 0.001).

**Table 9.4** The proportion of forest types sampled from transects at study sites (no. counts of forest type within a site/total number of counts per site). No transects exist for the Nkinki and Bekongo sites

| NEST FOREST TYPE | LOKOFA | ETATE | BEMINYO | ISANGA | YONGO | IKOLO | BONIMA | BIONDO | LOTULO |
|---|---|---|---|---|---|---|---|---|---|
| mm/M | 0.659 | 0.413 | 0.844 | 0.438 | 0.021 | 0.413 | 0.184 | 0.125 | 0.363 |
| mm/w | 0.091 | 0.225 | 0.156 | 0.229 | 0.375 | 0.231 | 0.243 | 0.250 | 0.275 |
| os/M | 0.193 | 0.163 | 0.000 | 0.000 | 0.292 | 0.000 | 0.146 | 0.063 | 0.000 |
| Other | 0.057 | 0.199 | 0.000 | 0.333 | 0.312 | 0.356 | 0.427 | 0.562 | 0.362 |

# Color Plates

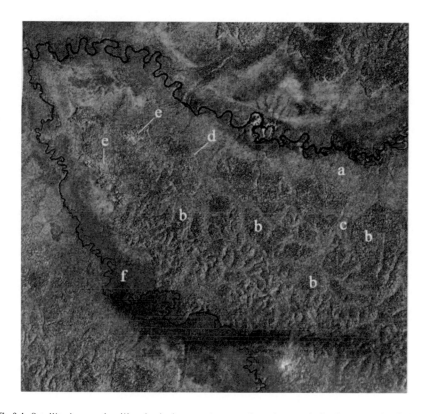

**Fig. 9.4** Satellite image detailing hydrology and vegetation characteristics between the Salonga (northeast) and Yenge (southwest) Rivers. a) inundated or seasonally inundated lowland; b) blocks of mixed mature forest; c) river with swamplands; d) *terra firma* edge; e) area with relatively open canopy (often associated with secondary forests and old village sites); and, f) inundated floodplain next to large rivers

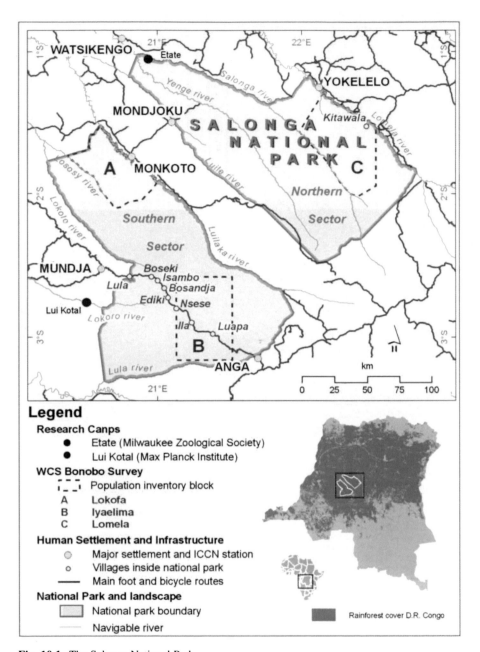

**Fig. 10.1** The Salonga National Park

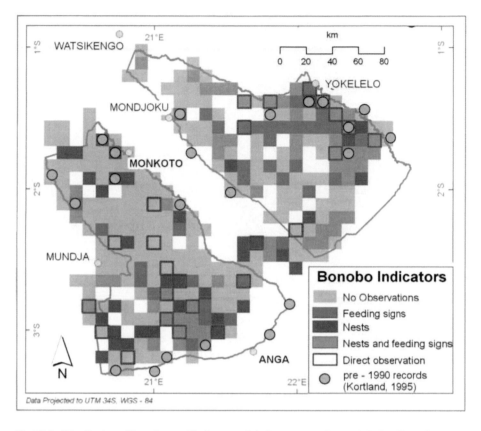

**Fig. 10.6** Distribution of bonobos and indicators of their presence observed during Phase I survey in the Salonga National Park. Locations of bonobo records predating creation of the park in 1970 are shown

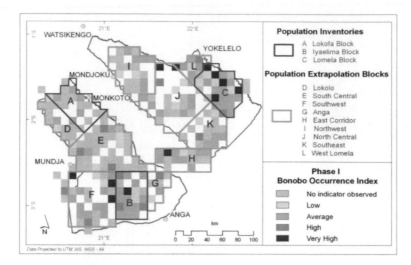

**Fig. 10.7** Bonobo occurrence indices integrate Phase I encounter rates of weighted field indicators for 10 × 10 km quadrats with ≥ 5 km reconnaissance coverage. Contiguous quadrats are combined into 12 population extrapolation blocks to calculate an estimate of bonobo populations for the total park area. Three extrapolation blocks cover the Phase II population inventory blocks

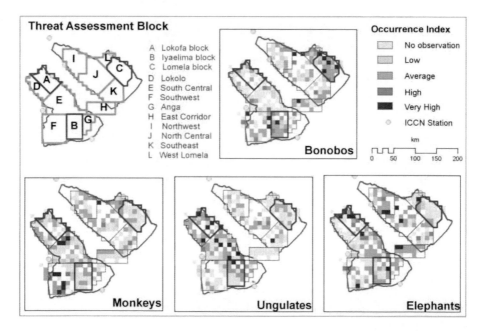

**Fig. 12.1** Occurrence of bonobos, ungulates, monkeys and elephants in the Salonga National Park, and surveyed corridor integrating weighted Phase I encounter rates of field indicators for 10 × 10 km quadrats having ≥ 5 km recce survey coverage. Contiguous quadrats are grouped into Threat Assessment Blocks to evaluate impact of hunting on bonobos and other fauna. Lokofa, Iyaelima, and Lomela Threat Assessment Blocks cover Phase II Population Inventory Blocks described in Grossmann et al. (2008)

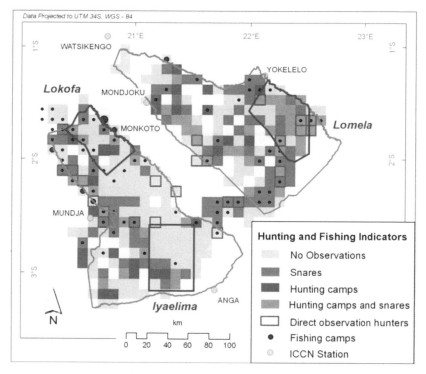

**Fig. 12.2** Hunting and fishing indicators recorded on Phase I surveys in the Salonga National Park and surveyed corridor. Phase II inventory blocks are shown in outline

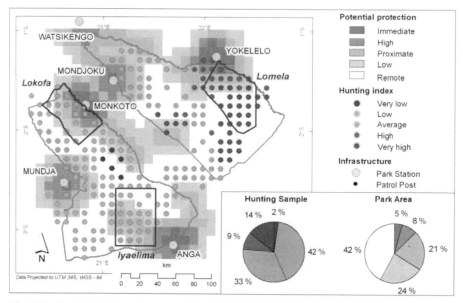

**Fig. 12.8** Getis-Ord G, hunting hot spots, and potential protection indices computed as a function of distance from established ICCN infrastructure (patrol posts and park stations) weighted by number of park guards based at the site

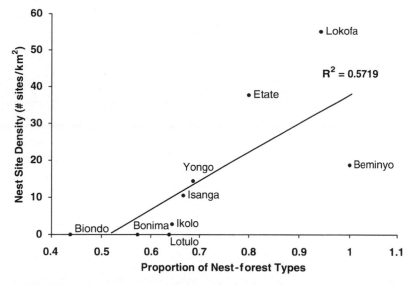

**Fig. 9.3** Relationship between the proportion of nest-forest types and nest site density at 9 survey locations. Nest-forest types are mixed mature/woody (mm/w), mixed mature/Marantaceae (mm/M), and old secondary/Marantaceae (os/M).

## Human Sign Encounter Rate and Bonobo Abundance

We encountered human signs at every location, either on transects or during reconnaissance exploration (Table 9.1), but encounter rates ranged widely and differed among sites (Kruskal-Wallis $H_{adj}$ = 23.469, d.f. = 9, P < 0.005, Table 9.1). In general, human signs that were associated directly with hunting, i.e., snares and hunting camps, appeared more commonly in sites with 0-low bonobo density (Kruskal-Wallis $H_{adj}$ = 9.7767, d.f. = 2, P<0.01, Table 9.1).

## *Phase 2*

Ground-truthing surveys during Phases 1 and 2 revealed that mm forests in Salonga occur in different soil types, but the mm/M forest type (*Haumania leonardiana*, *H. liebrechtsiana*, and *Megaphrynium macrostachyum*, being the most common Marantaceae species) tends to occur on *terra firma* soils on the plateau of forest blocks with varying degrees of canopy cover (ranging from 0% or open to 75% canopy cover). Forests of mm/M rarely occur in seasonally inundated soils. In contrast, mm/w occurs for large expanses in seasonally inundated soils (in lowlands near rivers and streams) and on drier soils on *terra firma*. The mm/w often occurs at a transition of waterways and swamps to plateaus. Forests of mm/l typically associate with wetter conditions, although large thickets of lianas can occur on dry soils usually coupled with low to sparse canopy cover. Monodominant forests (*Gilbertiodendron*), uncommon in the Etate sector, are most often confined to slopes approaching waterways (near the Yenge River).

Mapping forest type waypoint data onto satellite images confirmed that we could discern plateaus of *terra firma* and ridge tops, seasonally inundated lowlands, swamp forests, and open canopy on *terra firma* (Fig. 9.4). Within these plateaus, our experience predicted forest types to be primarily mm/w and mm/M, with the latter more prevalent towards the central portion (drier conditions). However, we could not reliably differentiate between mm/w and mm/M on satellite images, most likely because of the variance in forest canopy associated with mm/M.

To test this prediction, we sampled a total of 19 intentional transects and 20 randomly placed transects and analyzed only the proportion of mm/M and mm/w as nest-forest types. We found 222 nests (94 nest sites) in the 3 study sectors combined.

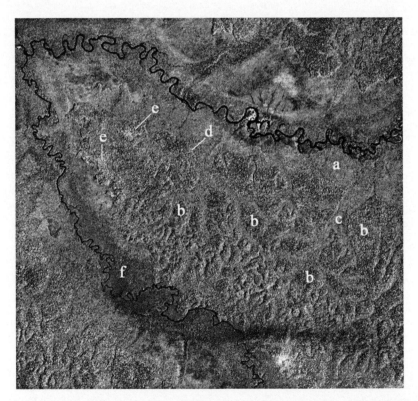

**Fig. 9.4** Satellite image detailing hydrology and vegetation characteristics between the Salonga (northeast) and Yenge (southwest) Rivers. a) inundated or seasonally inundated lowland; b) blocks of mixed mature forest; c) river with swamplands; d) *terra firma* edge; e) area with relatively open canopy (often associated with secondary forests and old village sites); and, f) inundated floodplain next to large rivers (*See Color Plates*).

We encountered 9 forest types across the Etate study area. Using only randomly placed transects and recces to assess representative forest types, mm/w (36.2%), mm/M (35.3%), and mm/l (10.8%) were the most common forest types. The proportions of mm/w and mm/M forest types were nearly equal, representing 71% of total forest types. The proportions of all other forest types were <11% (mm/o, os/M, os/w, os/l, o/m, o/l).

In contrast to representative forest types, 99% of intentional transects and recces were composed of mm forests, indicating that blocks of mm forest types can be located *a priori* via satellite images. The proportion of mm/M was twice as high (68.5%) in intentional than in random samples, while the proportions of mm/w and other forest types dropped to 24% and 6.8%, respectively (Table 9.6).

As in Phase 1, over all transects in the study area, bonobo nests were not evenly distributed between mm/M and mm/w forest types; mm/M forests contained a significantly higher proportion of nests ($X^2 = 29.77$, d.f.=1, p < 0.001). In all, 84% of nests occurred in mm/M, 15% in mm/w, and 1% in other forest types. Nest density was twice as high in the mm/M (221nests/km$^2$) as in the mm/w (113 nests/km$^2$). Only 0.5% of nests were in os/M.

Over the whole Etate study area, we estimated bonobo nest density at 139 nests/km$^2$, yielding 1.4 adult bonobos/km$^2$ (Table 9.5). However, Etate 1 appeared to be the core of the bonobo population. Composed of only randomly placed transects, Etate 1 had a higher adult bonobo density (1.9 adults/km$^2$) than the Etate 2 and Bofoku Mai sectors (1.2 adults/km$^2$ and 1.1 adults/km$^2$, respectively).

**Table 9.5** Transect type, number of nests, and nest density by sampling sector across the study area

| Transect Type | | Region | | | |
|---|---|---|---|---|---|
| | | Etate 1 | Etate 2 | Bofoku Mai | Study area |
| Random | No. transects | 11 | 8 | 1 | 20 |
| | Effort (km) | 19.5 | 14 | 0.5 | 34 |
| | No. nests | 101 | 37 | 0 | 138 |
| | Nest density (no. nests/km$^2$) (CV%) | 186.0 (20.8) | 100.0 (30.0) | – | 148.0 (18.8) |
| | % mm/M | 41 | 33 | – | |
| Intentional | No. transects | – | 12 | 7 | 19 |
| | Effort (km) | – | 17.4 | 6.1 | 23.5 |
| | No. nests | – | 65 | 19 | 84 |
| | Nest density (no. nests/km$^2$) (CV %) | – | 141.0 (32.0) | 104.0 (54.6) | 118.0 (26.9) |
| | %mm/M | – | 86 | 75 | |
| Total | No. transects | | | | 39 |
| | Effort (km) | | | | 57.5 |
| | No. nests | | | | 222 |
| | Nest density (no. nests/km$^2$) (%CV) | | | | 139.0 (15.4) |

**Table 9.6** Percentage of forest types on random and intentional samples across the study area

| Forest types | | Random/Representative samples (20 km) | | | Intentional samples (19 km) | | |
|---|---|---|---|---|---|---|---|
| | | mm | os | o | mm | os | o |
| *Understory* | M | 35.3 | 5.7 | 0.8 | 68.5 | – | – |
| | w | 36.2 | 5.1 | – | 24.6 | 0.7 | – |
| | l | 10.8 | 4.1 | 0.3 | 5.9 | – | 0.3 |
| | o | 1.0 | – | – | – | – | – |

Comparing data from the 2 transect types within and among sectors, we observed the following trends: For random transects, Etate 1 had a higher nest density and slightly higher proportion of mm/M forest (186 nests/km$^2$; 41%) than Etate 2 (100 nests/km$^2$; 33%). For intentional transects, nest density and the proportion of mm/M forest at Etate 2 (141 nests/km$^2$; 86%) were higher than at Bofoku Mai (104 nests/km$^2$; 75%) (Table 9.5). The sector density estimates, having relatively high coefficients of variation (CV = 20%–55%), are preliminary. Nest density was higher for Etate 2 intentional transects as compared to random, but the CVs for the density estimates are nearly equal (CV = 32% vs. 30%). To detect a difference in the precision of the density estimates for both sampling methods, larger sample sizes are needed (Plumptre 2000).

## Discussion

Our study reveals several findings important for understanding the ecology and conservation biology of the bonobo. Knowing where bonobos occur and what eco-logical factors influence density are the cornerstones for developing a conservation strategy for the species (Thompson-Handler et al. 1995). In SNP, bonobo nesting distribution and density are determined largely by the proportion of mixed mature, semideciduous forests on *terra firma*, most strongly associated with a Marantaceae understory. Secondarily, human presence related to hunting severely impacted bonobo distribution and abundance. Thus, the proportion of preferred nesting forest types and hunting partially explain the nonuniform distribution of bonobos in Salonga (Alers et al. 1992). By identifying forest types associated with nesting and those in which nesting is rare, we furthermore provide a prospective method by which survey design can be stratified by forest types in order to increase survey efficiency and precision and to compare density estimates among sites.

Contrary to early reports (Badrian and Badrian 1977, Kano 1979, 1984, IUCN/UNEP 1987), bonobos are present in SNP, and in certain locations they occur in relatively high densities. However, the species' distribution within the park is patchy. Our findings support Kortlandt's (1995) hypothesis that the existence of mature dry forests and edible terrestrial vegetation correlates with bonobo occurrence and that vegetation patterns play a central role in determining the bonobo's patchy distribution (Kano 1983, 1984, 1992).

Earlier studies used various forest classification systems to describe bonobo habitat and forest composition. However, the inconsistencies in terminology impede intersite comparison. In order to more finely differentiate habitat characteristics while respecting seminal forest classifications, we offer a more detailed system to predict bonobo occurrence whereby forest types are discretely categorized by forest structure: combinations of dominant tree species, understory, and soil hydrology. Because bonobos feed in and utilize the lower strata of the forest, we include the dominant understory characteristics as an integral part of the forest type designation. Conversely, other authors have broadly classified the forests of the Wamba and Lomako study sites into primary, secondary (old and young), swamp forests, and agricultural lands (Kano and Mulavwa 1984, White 1992, Idani et al. 1994, Fruth 1995, Hashimoto et al. 1998). The primary forest described for Wamba by Kano and Mulavwa (1984) and Idani et al. (1994) or the dry forest by Hashimoto et al. (1998) is synonymous with the mixed mature forest and potentially includes 4 forest types that we further differentiate by understory: woody, Marantaceae, liana, and open.

The results of our Salonga study are consistent with studies of nesting habitat in other areas. Bonobos at Wamba nest predominantly in the primary/dry forest (Kano 1992, Idani et al. 1994). In Lomako, Fruth (1995, p. 106) reports that the nest sites are restricted to the primary forest, defined as "*polyspecific* evergreen forest and slope forest [of] *Gilbertiodendron dewevrei*...," (p. 72). Mohneke and Fruth (2008) state that in the southwest corner of the SNP, 97.4 % of bonobo nests occur in "heterogenous primary forest on terra firma soils." Our analysis of understory and the predominance of nests in mixed mature Marantaceae and woody forest types suggest that only a subset of primary forests appears to be associated with nest-building.

Owing to the presence of preferred nesting tree species and edible herbaceous vegetation, bonobos are refined in their selection of nest forest types (Fruth 1995). However, bonobos use a wider range of habitats during the day (as confirmed by direct observation and food remains in most other forest types). Secondary forests, swamp forests, and forests with liana understory are important for feeding (Kano 1983, Badrian and Malenky 1984, Idani et al. 1994, Hashimoto et al. 1998), though bonobos appear to avoid nesting in swamp forests (Kano 1984, 1992, Fruth 1995, Hashimoto et al. 1998, but see Kano 1983) and possibly secondary forests when other choices are available (Kano 1984, Fruth 1995, Van Krunkelsven 2001, but see Sabater Pi and Vea 1990). We found that 6 % of bonobo nests occurred in inundated or seasonally inundated soils and that 3.5% occurred in secondary forests.

The density of nest sites and nests is at least twice as high in the mm/M forest compared to all other forest types. Nest group size also tends to be higher in the mm/M. These findings emphasize the importance of Marantaceae, in combination with nesting tree availability, as a determinant of bonobo nest distribution and abundance. Species of Marantaceae are important year-round food sources for the bonobo (Kuroda 1979, Horn 1980, Kano 1984, 1992, Malenky and Stiles 1991).

However, this does not necessarily imply a strict causal relationship between Marantaceae and nesting distribution. Despite its importance as a food source, it is likely that the same environmental and geophysical conditions which promote the growth of Marantaceae understory also correspond to the growth conditions for preferred nesting tree species and other nest habitat features (Fruth 1995). Moreover, the presence of Marantaceae alone is an inadequate predictor of bonobo nest occurrence because Marantaceae species occur under a myriad of forest conditions. Where Marantaceae are superabundant, climbing and overwhelming trees, bonobo density can be 0-low, such as in the old secondary forests at Yongo. Under these conditions, a low density of nesting trees exists, and the canopy becomes sparse (10–25%).

Concomitantly, bonobos may have more difficulty locating nest sites in areas having < 70% nest-forest type, where smaller patch size and frequent transitions to other forest types occur. Nest sites in mixed mature woody forest types often occur in the transition zone from woody to Marantaceae understory. We rarely encountered nests in large continuous blocks of woody understory.

Despite the existence of suitable habitat, we encountered large areas, e.g., Bekongo and Nkinki, where there was no evidence of bonobos. Hunting is the largest and most immediate threat to bonobo and wildlife survival in the Salonga (D'Huart 1988, Blom and Tshobo 1989, Alers et al. 1992, Fotso 1996, Van Krunkelsven et al. 2000, Ilambu and Grossmann 2004, Hart et al. 2008). In low bonobo density areas such as Ikolo, Bonima, Biondo Biondo, Nkinki and Bekongo (Fig. 9.1), which are near large settlements or near trails connecting major human populations, the frequency of hunting signs is highest. At Ikolo, 71 metallic snares occurred over 9 km of transects. The Beminyo reconnaissance (vs. transects) had the third highest snare encounter rate and also one of the highest densities of bonobos (and other large mammals); however, we did not find bonobo signs and nests until we had advanced beyond the limits of hunters' snare lines. We found evidence of either past or ongoing hunting at all study locations. Even at low human encounter rates, human presence had a profoundly low threshold effect on bonobo abundance. Within the Etate sector, bonobo density tended to decrease with distance from the patrol post - in areas that have higher hunting sign encounter rates - despite the continuation of appropriate bonobo habitat.

A priority for bonobo conservation in the Salonga is to identify bonobo high density areas and to develop means to protect and monitor them. Density estimates of a specified area can be obtained either by placement of transects across all forest types (Grossmann et al. 2008) or by stratifying the survey and more intensively sampling nesting habitat systematically. Grossmann et al. (2008) acknowledge that allocation of survey effort is "a question of major concern." They choose to allocate effort to regions of higher bonobo density. We advocate adding a survey stratification component based on nest forest types. If nesting habitat, as identified in Phase 1, can be located *a priori*, a portion of the survey effort that would sample marginal habitat (in nonstratified surveys) could be reallocated to habitat areas, thus improving the precision of the density estimate (Buckland et al. 2001). In Phase 2 we present a model that predicts the likelihood of bonobo nest occurrence, so that nest habitat areas can be more efficiently identified and periodically reassessed. Furthermore, accounting for density within the area of nest habitat provides a basis by which to standardize and to compare survey results among sites. For example, 2 survey locations may differ considerably in the amount (area) of available nesting habitat; yet, 2 nonstratified surveys may yield similar bonobo nest densities for both sites even when the bonobos in fact are more concentrated in a smaller area of habitat at one site versus the other. In failing to stratify the sampling design *a priori*, the habitat area may be undersampled and the actual density within it, underestimated. Moreover, ecological reasons for any possible difference between the 2 sites can not be ascertained. Therefore, we caution against prioritizing sites based on density estimates that have not accounted for habitat conditions.

In Phase 1, we randomly placed transects and then poststratified site surveys to estimate the forest type-specific detection functions and densities. Phase 1 results showed that bonobos nested almost entirely in the mm/M and mm/w (in Salonga's northern latitudes and confirmed in the southwestern sector by Mohneke and Fruth (2008). The Phase 2 experiment confirmed that it was possible to *a priori* stratify the survey design.

To locate probable blocks of nest forest types, satellite images were essential. Landsat images of this region depicted changes in hydrology, canopy cover, and elevation (Fig. 9.4). Ground-truthing observations confirmed that elevated patches were plateaus of *terra firma* surrounded by small rivers and streams. Furthermore, we found that most *terra firma* plateaus contained a high proportion of mm/w and mm/M. Within the preselected forest blocks, 93% of forest sampled was mm/M and mm/w, whereas these forest types accounted for only 71% of the representative samples. Furthermore, 68.5% were mm/M versus 35.3% mm/M in random samples. Other forest types existed within the *terra firma* blocks, but they generally corresponded to changes in hydrology too fine to be resolved on the image. While these results are still preliminary, we demonstrated that satellite images can guide survey stratification and confine the sampling area to *terra firma*, where nest forest types and nests are most likely to occur. Concomitantly, preselected forests tended to have higher bonobo density than randomly selected ones. These results are promising for developing a method to improve survey efficiency and to compare bonobo densities among sites that may vary in nest habitat availability. In the future, we will test this model in forest outside of the Salonga where forest types may differ. A thorough analysis of satellite images can be used to study population distribution (as related to ecological/habitat factors), to assess population fragmentation, and to identify optimal placement of corridors between populations. A remote sensing study of canopy cover and understory may help distinguish between different mixed mature forest types and aid in the calculation of nest forest habitat area.

**Acknowledgements**  This work was carried out in partnership with the Institut Congolais pour la Conservation de la Nature and the Ministère de l'Environnement, Conservation de la Nature, Eaux et Fôrets. The US Embassy-Kinshasa, USAID- Kinshasa Mission, the Governor of Equateur, and the Commandement de 3ème Région Militaire, Mbandaka, provided logistical support. We pay a special tribute to those who generously gave their time: Stefanie McLaughlin assisted with every aspect of the project. S. Buckland, K. Anderson, P. Dunn, J. Reinartz, J. Witebsky, T. Butynski,

N. Thompson-Handler and R. Malenky, B. Konstant, E. Van Krunkelsven, and J. Hall provided expert technical advice. We thank them for years of support. For logistical support, we owe unending gratitude to Steven and Julia Weeks, Elizabeth and James Williamson, M.K. Koerner, Lisa Steel, Ntuntani Etienne, Bokitsi Bunda, the Conservateurs and guards of the Salonga National Park, and the staff of TRAWA, Mbandaka. We also thank M. Hamadi, R. Carroll, N. Laporte, S. Sheper, D. Messinger, G. Boese, and the Royal Zoological Society of Antwerp. Financial support was graciously provided by the American people through the United States Agency for International Development – Central Africa Regional Program for the Environment, World Wildlife Fund International, the American Zoo and Aquarium Conservation Endowment Fund, the Margot Marsh Biodiversity Foundation, the Wildlife Conservation Network, Jones Family Trust, F. & S. Young family, D. Kern, Laacke and Joys-Milwaukee, World Wildlife Fund-US, Conservation Food and Health Foundation, Beneficia Foundation, San Diego Zoological Society, Columbus Zoo, Milwaukee County Zoo, and private donors. Landsat TM images were provided courtesy of University of Maryland (N. Laporte).

# References

Alers MPT, Blom A, Sikubwabo Kiyengo C, Masunda T, Barnes RFW (1992) Preliminary assessment of the status of the forest elephant in Zaire. Afr J Ecol 30:279–291

Badrian A, Badrian N (1977) Pygmy chimpanzees. Oryx 13:463–468

Badrian NL, Malenky RK (1984) Feeding ecology of *Pan paniscus* in the Lomako Forest, Zaire. In: Susman R (ed) The pygmy chimpanzee: evolutionary biology and behavior. Plenum Press, New York, pp 275–299

Blake S (2005) Central African Forests: final report on population surveys (2003–2004). Long Term System for Monitoring The Illegal Killing Of Elephants (MIKE), Wildlife Conservation Society, USA, http://www.cites.org/common/prog/mike/survey/central_africa_survey03–04. pdf, accessed 31 May 2007

Blom IA, Tshobo M (1989) Rapport provisoire sur la situation actuelle : activites humaines, conservation, gestion et developpment. Wildlife Conservation International, World Wide Fund for Nature

Buckland ST, Anderson DR, Burnham KP, Laake JL, Borchers DL, Thomas L (2001) Introduction to distance sampling: estimating abundance of biological populations. Oxford University Press, Oxford

Butynski TM (2001) Africa's great apes. In: Beck BB, Stoinski TS, Hutchins M, Maple TL, Norton B, Rowan A, Stevens EF, and Arluke A (eds) Great apes and humans: the Ethics of Coexistence, Smithsonian Institution Press, Washington DC, pp 3–56

Coxe S, Rosen N, Miller P, Seal U (2000) Bonobo conservation assessment-Nov. 21–22, 1999. Unpublished report of pre-PHVA workshop, Kyoto University Primate Research Institute, Inuyama, Japan, IUCN/SCC Conservation Breeding Specialist Group, Apple Valley

D'Huart JP (1988) Parc national de la Salonga (Equateur, Zaire): conservation et gestion. developpement des collectives locales. Unpublished Report. IUCN, Gland, Switzerland

Dupain J, Van Elsacker L (2001) The status of the bonobo in the democratic republic of Congo. In: Galdikas BMF, Briggs NE, Sheeran LK, Shapiro GL and Goodall J (eds) all apes great and small, volume 1: african apes. Kluwar Academic/Plenum, New York. pp 57–74

Dupain J, Van Krunkelsven E, Van Elsacker L, Verheyen RF (2000) Current status of the bonobo (*Pan paniscus*) in the proposed Lomako Reserve (Democratic Republic of Congo). Biol Cons 94:265–272

Evrard C (1968) Recherches ecologiques sur le peuplement forestier des sols hydromorphes de la Cuvette Centrale Congolaise. Série Scientifique de l'Institut National des Etudes Agronomiques en Congo Belge, Brussels, pp 11–295

Fotso R (1996) Examen du status, etude de la distribution et de l'utilisation du perroquet gris (Psittacus erythacus) au Zaire. CITES, Lausanne

Fruth B (1995) Nests and nest groups in wild bonobos (*Pan paniscus*): ecological and behavioural correlates. Verlag Shaker, Aachen

Fruth B, Hohmann G (1993) Ecological and behavioral aspects of nest building in wild bonobos (*Pan paniscus*). Ethology 94:113–126

Grossmann F, Hart JA, Vosper A, Ilambu O (2008) Range occupation and population estimates of bonobo in the Salonga National Park: results of a large scale, multi-phase inventory. In: Furuichi T and Thompson J (eds) The bonobos: ecology, behavior, and conservation. Springer, New York, pp 189–216

Hall JS, White LJT, Inogwabini BI, Omari I, Simons Mooreland H, Williamson EA, Saltonstall K, Walsh P, Sikubwabo C, Bonny D, Kiswele KP, Vedder A, Freeman K (1998) Survey of Grauer's gorillas (*Gorilla gorilla gaueri*) and eastern chimpanzees (*Pan schweinfurthi*) in the Kahuzi-Biega National Park Lowland Sector and adjacent forest in eastern Democratic Republic of Congo. Int J Primatol 19:207–234

Hart JA, Grossmann F, Vosper A, Ilanga J (2008) Human hunting and its impact on bonobo in the Salonga National Park, D.R. Congo. In: Furuichi T and Thompson J (eds) The bonobos: ecology, behavior, and conservation. Springer, New York, pp 245–271

Hashimoto C, Furichi T (2001) Current situation of bonobos in the Luo Reserve, Equateur, Democratic Republic of Congo. In: Galdikas BMF, Briggs NE, Sheeran LK, Shapiro GL, and Goodall J (eds) All apes great and small, vol 1: african apes. Kluwer Academic/Plenum, New York, pp 83–90

Hashimoto C, Tashiro Y, Kimura D, Enomoto T, Ingmanson E, Idani G, Furuichi T (1998) Habitat use and range of wild bonobos (*Pan paniscus*) at Wamba. Int J Primatol 19:1045–1049

Horn AD (1980) Some observations on the ecology of the bonobo chimpanzee (*Pan paniscus*, Schwarz 1929) near Lake Tumba, Zaire. Folia Primatol 34:145–169

Hurlbert SH (1984) Pseudoreplication and the design of ecological field experiments. Ecol Monogr 54:187–211

Idani G, Kuroda S, Kano T, Asato R (1994) Flora and vegetation of Wamba Forest, Central Zaire with reference to bonobo (*Pan paniscus*) foods. Tropics 3:309–332

Ilambu O, Grossmann F (2004) Inventaires MIKE/WCS a la Salonga: resultats preliminaires. Unpublished Report to ICCN and Wildlife Conservation Society, Kinshasa

Inogwabini BI, Ilambu O (2005) A landscape-wide distribution of *Pan paniscus* in the Salonga National Park, Democratic Republic of Congo. Endangered Species Update 22:116–123

IUCN/UNEP (1987) Parc National de la Salonga. In: The IUCN Directory of Afrotropical Protected Areas. IUCN, Gland Switzerland and Cambridge, pp 918–919

Kano T (1979) A pilot study on the ecology of pygmy chimpanzees (*Pan paniscus*). In: Hamburg DA, McCown ER (eds) The great apes. Benjamin/Cummings, Menlo Park, pp 123 135

Kano T (1983) An ecological study of the pygmy chimpanzees (*Pan paniscus*) of Yalosidi, Republic of Zaire. Int J Primatol 4: 1–31

Kano T (1984) Distribution of pygmy chimpanzees (*Pan paniscus*) in the Central Zaire basin. Folia Primatol 43:36–52

Kano T (1992) The Last Ape: pygmy chimpanzee behavior and ecology. Stanford University Press, Stanford

Kano T, Mulavwa M (1984) Feeding ecology of the pygmy chimpanzees (*Pan paniscus*) of Wamba. In: Susman RL (ed) The pygmy chimpanzee: evolutionary biology and behavior. Plenum Press, New York, pp 233–274

Kempf E, Wilson A (1997) Great apes in the wild. World Wildlife Fund for Nature, Gland

Kortlandt A (1995) A survey of the geographical range, habitats and conservation of the pygmy chimpanzee (*Pan paniscus*): An ecological perspective. Primate Conserv 16:21–36

Kuroda S (1979) Grouping of the pygmy chimpanzees. Primates 20:161–183

Kuroda S (1980) Social behavior of the pygmy chimpanzees. Primates 21:181–197

Malenky RK, Stiles EW (1991) Distribution of terrestrial herbaceous vegetation and its consumption by *Pan paniscus* in the Lomako Forest, Zaire. Am J Primatol 23:153–169

Malenky RK, Thompson-Handler N, Susman R (1989) Conservation status of *Pan paniscus*. In: Heltne PG and Marquardt LA (eds) Understanding chimpanzees. Harvard University Press, Cambridge, pp 362–368

Marques FFC, Buckland ST (2003) Incorporating covariates into standard line transects analyses. Biometrics 59:924–935

Mohneke M, Fruth B (2008) Bonobo (*Pan paniscus*) density estimation in the SW-Salonga National Park, DRC: common methodology revisited. In: Furuichi T and Thompson J (eds) The bonobos: ecology, behavior, and conservation. Springer, New York, pp 151–166

Plumptre A (2000) Monitoring mammal populations with line transect techniques in African forests. J Appl Ecol 37:356–368

Reinartz GE, Inogwabini BI, Ngamankosi M, Wema Wema L (2006) Effects of forest type and human presence on bonobo (*Pan paniscus*) density in the Salonga National Park. Int J Primatol 27:603–634

Remis MJ (2000) Preliminary assessment of the impacts of human activities on gorillas (*Gorilla gorilla gorilla*) and other wildlife at Dzanga-Sangha Reserve, Central African Republic. Oryx 34:56–65

Sabater Pi J, Vea JJ (1990) Nest-building and population estimates of the bonobo (*Pan paniscus*) from the Lokofe-Lilungu-Ikomaloki region of Zaire. Primate Conserv 11:43–48

Sokal RR, Rohlf FJ (1995) Biometry, 3rd edition. WH Freeman and Company, New York

Susman RL (1995) The only way to determine the conservation status of the pygmy chimpanzee is to conduct a survey in the Zaire Basin: a reply to Dr. Kortlandt. Primate Conserv 16:37–39

Thomas L, Laake JL, Strindberg S, Marques FFC, Buckland ST, Borchers DL, Anderson DR, Burnham KP, Hedley SL, Pollard JH, Bishop JRB (2003) Distance 4.1, Release 2. Research Unit for Wildlife Population Assessment, University of St. Andrews, UK. (http://www.ruwpa.st-and.ac.uk/distance/)

Thompson J Myers (1997) The history, taxonomy, and ecology of the bonobo (*Pan paniscus*, Schwarz, 1929) with a first description of a wild population living in a forest/savanna mosaic habitat. Doctoral Dissertation, University of Oxford, Oxford

Thompson-Handler N, Malenky RK, Reinartz G (1995) Action plan for *Pan paniscus*: report on free ranging populations and proposals for their preservation. Zoological Society of Milwaukee County, Milwaukee

Van Krunkelsven E (2001) Density estimation of bonobos (*Pan paniscus*) in Salonga National Park, Congo. Biol Cons 99:387–391

Van Krunkelsven E, Inogwabini BI, Draulans D (2000) A survey of bonobos and other large mammals in the Salonga National Park, Democratic Republic of Congo. Oryx 34:180–187

White FJ (1989) Ecological correlates of pygmy chimpanzee social behavior. In: Standen V, Foley RF (eds) Comparative socioecology: the behavioral ecology of humans and other mammals. Special Publication of the British Ecological Society, Blackwell Science, Oxford, pp 151–164

White FJ (1992) Pygmy chimpanzee social organization: variation with party size and between study sites. Am J Primatol 26:203–214

White L, Abernethy K (1997) A guide to the vegetation of Lopé Reserve. Wildlife Conservation Society, Bronx

White L, Edwards A (2000) Vegetation inventory and description. In: White, L, Edwards, A (eds) Conservation research in the african rain forest: a technical handbook. Wildlife Conservation Society, Bronx, pp 119–155

Wolfheim JH (1983) Primates of the world: distribution, abundance, and conservation. University of Washington Press, Seattle

# Range Occupation and Population Estimates of Bonobos in the Salonga National Park: Application to Large-scale Surveys of Bonobos in the Democratic Republic of Congo

Falk Grossmann[1], John A. Hart[2], Ashley Vosper[2], and Omari Ilambu[3]

## Introduction

Conservation of the bonobo, Congo's endemic ape, is one of the most important conservation priorities in the Democratic Republic of Congo (DRC). Bonobos are classified as endangered by both the IUCN (1996) and CITES (2001). In determining where bonobos occur, their population numbers and the threats to them are critical for development of a range-wide conservation strategy for the species. The need for information on the bonobo's status is all the more urgent given the imminent opening of their range to logging and other extractive activities following the end of DRC's conflict.

The potential bonobo range, variably estimated from $341,000 - 472,000$ km$^2$, is restricted to DRC's central cuvette; however, occupation of this area by bonobos is not contiguous (Butynski 2001, Meyers Thompson 1997). Large areas of forest contain few or no bonobos, while the species occurs in relatively high numbers in other areas. Most studies of bonobos have been conducted in very small study areas, widely dispersed within the range. Even at this scale, researchers report wide variability in occurrence and population size. Speculative estimates of the bonobo's global population range from $13,500 - 100,000$, though figures from $20,000 - 50,000$ are the most widely cited (Butynski 2001). Some authors suggested that 50% of the bonobo's range might have been lost over the past several decades (Dupain and Van Elsacker 2001, Thompson-Handler et al. 1995), though early records of bonobos suggest that there were major discontinuities in their distribution over 80 years ago (Kortland 1995).

Until recently, much of the bonobo's range was isolated from major settlements and had historically low human population density. This has likely ensured the protection of many bonobo populations. Passive protection, however, may no

---

[1] Wildlife Conservation Society, Africa Program, c/o K.C.M.C., Private Bag, Moshi, Tanzania

[2] Lukuru Project, Tshuapa-Lomami-Lualaba Landscape, D.R. Congo

[3] World Wide Fund for Nature, WWF-DRC Program, D.R. Congo

longer be adequate. A decade of conflict and political instability (1996 – 2006) has weakened the national parks service (Institut Congolais pour la Conservation de la Nature, ICCN) and favored widespread access to firearms. Artisanal scale extraction of natural resources, including bushmeat, increased in many areas in the bonobo's range during the period of conflict (Draulans and van Krunkelsven 2002, Dupain et al. 2000). As Congo's human population and economy grow, even the most remote forests occupied by bonobos will be opened to exploitation. Threats to bonobos and other wildlife will intensify. Active protection and conservation will become increasingly more important.

Mobilizing the financial resources and creating the political will to protect bonobos and conserve key areas of their range will require strategies that are well-informed and focused if they are to have any chance of success. This will require updated information on the distribution and abundance of bonobos, and a well-founded evaluation of the impact of the threats they face. Developing conservation priorities and monitoring the status of bonobos will require large-scale surveys over important areas of their range (Mohneke and Fruth 2008, Reinartz et al. 2008, Thompson-Handler et al. 1995). An important question is how these surveys should be completed?

## The Salonga National Park and its Bonobos

The Salonga National Park is the largest, and until recently the only, protected area within the bonobo's range. The park was established in 1970 and enrolled as a World Heritage Site in 1984. The park is composed of two sectors, a northern and a southern, separated by a corridor buffer zone between them (Fig. 10.1). It covers ca. 33,346 km², about 10% of the bonobo's range, and represents one of the most intact blocks of tropical forest in DRC (Siegert 2003, Sanderson et al. 2002). Closed mixed tropical forests cover > 90% of the park and ca. one third are permanently or seasonally inundated. Recent clearings, regenerating forests, and natural savannas represent a small percentage of the park area (Siegert 2003).

Current human occupation averages less than 3 inhabitants per km² over the area within 15 km of the park limits. About 215 villages are within 15 km of the park borders. Most of them are small with < 500 individuals. There are nine villages within the park. Under current legislation they are all illegal. Kitawala, located just inside the park border in the northern sector, has a total population of 5,000 – 7,000 people, many of whom belong to a syncretic religious sect of the same name that retreated into what was to become the park in the1960s to avoid contact with other groups. Eight settlements of the Iyaelima people, with a total population of ca. 2500 and comprising the entire population of this ethnic group, are located along a major footpath bisecting the southern sector of the park (Thompson et al. 2008). Almost half of the park area is located > 15 km from a permanent human settlement (Fig. 10.2).

Despite a low level of permanent human occupation and distance from major settlements, the park remains relatively accessible along a network of rivers that can be navigated by dugout canoe. Less than one-third of the park area is located more

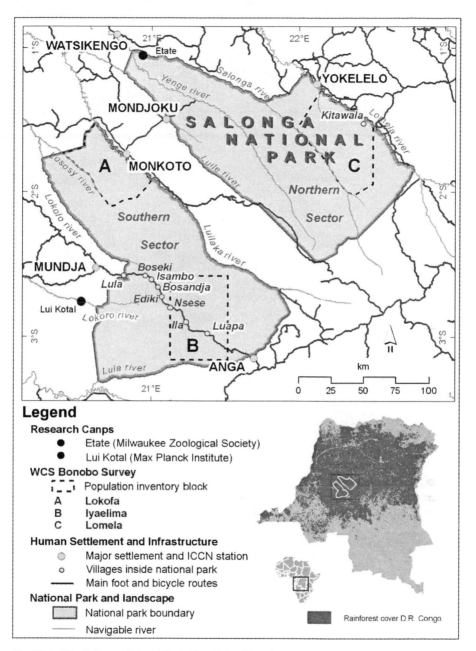

**Fig. 10.1** The Salonga National Park (*See Color Plates*).

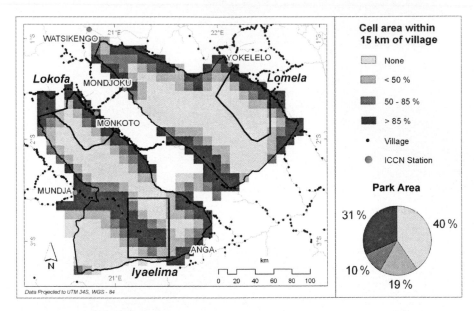

**Fig. 10.2** Human settlement in the Salonga National Park and vicinity. Bonobo population inventory blocks are indicated in outline.

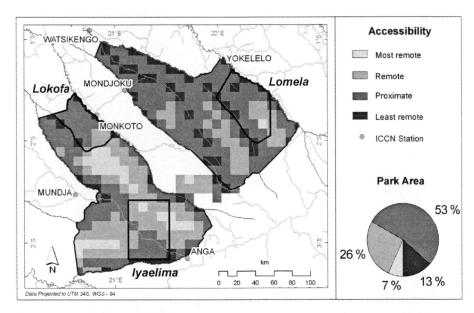

**Fig. 10.3** Accessibility of the Salonga National Park and eastern corridor survey zone is calculated as the percentage of each 10 × 10 km quadrat that is within 15 km of human access (roads and rivers accessible by dugout canoe). Percentage quadrat area > 15 km from access: Most remote, 100; remote, 75 – 99; proximate, 50 – 74; least remote < 50. Bonobo population inventory blocks are indicated in outline.

than 15 km from a navigable river. Nearly 85% of the northern sector of the park is within 15 km of a river navigable by dugout canoe (Fig. 10.3).

The status of bonobos in the Salonga National Park was poorly known throughout the early years of the park's history. As recently as the 1980s, it was uncertain whether the park even contained bonobos (Susman et al. 1981). Van Krunkelsven (2001) and van Krunkelsven et al. (2000) reported the first population surveys of bonobos in the Salonga National Park in the late 1990s. Other surveys followed, including Lui Kotal, just outside the park (Mohneke and Fruth 2008, Mohneke 2004) and at a number of sites in both the northern and southern sectors (Reinartz et al. 2008, 2006). Intensive studies of semi-habituated bonobos were initiated at Lui Kotal in 2000 (Hohmann and Fruth 2003) and Etate in 2004 (Reinartz et al. 2008).

Taken as a whole, these surveys confirmed that at least some areas of the Salonga National Park contained important numbers of bonobos and provided a useful comparison of the abundance of different bonobo communities. However, most of the survey sites covered relatively small areas, < 300 km$^2$, and often much smaller. Direct extrapolation of the results to larger areas is problematical because there is little basis to determine how representative these study areas are of communities and populations elsewhere. A comprehensive picture of the park's bonobos and factors affecting them was lacking.

In the late 1990s, following the outbreak of conflict, ICCN patrols were reduced leaving the park mostly untended for much of the past decade. Just how badly the park and its bonobos were threatened by the conflict remained uncertain (Draulans and van Krunkelsven 2002). Surveys were urgently needed to provide an up-to-date status of bonobos and to identify needs for protection.

## Objectives

The current chapter and Chapter 12 in this volume present the results of a multiphase, spatially nested survey to develop the first park-wide estimate of the distribution and abundance of bonobos and an evaluation of the impact of human activities, in particular hunting, in the largest protected area within their range. We first present the design and results of the surveys, which at the largest spatial extent cover an area >30,000 km$^2$. We then integrate the results of the different survey phases to provide an estimate of the population of bonobos in the park. The paper concludes with recommendations for the use of multiphase surveys to determine the occurrence and abundance of bonobos in other areas of their range.

## Survey Design and Data Collection

Two major challenges face large-scale surveys of bonobos. First, is the impossibility of using direct counts for the census. Bonobos are shy, and visibility in the forests they occupy is limited. Thus, inventories depend upon counts of their sign, in particular nests, and their conversion to estimates of bonobo density. The second challenge is to develop a survey design that will provide a representative sample of observations of bonobos and their sign, the habitats they occupy, and human activity in the forest at an appropriate degree of spatial resolution.

Field data for forest surveys, including those reported here, are generally collected over relatively small areas and at a fine spatial resolution, with observations made from line transects and reconnaissance walks usually at distances of a few tens of meters or less from the observer and line of travel. Yet the total extent of the area to be surveyed is much larger, and in the case of landscape-scale surveys, such as the Salonga National Park, tens of thousands of square kilometers must be evaluated. Costs and logistical difficulties preclude covering all areas of the landscape with the same degree of survey resolution. Determining what will be measured for each observation and where observations will be made (allocation of survey effort) are questions of major concern.

To resolve these problems we used a multiphase survey design. In multiphase designs, an initial area is surveyed for readily measured, coarse-resolution variables. In subsequent phases, subsets of the overall survey area are selected based on the results of the initial survey and resurveyed for the same and new variables at a finer spatial resolution (Urban 2002). Multiphase designs provide a means to allocate survey effort efficiently and optimally across a range of spatial scales. They also provide a statistically sound framework for the extrapolation of the results of smaller scale surveys to larger areas.

The Salonga survey program used a three-phase design spanning a twenty-fold range of spatial resolution. The largest survey zone was > 2500 times the area of the smallest. The surveys shared the same overall goals. However, the objectives of each survey phase were specific to the spatial scale of the design. Figure 10.4 provides an overview of the survey objectives and associated spatial design and data collection of each phase.

In Phase I, at the largest spatial extent, the survey area covered most of the park and portions of the immediate buffer zone and corridor between the two park sectors. We made field observations from compass-directed reconnaissance walks, termed "recces," placed systematically at a spatial grain of ca. $10 \times 10$ km (quadrats of $100$ km$^2$). In Phase II, we surveyed three subsets of the landscape covered in Phase I, termed inventory blocks, which cover $2000$–$3000$ km$^2$ each, using quadrats of ca. $5 \times 5$ km ($25$ km$^2$) to allocate survey effort. We collected field data from both recces and formal line transects, allocated spatially using the DISTANCE software (Thomas et al. 2001a) and using data collection methods and analytical protocols described in Buckland et al. (2001). In Phase III surveys, we evaluated the persistence of bonobo nest site use in spatial units termed monitoring zones covering ca. $12.5$ km$^2$ each. We collected Phase III data from line transects at a spatial grain of $0.5 \times 0.5$ km ($0.25$ km$^2$). Phase III surveys were repeated in a sample of 8 monitoring zones at intervals of $5 - 7$ months.

At the outset of the surveys we had very little information on the distribution and abundance of bonobos across the park. We used a nonstratified, systematic placement of recces and transects. This is recommended to ensure unbiased and representative samples of observations where little antecedent information is available to stratify or otherwise model survey design (Thomas et al. 2001b).

## A. Survey objectives specific to spatial scale.

| GOALS: | Determine bonobo occurrence | Identify key habitats | Assess impact hunting |
|---|---|---|---|
| **OBJECTIVES** | | | |
| **Phase I** | Occupancy | Geographic clines | Landscape-scale patterns |
| **Phase II** | Density | Nesting habitats | Relationship with bonobo density |
| **Phase III** | Nesting site use over time | Characteristics of intensively used nesting sites | Impact on nesting site use |

## B. Survey effort and data collection specific to spatial scale.

| Phase | Spatial class (sample size) | Survey unit area (km$^2$) | Survey grid (km) | Data collection | Sampling placement |
|---|---|---|---|---|---|
| I. Exploratory | Landscape ( n=1) | 33,000 | 10 x 10 | Recces | Systematic |
| II. Inventory | Block (n = 3) | 2,000 - 2,750 | 5 x 5 | Recces, transects | Systematic |
| III. Monitoring | Zone (n = 8) | 12.5 | 0.5 x 0.5 | Transects | Systematic |

**Fig. 10.4** Objectives and design of the Salonga National Park multiphase survey. A) Survey objectives are based on the same survey goals but are specific to each survey phase determined by the spatial scale and resolution of the data collection. B) Survey zone area, the spatial grain and placement of survey effort and data collection methods are specific to each survey phase. Phase I results inform design and data collection of Phase II inventories. Phase I and Phase II results are used to design Phase III monitoring data collection. Phase I and Phase II are single data collection designs. Phase III data collection is repeated over intervals of 4–6 months.

We conducted Phase I field work from 2003 – 2005. Field teams covered the southern sector and about half the northern sector of the park from 2003–2004 during the CITES-MIKE project (Blake 2005). They completed the remaining half of the northern sector and the eastern corridor in 2005. We conducted Phase II surveys in the Lokofa, Iyaelima and Lomela blocks from 2005 – 2006. We initiated Phase III surveys in 2005 in eight monitoring zones in the Lokofa block. Six monitoring zones covered bonobo nesting areas discovered during Phase I and Phase II surveys. Two monitoring zones covered areas where nests were not previously encountered. We resurveyed the monitoring zones two to three additional times each from 2005 – 2006.

We used GIS to determine geographic coordinates of line transects (start and end points) and to plot quadrat centroids used to orient recces. We used GPS units and compasses to locate recce and transect positions in the field and to orient in the forest. We measured distances along line transects with topofils.

Teams composed of a team leader, assistant leader, compass man, two observers, and supported by 6 – 8 porters and local guides collected the field data. We used GPS track logs to document the geographic position of survey teams as they moved across the survey zone. Field teams recorded geographic coordinates (waypoints) for all observations, and measured perpendicular distances from the line of travel to the center of the observed object (bonobo nest, snare, etc) on line transects (but not on recces).

Indicators of bonobo occurrence recorded in the field include direct observations of subjects (seen, heard, or both), feeding signs, and nests. We recorded tree species containing nests, nest height, nest age class (fresh, recent, old, disappearing) based on criteria established for this study and photographed each nest. After completing nest measures from the recce or line transect, field teams located additional nests not seen from the line of travel and produced a field map of each nest aggregation showing nest locations.

Habitat indicators recorded in the field include substrate and vegetation type and under-story class. We classified the habitat for all observations at every 100 m along line transects. Substrate types were: permanently inundated, seasonally inundated, and *terra firma* (non-inundated). Vegetation types included permanently flooded forest, seasonally inundated forest, mixed *terra firma* forest, monodominant *terra firma* forest, open canopy Marantaceae forest, recent regeneration, secondary forest, and savanna. Understory classes included: open shrub, closed liana/shrub and herbaceous dominated. Observers classified habitats as the dominant types covering a circle of ca. 10m surrounding their position or the position of the observation.

Indicators of human hunting included encounters with hunters, snares (classed as active or inactive and by the size of the sapling anchor) and hunting camps. We recorded hunting camp activity (occupied, recently abandoned, long-abandoned), the number of shelters and beds, and the presence and size of meat drying racks. We also recorded fishing camps and other fishing signs, trail crossings, machete cuts, and evidence of other extractive activities. We photographed most of the illegal hunting and fishing camps encountered in the park.

We recorded field observations with associated geo-referencing data (GPS waypoints and tracklogs) on Excel spread sheets for data analyses. We used DISTANCE software (Thomas et al. 2001a, Southwell and Weaver 1993) to determine nest densities based on line transect nest counts. We mapped encounter rates (number of observations per km surveyed) of bonobo and human activity indicators to each survey quadrat using ARC GIS and conducted further spatial analyses via ESRI statistical packages (Mitchell 2005) and other sources. Table 10.1 is a summary of data collection and analytical methods for the three survey phases.

**Table 10.1** Data collection and analysis of the multiphase bonobo surveys in the Salonga National Park

| Survey phase | Dates | Location | Survey units | Indicators | Analysis |
|---|---|---|---|---|---|
| I. Landscape exploration | 2003 – 2005 | Southern and northern sectors of park; eastern corridor | Pilot transects (1 km each) linked by recces for MIKE surveys. Recces linking centroids of $10 \times 10$ km quadrats for northern sector and corridor surveys. | • Bonobo: direct observation, nest count, feeding sign. <br> • Habitat: substrate, vegetation type, sub-canopy <br> • Hunting: snares, camps, encounters with hunter | Spatial analysis of indicator encounters. |
| II. Inventory block | 2005 – 2006 | Lokofa, Iyaelima, Lomela | Systematically placed line transects (1.4 km each) linked by recces. | • Bonobo: nest count, direct observation <br> • Habitat: substrate, vegetation type, sub-canopy <br> • Hunting: snares, camps, encounters with hunter | DISTANCE estimates of nest density from line transect encounters. Spatial analysis of indicator encounters. |
| III. Monitoring zone | 2005 – 2006 | Lokofa | Five Systematically placed line transects; 500 m, each, separated by 500 m, linked by recces. | • Bonobo: nest count, direct observation <br> • Habitat: substrate, vegetation type, sub-canopy <br> • Hunting: snares, camps, encounters with hunter | Spatial analysis of i ndicator encounters. |

## Estimating Bonobo Densities

We used standing crop nest counts (Mohneke and Fruth 2008, Mohneke 2004) to estimate bonobo density as follows:

Bonobo density (number/km$^2$) = [Density of bonobo nests (number/km$^2$) / mean decay rate of nests (day)] / daily nest production per individual (number/day).

This estimate of the population of bonobos is actually an estimate of nest building individuals. Infant bonobos nest with their mothers, and even older individuals may share nests (Fruth 1995). Hashimoto and Furuichi (2001) estimated that two-thirds of nest-building individuals built nests at Wamba. We have no basis to estimate the proportion of the population that does not build its own nests. Our estimates of nest building individuals are conservative and underestimate total population.

To convert estimates of standing crop nest density to bonobo density requires estimates of two additional parameters: daily nest production rates and nest decay rates. We used an estimate of daily nest production per individual of 1.37 nests/day based on observations in the Lomako forest (Fruth 1995). We used mean nest decay estimates of 78 days with a 95% confidence interval upper and lower range of 68 and 83 days, based on nest decay studies at Lui Kotal (Mohneke and Fruth 2008, Mohneke 2004).

## Phase I: Bonobo Occurrence in the Salonga Park

### *Phase I: Coverage*

Field teams conducted 2,900 km of reconnaissance during the Phase I survey, covering an area of 33,000 km$^2$, including 2,100 km$^2$ of the eastern corridor between the two park sectors. We did not survey the eastern limits of the northern and southern sectors, as these areas were thought to be occupied by rebels and not safe when the survey was designed. The excluded areas represent < 8% and < 15% of the southern and northern sectors respectively.

Survey teams sampled 325, 10 × 10 km quadrats, of which 233 quadrats had ≥ 5 km reconnaissance coverage. We considered cells with < 5 km reconnaissance coverage insufficiently sampled and excluded them from statistical analyses. Figure 10.5 shows the Phase I survey coverage.

### *Phase I: Results*

Figure 10-6 shows the distribution of all 10 × 10 km grid cells wherein we observed bonobo indicators. We also include locations of historic records provided by Kortland (1995). We recorded evidence of bonobos in 173 (53%) of the 325 quadrats

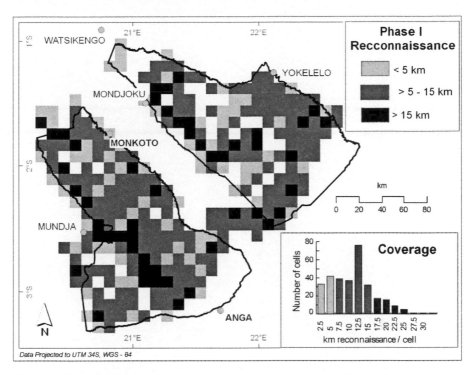

**Fig. 10.5** Phase I survey coverage. Survey coverage classes determined by the distance of reconnaissance (recce) track traversed within each 10 × 10 km quadrat. Grid cells with < 5 km of recce are not included in the statistical analyses.

sampled. This included 31 direct encounters with bonobos in 25 quadrats. Bonobo nests were present in 93 quadrats.

We integrated the field indicators of bonobos into a composite index of occurrence for each grid cell by summing the encounter rates of each indicator weighted by a score based on the indicator's probability of detection, the certainty of its identity, possible time lapse between the detection of the indicator and the occurrence of bonobos, and the production and decay rates of the indicator. The criteria for the scoring are in Table 10.2. The weighting scores of the indicators are in Table 10.3.

Bonobo occurrence indices have a log normal distribution. We log transformed and classed them on an ordinal scale as low, average and high with the mean +/− one standard deviation of the log transform values classed as average. We classed the quadrats with the highest 12 occurrence values as very high.

The distribution of bonobo occurrence for 233 quadrats with ≥ 5 km of survey effort is illustrated in Fig. 10.7. We grouped the sampled survey quadrats into larger contiguous areas, termed population extrapolation blocks, to be used for developing a park-wide estimation of bonobo numbers. Bonobo occurrence increases significantly in a cline from west to east across the park using Rosenberg's (2000) bearing correlogram. The spatial trend is correlated with the geography of the hydrological network and relative elevation of the park. Bonobos are less abundant and more

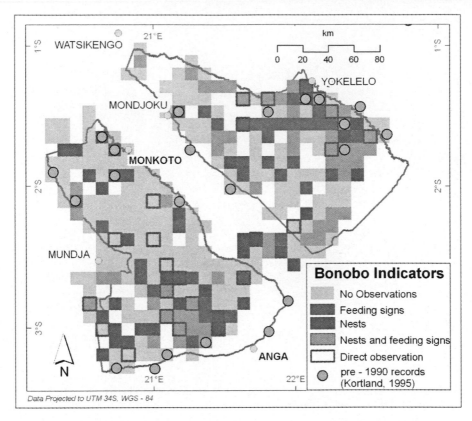

**Fig. 10.6** Distribution of bonobos and indicators of their presence observed during Phase I survey in the Salonga National Park. Locations of bonobo records predating creation of the park in 1970 are shown (*See Color Plates*).

localized in the western area of the park, a lower-lying region of river confluences. They are more abundant and widespread in the eastern area of the park, a region of higher-lying plateau forests and river headwaters.

Bonobos are not adverse to proximity of human settlement. They are consistently associated with areas that are accessible to and used by humans (Fig. 10.8). Some of the highest bonobo indices were found near villages in the Iyaelima and Lomela inventory blocks.

## Phase II: Population Estimation in Inventory Blocks

### *Phase II: Coverage*

Two of the inventory blocks – Iyaelima and Lomela – had high mean Phase I bonobo occurrence indices, while the Lokofa block had one of the lowest mean index values. We allocated line transects of 1.4 km systematically across each inventory block at

**Table 10.2** Criteria for determining field indicator weighting score used to calculate composite indices of faunal occurrence and human activity for survey girds

| Score | Certainty | Detection | Criteria Time lapse | Production rate | Decay rate |
|---|---|---|---|---|---|
| 0 | *Ambiguous association.* Indicator may be confused with more than one faunal species or human activity. | *Difficult.* Unpredictable and not consistent; uncertain classification by age or category. | *Large time lapse.* Long delay possible between the observation of the indicator and the occurrence of the fauna or human activity producing it. | *Unpredictable.* Rarely produced, highly variable by season or habitat or variable for unknown reasons. | *Unpredictable.* Rapid decay, highly variable by season or habitat or variable for unknown reasons. |
| 1 | *Indirect evidence.* Dung, feeding remains, constructions, evidence of tool use or other sign produced by animals or humans. | *Seasonally variable.* Observation and classification affected by habitat, season or other external factors. | *Small time lapse.* Small delay between the observation of the indicator and the occurrence of the fauna or human activity producing it. | *Known variability.* Consistent known relationship between occurrence of indicator and fauna or activity. | *Known variability.* Consistent known relationship with habitat and season. |
| 2 | *Direct observation.* Fauna or human activity observed directly (seen or heard). | *Consistent.* Limited variability; classification based on well defined criteria. | *Immediate.* Indicator immediately associated with the fauna or human activity producing it. | *Stable.* Consistent relationship between indicator and occurrence of fauna or human activity. | *Stable.* Consistent across habitat and season; variability associated with known factors. |

**Table 10.3**  Weighting values for indicators of bonobo occurrence

|  | | | Criteria | | | |
|---|---|---|---|---|---|---|
| Indicator | Certainty | Detection | Time lapse | Production rate | Decay rate | Total score |
| Feeding sign | 0 | 1 | 1 | 0 | 0 | 2 |
| Nest | 1 | 2 | 0 | 1 | 1 | 5 |
| Bonobo encounter | 2 | 0 | 2 | 0 | 0 | 4 |

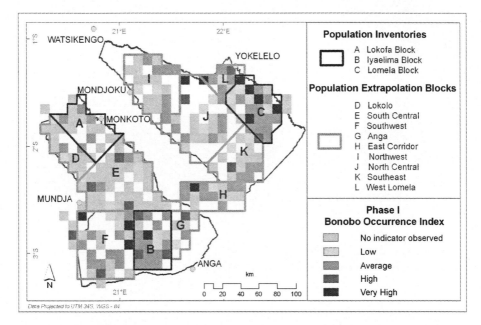

**Fig. 10.7**  Bonobo occurrence indices integrate Phase I encounter rates of weighted field indicators for 10 × 10 km quadrats with ≥ 5 km reconnaissance coverage. Contiguous quadrats are combined into 12 population extrapolation blocks to calculate an estimate of bonobo populations for the total park area. Three extrapolation blocks cover the Phase II population inventory blocks (*See Color Plates*).

a rate of about one transect per 6 km². In addition, we conducted between 511 and 583 km of recce in each block. We completed full surveys for the Lokofa and Iyaelima blocks; however, 11 transects and ca. 400 km² of the planned Lomela block were truncated when field teams were threatened by residents of the Kitawala village. In total, we inventoried 7,250 km² for standing crop nest counts in all three blocks, using 186 transects totaling 260 km and an additional 1,609 km of recce.

## Phase II: Results

Table 10.4 is a summary of survey effort and nest encounter rates for the three inventory blocks. Figure 10.9 presents the spatial distribution of the line transect nest encounter rates within each block. The spatial distribution of Phase II nest

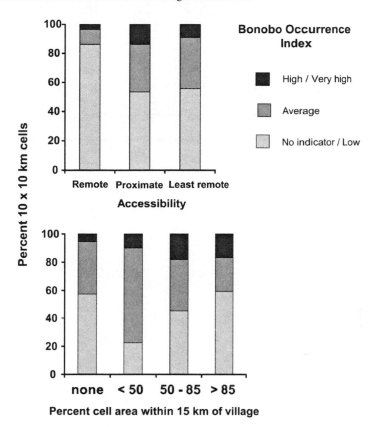

**Fig. 10.8** Bonobo occurrence in relation to (A) human settlement and (B) accessibility. Bonobo occurrence varied significantly between classes for both human settlement and accessibility (Chi Square probability = 0.0002 and 0.0194 respectively). Bonobos consistently occur in accessible areas and in proximity to human settlement in the park.

encounters reflects the distribution of bonobo occurrence determined during Phase I surveys in all three inventory blocks. Mean transect nest encounter rates varied from 1.15 to 2.29 nests per km for the three blocks. Nest encounter rates on recces averaged 82% of encounter rates recorded on transects. In total, we recorded 2,941 nests in 1,032 nest aggregations on the three inventories.

Table 10.5 is a summary of nest density estimates for each inventory block calculated via the DISTANCE software. Nest detections exhibited a strong shoulder at distances near the transect line for all three inventories, thus facilitating the fit of the detection curves. Effective strip width varied from 14.9 to 16.2 m across the three blocks.

Mean nest densities ranged from 29.9 – 90.2 nests per km$^2$ for the three blocks. These are equivalent to 0.27 – 0.84 nest-building bonobos/km$^2$, via a mean nest decay value of 78 days and nest production rate of 1.37 nests per day per bonobo. We estimated upper and lower densities using 95% confidence limits generated by

**Table 10.4** Survey effort and nest count data for Phase II inventory blocks in Salonga National Park

| | | Line Transects | | | | | | Recces | | | | Total survey | | |
|---|---|---|---|---|---|---|---|---|---|---|---|---|---|---|
| Block | Area (km²) | Transects | Effort (km) | Nests observed | Transects with nests | Nest aggregations | Nest encounter (per km) | Effort (km) | Nests observed | Nest aggregations | Nest encounter (per km) | Effort (km) | Nests observed | Nest aggregations |
| Lokofa | 2000 | 55 | 76.6 | 88 | 11 | 23 | 1.15 | 583 | 542 | 156 | 0.93 | 660 | 630 | 179 |
| Iyaelima | 2500 | 63 | 88.2 | 194 | 34 | 67 | 2.20 | 511 | 974 | 378 | 1.91 | 599 | 1168 | 445 |
| Lomela | 2750 | 68 | 95.2 | 218 | 31 | 49 | 2.29 | 515 | 925 | 359 | 1.80 | 610 | 1143 | 408 |
| TOTAL | 7250 | 186 | 260 | 500 | 79 | 139 | 1.93 | 1609 | 2441 | 893 | 1.52 | 1869 | 2941 | 1032 |

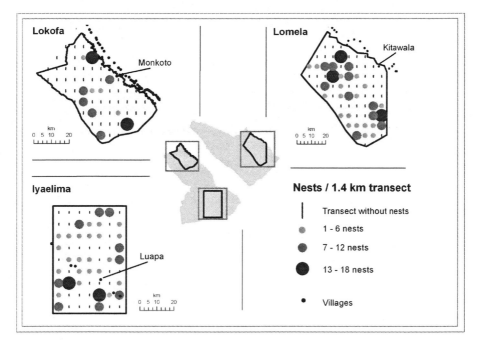

**Fig. 10.9** Encounter rates of the standing crop of bonobo nests recorded on line transects in Phase II inventories.

**Table 10.5** Estimates of bonobo numbers in Phase II inventory blocks

| | Nest density (per km²) | | | | | Nest building bonobos | | |
|---|---|---|---|---|---|---|---|---|
| | | Standard | Coefficient of variation | (95% CI) | | Mean density | Population in block [a] | |
| Block | Mean | error | (percent) | Low | High | (per km²) | Mean | Range |
| Lokofa | 28.9 | 8.74 | 30.3 | 16.0 | 52.1 | 0.27 | 541 | 281 – 1119 |
| Iyaelima | 54.8 | 15.41 | 28.1 | 31.8 | 94.4 | 0.51 | 1282 | 699 – 2533 |
| Lomela | 90.2 | 17.2 | 19.1 | 61.9 | 131.3 | 0.84 | 2321 | 1497 – 3876 |
| Mean or total | Mean 58.0 | | | | | Mean 0.54 | Total 4144 | 2477 – 7528 |

[a] Parameters used to estimate bonobo populations in blocks:
Mean estimate: mean nest density from DISTANCE, nest decay 78 days, nest production 1.37 nests / day.
Low estimate: lower 95 % confidence interval nest density from DISTANCE, nest decay 83 days, nest production 1.37 nests / day.
High estimate: upper 95 % confidence interval nest density from DISTANCE, nest decay 68 days, nest production 1.37 nests / day.

DISTANCE, with lower and upper nest decay estimates used in the conversion of nest densities to bonobo densities. Together, the three inventory blocks, covering 7,250 km$^2$, contain an estimated 4,144 (2,477 – 7,528) nest building bonobos, an average density of 0.54 individuals/km$^2$.

# Phase III: Spatial-Temporal Use of Nesting Zones

## *Phase III: Coverage*

We selected the Lokofa block for Phase III surveys since it had the longest survey history (dating to 2003), and the low bonobo indices were typical of many other areas in the park. We delimited 8 circular monitoring zones (2 km radius each) following completion of the Phase II inventories in April 2005. We centered 6 zones on areas that contained bonobo nests during the Phase I or Phase II surveys. We centered 2 zones on areas where no nests had been detected. In each zone we laid out 5 transects, 2.5 km each separated by 500 m (12.5 km total) (Fig. 10-10). We visited each zone and conducted a standard line transect nest count once or twice over the subsequent 14 months. The average interval separating visits was 5 months (range

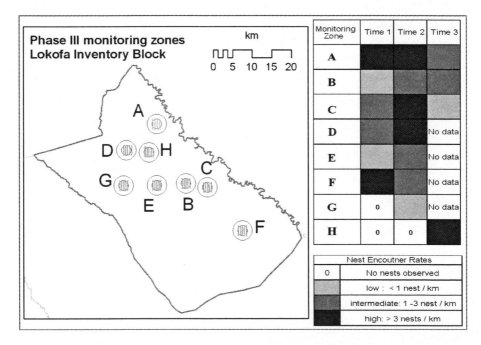

**Fig. 10.10** Phase III encounter rates of bonobo nests in 12.5 km$^2$ monitoring zones in the Lokofa block.

4 – 7 months). This interval ensured that nests counted at one period would have a low probability of persisting to the next period.

## *Phase III: Results*

Nest counts in individual monitoring zones were often too low to estimate nest density. Thus, we used nest encounter rates for comparisons. Following each visit we classed nest encounter rates in four relative abundance classes from no nest to high nest encounter rates (Fig. 10.10).

Six of the eight monitoring zones had nests initially. We inspected 3 zones three times and 3 zones twice. All 6 monitoring zones had nests on all inspection visits. Nest encounter rates varied within monitoring zones over time, but most changes were not large. Two zones did not have nests initially, but eventually contained them. One zone that was empty on the first and second visit had high nest encounter rates on the third visit. Of the nine pair-wise comparisons from one inspection visit to the next, encounter rates remained in the same frequency class only twice. However, 6 changes were relatively small, shifting one frequency class. Only one change was of a large magnitude. Zone C dropped from a high nest encounter rate at inspection 2 to a low rate at inspection 3, 5 months later.

## Estimating the Bonobo Poulation of Salonga National Park

We regressed Phase I bonobo occurrence indices with nest densities determined in Phase II inventories to estimate bonobo populations for 12 extrapolation blocks covering the park and eastern corridor using the following equation:

$$y = ax + b$$

wherein y = mean nest density for the block, x = mean bonobo occurrence for the block, a = slope, and b = intercept value, both determined from the regression equation. We calculated mean and upper and lower estimator equations separately using the mean, and the upper and lower 95 percent confidence interval nest density values produced by the DISTANCE analysis of the Phase II surveys (Fig. 10.11).

We delimited 3 extrapolation blocks to cover the inventory blocks surveyed in Phase II. These represent ca. 20% of the total park area and span the range of bonobo occurrence from lowest to highest. We delimited 9 blocks to include contiguous areas of comparable extent to the Phase II blocks and with relatively homogenous bonobo occurrence indices for the 10 × 10km grid cells within their limits. We calculated a mean occurrence index for each block. Most blocks had > 75% Phase I coverage, and all blocks had ≥ 50% Phase I coverage (Fig. 10.7).

Estimates of bonobo populations for the 12 blocks are in Table 10.6. We used the mean nest decay period of 78 days to convert mean nest densities to an estimate

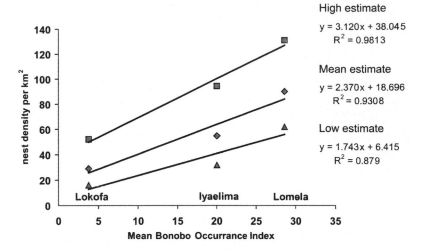

**Fig. 10.11** Equations for estimating bonobo nest densities in the Salonga National Park based on the relationship between nest densities determined in Phase II inventories and mean Phase I occurrence indices for population extrapolation blocks. Mean, low and high estimating equations are based on mean and 95% confidence interval values for nest densities determined by the DISTANCE analysis of the Phase II nest encounters for the three inventory blocks.

of nest-building bonobos for each estimator block. We used the lower nest decay value of 68 days to estimate the upper population values and the upper nest decay value of 83 days to estimate the lower population values.

Via these parameters, we estimated 14,883 nest-building bonobos in the park (0.42/km$^2$), with lower and upper estimates of 7,119 (0.20/km$^2$) and 20,434 (0.57/km$^2$) respectively. Estimates of mean bonobo density for the different blocks within the park range from 0.26 to 0.84 nest building individuals/km$^2$. We estimated the eastern corridor block, covering 2,100 km$^2$, to contain 809 nest building bonobos (377 – 1,128) with a mean density of 0.39/km$^2$.

## Discussion

The results of the surveys confirm that the Salonga National Park contains a globally significant population of bonobos. They are numerous and widespread within the park, and there are important populations in at least some areas of the corridor linking the northern and southern sectors of the park. Our estimates of mean bonobo densities are lower than density estimates of 0.72/km$^2$ and 1.15/km$^2$ given in two earlier limited surveys of the park (Reinartz et al. 2006, van Krunklesven 2001). They are lower than the densities of 0.83 – 1.04/km$^2$ for bonobos in the forest-savanna ecotone at Lukuru (Meyrs Thompson 1997) and less than one third the estimate of 1.3 – 1.4/km$^2$ for the Lomako (Eriksson, 1999). They are consistent

Table 10.6 Population estimates of nest building bonobos in the Salonga National Park

| Sector | Population extrapolation block[a] | Area (km²) | Mean occurrence index | Low estimate | | | Mean estimate | | | High estimate | | |
|---|---|---|---|---|---|---|---|---|---|---|---|---|
| | | | | nest density | bonobo density | bonobo population | nest density | bonobo density | bonobo population | nest density | bonobo density | bonobo population |
| Southern | A. Lokofa | 2300 | 3.79 | 13.0 | 0.106 | 243 | 27.7 | 0.266 | 611 | 49.9 | 0.404 | 930 |
| | B. Iyaelima | 2400 | 20.04 | 41.3 | 0.335 | 805 | 66.2 | 0.636 | 1526 | 100.6 | 0.816 | 1958 |
| | D. Lokolo | 1550 | 3.87 | 13.2 | 0.107 | 165 | 27.9 | 0.268 | 415 | 50.1 | 0.406 | 630 |
| | E. South Central | 4750 | 3.33 | 12.2 | 0.099 | 471 | 26.6 | 0.255 | 1213 | 48.4 | 0.393 | 1866 |
| | F. Southwest | 4200 | 15.25 | 33.0 | 0.268 | 1124 | 54.8 | 0.527 | 2212 | 85.6 | 0.694 | 2917 |
| | G. Anga | 1100 | 8.57 | 21.4 | 0.173 | 190 | 39.0 | 0.375 | 412 | 64.8 | 0.525 | 578 |
| | Unsampled zone | 1800 | 8.40 | 21.1 | 0.171 | 307 | 38.6 | 0.371 | 667 | 64.3 | 0.521 | 938 |
| | TOTAL | 18100 | | | | 3306 | | | 7057 | | | 9816 |
| Northern | C. Lomela | 2700 | 28.59 | 56.2 | 0.456 | 1232 | 86.5 | 0.830 | 2242 | 127.2 | 1.032 | 2786 |
| | I. Northwest | 3100 | 7.19 | 18.9 | 0.154 | 476 | 35.7 | 0.343 | 1064 | 60.5 | 0.490 | 1521 |
| | J. North Central | 4700 | 5.11 | 15.3 | 0.124 | 584 | 30.8 | 0.296 | 1391 | 54.0 | 0.438 | 2058 |
| | K. Southeast | 2500 | 11.55 | 26.5 | 0.215 | 538 | 46.1 | 0.442 | 1106 | 74.1 | 0.601 | 1502 |
| | L. West Lomela | 900 | 23.51 | 47.4 | 0.384 | 346 | 74.4 | 0.715 | 643 | 111.4 | 0.903 | 813 |
| | Unsampled zone | 3700 | 8.50 | 21.2 | 0.172 | 637 | 38.8 | 0.373 | 1380 | 64.6 | 0.524 | 1937 |
| | TOTAL | 17600 | | | | 3813 | | | 7826 | | | 10618 |
| Park | **Block Totals** | **35700** | | | **0.199** | **7119** | | **0.417** | **14883** | | **0.572** | **20434** |
| Corridor | H. East Corridor | 2100 | 9.03 | 22.2 | 0.180 | 377 | 40.1 | 0.385 | 809 | 66.2 | 0.537 | 1128 |

[a] See Fig. 10-7 for delimitation of blocks.

with densities of 0.54/km² reported from the northern sector of the Luo Reserve (Hashimoto and Furuichi 2001) and 0.52/km² for the Lui-Kotal study area, using the same standing crop nest count methods as this study (Mohneke and Fruth 2008). They are similar to an estimate of 0.4/km² presented more than 20 years ago by Kano (1984) for the overall bonobo range. All of the high estimates cited above are based on survey areas that are considerably smaller than the 3 inventory blocks we surveyed, and they can not be extrapolated to larger areas.

We have no evidence of major population declines of bonobos in the Salonga National Park in the recent past. Historic records mapped by Kortland (1995) are mostly located in areas where we found bonobos. Of the 21 bonobo records before 1990, 11 were located in Phase I quadrats with ≥ 5 km reconnaissance coverage. We confirmed occurrence of bonobos in 9 of the 11 sites (Fig. 10.5). We recorded no evidence that the bonobos in the Salonga National Park have been reduced by recent widespread epidemic diseases, as has occurred in some populations of western gorilla (*gorilla gorilla*) and chimpanzees (*Pan troglodytes*) (Lahm et al., 2006; Formenty et al.,1999), though we can not eliminate the possibility. A further analysis of the impact of human hunting on bonobos is presented in Hart et al. (2008).

At a landscape scale, bonobos increase in abundance and occurrence from west to east across the park. Extensive areas of the western Salonga landscape are covered by black water swamp forests not favored by bonobos, and the entire region is underlain by highly leached white sand soils. White sand soils often have reduced primary productivity and are dominated by chemically protected plants less palatable to many primary consumers, including primates (Oates et al. 1990, Freeland and Janzen 1974). The bonobo's localized distribution in some areas of Salonga National Park may be determined by the limited availability of areas of marginally higher productivity suitable for their needs. The repeated use of the same nesting areas observed in the Lokofa block, an area of extensive white sand substrate, indicates intensive use of limited areas by bonobos there. A further investigation of the relationship of habitat productivity and bonobo socioecology may provide a useful basis for developing conservation programs appropriate to their varied ecological context.

Reinartz et al. (2008, 2006) report that bonobos avoid areas of human settlement and activity in the Salonga landscape. We did not confirm their conclusion. While bonobos occur in remote areas of the park, we found some of the highest concentrations of bonobo nests near villages, particularly in the Iyaelima block. The Iyaelima people traditionally avoid contact with bonobos, which they consider to be highly capable fighters (Thompson et al. 2008). Bonobos and humans coexist in other areas as well, such as Wamba, Yasa and Lilungu, (Thompson et al. 2008, Kano et al. 1996). Our surveys show that the relationship between humans and bonobos must be evaluated in a site-specific context and that the relationship is dynamic (Hart et al. 2008). The possibility that humans and bonobos selectively occupy the same localized areas of marginally higher productivity within an overall nutrient constrained environment needs to be evaluated. Human modification of the forest may also attract bonobos. A better understanding of the relationship between human settlement and use of the forest and bonobo occurrence is important as human populations grow and disperse within the bonobo's range.

## The Use of Nest Counts to Estimate Bonobo Occurrence and Density

All great apes make nests, and nest counts can be used to estimate populations where animals are difficult to detect, and where large areas must be covered. Researchers have used nest counts to develop landscape estimates of chimpanzees and gorillas, including nation-wide surveys of apes in Gabon (Tutin and Fernandez 1984), Cote d'Ivoire (Marchesi et al. 1995) and Uganda (Plumptre et al. 2003). We also confirm their utility for large-scale bonobo surveys.

Apes live in groups and often nest together, so many nest counts use nest groups as the observational unit with line transect measures to the center of the group. Counts of nest groups assume that each one represents a single nesting event, that separate nest groups can be distinguished, and that all nests in a group are constructed at the same time. Early in the Salonga survey, we found that some aggregations of bonobo nests observed in the field represented different nesting events in close spatial proximity. As nests aged, it was impossible to distinguish one nesting event from another, especially if different nesting events were separated by short periods of time. We opted to use the individual nests instead of nest groups as the observational unit.

The Phase III surveys confirmed that bonobos consistently use at least some nesting zones over time, and that nest site fidelity can be very specific. In 9 of 15 pair-wise comparisons from one time period to the next, bonobo nesting events occurred on the same transect within the monitoring zone and in all but two cases, within the same 250m of transect line. Mohneke (2004), Fruth (1995) and Fruth and Hohmann (1993) observed in Lomako and Lui Kotal that nesting is concentrated in selected areas of a bonobo community's home range, and that nesting events accumulate in the same locations. They also opted to count individual nests instead of nest groups in population surveys. Len Thomas (personal communication, 2005) reports that the potential bias due to non-independence of observations in counting individual nests from clusters is small, and also recommends counts of nests rather than nest groups on line transects where there is inability to distinguish the groups.

The phenomenon of repeated nesting in limited areas is not restricted to bonobos. Furuichi et al. (2001) compared counts of individual nests versus nest groups for chimpanzees in Uganda and justified the use of individual nests counts. While it has rarely been evaluated, it is likely that at least some of the large aggregations of nests reported for bonobos, and possibly chimpanzees, represent multiple nesting events in the same location instead of a large single nesting event (Kuroda 1979).

We observed that bonobos sometimes refurbished and reused older nests, especially in repeatedly used nesting areas. Therefore, estimates of daily nest production rates should be given with estimates of percentage reuse of old nests. This allows for the calculation of a correction factor in estimating bonobo densities from nests estimates (Plumptre and Reynolds 1997). If nest production rates are not corrected for reuse, estimates of bonobo density from nest counts might be biased downward.

Our surveys used standing crop nest counts. A second nest count method, using marked nests, requires repeated visits of the same transects. This has the advantage that it does not require independent estimates of nest decay (Plumptre and Reynolds 1996). However, the method is not feasible for large surveys where each sampling unit is visited once. It may not be feasible where nest accumulation rates are very low. In one study comparing marked and standing crop nest counts, the marked nest methods led to a higher density estimate (Mohneke 2004). It is not known if this is likely to be a consistent trend.

## Evaluation of Multiphase Design and Recommendations for Large-Scale Surveys

Bonobo populations can be surveyed at a range of spatial scales; however, the spatial resolution and the spatial extent of survey effort will determine the precision of the results and how representative they are of larger areas. No single spatial scale will be appropriate for all questions concerning bonobo distribution and abundance. Thus, it is important to know at the outset what questions to ask and what conclusions to seek for surveys at different spatial scales. In the Salonga design, we identified patterns of bonobo occurrence over a large area and coarse spatial resolution, and correlated these with results of population inventories at a finer spatial resolution in subsequent survey phases to provide a basis for extrapolation of population estimates over large areas.

Our large scale surveys used direct encounters with bonobos, and observations of nests and feeding signs to confirm their presence. We integrated observations of field indicators into a composite index of occurrence that weighted each indicator by its relative utility, based on five criteria, and its frequency calculated by its encounter rate along recces and transects within each mapped grid cell. The composite index permits comparisons of bonobo occurrence across the landscape and is correlated with estimates of population density when averaged over contiguous blocks of grid cells. We recommend a similar approach to evaluating the occurrence of bonobos in other large scale surveys where different indicators are likely to be encountered and comparisons made between different sites and seasons. We also recommend that Phase II population inventories include blocks that span the range of bonobo occurrence from low to high when used to estimate population abundance over larger Phase I landscapes.

The single biggest challenge to large scale surveys is to ensure that the allocation of survey effort provides a representative sample of the spatial variation in bonobo occurrence across the survey zone. This is especially important in the Phase I surveys where results covering a large spatial extent inform finer resolution surveys over smaller areas. It is generally better to increase the number of sample sites rather than increasing sample coverage per site when allocating limited survey effort. Large scale spatial variation is generally the largest component of survey variance and the most important factor in determining survey design (Thomas et al. 2001b). We recommend non-stratified, systematic placement of reconnaissance walks and transects. This ensures that the results are unbiased and is the most efficient survey design where bonobo occurrence is variable or poorly known, and where factors contributing to differences in occurrence are likely to vary spatially in their importance. Stratification may increase survey efficiency; however, this is most likely to be the case at smaller spatial scales, as shown by Reinartz et al. (2008). We stratified survey effort in allocating bonobo nest monitoring sites in Phase III of this study.

We recommend line transect standing crop nest counts to estimate bonobo populations over large survey areas and counts of individual nests, as opposed to nest groups. This is especially recommended in areas where bonobos are likely to re-use the same nesting areas intensively and where nest clusters can not be separated into discrete nesting events with certainty. Differences in nest decay rates and, to an unknown extent, nest production rates can have a major impact on the conversions of standing crop nest densities to estimates of bonobo densities (Monheke and Fruth 2008). These parameters are often not measured directly, and their range of variability is often poorly known (Monheke 2004). We recommend that other large scale surveys gather further data on nest construction and decay rates when possible.

Nest encounter rates and estimates of nest densities are the most reliable indicators of relative bonobo abundance for standing crop nest counts covering large areas and where survey locations are visited only once. We recommend that all surveys present geo-referenced survey efforts and results separately for line transects and reconnaissance walks, and provide encounter rates of individual nests recorded, even if information on nest groups is also provided. This will facilitate comparisons across sites and between different surveys within the same site. It will also permit estimates of bonobo populations to be revised as new data on nest decay and nest production become available (Plumptre 2000).

**Acknowledgements** We acknowledge the support of the Institut Congolais pour la Conservation de la Nature (ICCN) for the Salonga National Park surveys. We wish to thank the CITES-MIKE program, Wildlife Conservation Society (WCS), USAID's Central African Regional Program for the Environment (CARPE), the Alexander Abraham Foundation and World Wide Fund for Nature (WWF) for financial support.

We are grateful to the over 100 individuals who participated in the surveys and assisted in providing the major logistical effort to support the survey teams in remote locations. Field team leaders deserve special mention: Maurice Emetshu, Aime Bonyenge, Bernard Ikembelo, Simeon Dino, Pupa Mbenzo, Samy Matungila, Menard Mbende and Pele Misenga. Inogwabini Bila Isia assisted in the training of the field teams. Jose Ilanga provided assistance for the teams operating from Monkoto.

We acknowledge the following colleagues who contributed both directly to the survey design and base maps, or who provided key information and supporting data based on their own experience with bonobos in the Salonga landscape: Steve Blake, Jonas Eriksson, Barbara Fruth, Gottfried Hohmann, Adrian Kortland, Menke Mohneke, Florian Siegert, Jo Thompson and Carlos de Wasseige. Terese Hart and two independent reviewers provided useful critiques of early versions of this chapter.

214 F. Grossmann et al.

# References

Blake S (2005) Forêts d'afrique centrale: rapport final sur les relevés démographiques d'éléphants (2003–2004). CITES–MIKE, Nairobi

Buckland ST, Anderson DR, Burnham KP, Laake JL, Borchers, DL, Thomas L (2001) Introduction to distance sampling: estimating abundance of biological populations (2$^{nd}$ edn). Chapman & Hall, London

Butynski TM (2001) Africa's great apes. In: Beck B, Stoinski TS, Hutchins M, Maple TL, Norton, B, Rowan A, Stevens EF, Arluke A (eds) Great apes and humans: the ethics of coexistence. Smithsonian Institution Press, Washington DC, pp 3–56

CITES (2001) Database of listed species. *http://www.cites.org/CITES/common/dbase/fauna/index. shtml*

Draulans D, Van Krunkelsven E (2002) The impact on war on forest areas in the Democratic Republic of Congo. Oryx 36(1):35–40

Dupain J, Van Elsacker L (2001) The status of the bonobo in the Democratic Republic of Congo. In: Galdikas BMF, Briggs NE, Sheeran LK, Shapiro GL, Goodall J (eds) All apes great and small, vol. 1: African apes. Kluwer Academic / Plenum, New York, pp 57–74

Dupain J, Van Krunkelsven E, Van Elsacker L, Verheyen RF (2000) Current status of the bonobo (*Pan paniscus*) in the proposed Lomako Reserve (Democratic Republic of Congo). Biological Conservation 94:265–272

Eriksson J (1999) A survey of the forest and census of the bonobo (*Pan paniscus*) population between the Lomako and the Yekokora rivers in the Equateur province, D R Congo. MSc Thesis, University of Uppsala

Formenty P, Boesch C, Wyers M, Steiner C, Donati F, Dind F, Walker F, Le Guenno B (1999) Ebola virus outbreak among wild chimpanzees living in a rain forest of Côte d'Ivoire. Journal of Infectious Disease 179 (Suppl 1): S120—S126

Freeland WJ, Janzen DH (1974) Strategies in herbivory by mammals: the role of plant secondary compounds. American Naturalist 108:269–289

Fruth B (1995) Nests and nest groups in wild bonobos (*Pan paniscus*) ecological and behavioral correlates. Shaker-Verlag, Aachen

Fruth B, Hohmann G (1993) Ecological and behavioral aspects of nest building in wild bonobos (*Pan paniscus*). Ethology 94: 113–126

Furuichi T, Hashimoto C, Tashiro Y (2001) Extended application of a marked–nest census method to examine seasonal changes in habitat use by chimpanzees. Int J Primatol 22: 913–928

Hart J, Grossmann F, Vosper A, Ilanga, J (2008) Human hunting and its impact on bonobos in the Salonga National Park, D.R. Congo. In: Furuichi T, Thompson J (eds) The bonobos: ecology, behavior, and conservation, Springer, New York, pp 245–271

Hashimoto C, Furuichi T (2001) Current situation of bonobos in the Luo Reserve, Equateur, Democratic Republic of Congo. In: Galdikas BMF, Briggs NE, Sheeran LK, Shapiro GL, Goodall, J (eds) All apes great and small, vol. 1: African apes. Kruwer, Academic/Plenum, New York, pp 83–93

Hohmann G, Fruth B (2003) Lui Kotal: a new site for field research on bonobos in the Salonga National Park. Pan Africa News 10:25–27

IUCN (1996) IUCN red list of threatened snimals. IUCN, Gland, Switzerland

Kano T (1984) Distribution of pygmy chimpanzees (*Pan paniscus*) in the Central Zaire Basin. Folia Primatologica 43:36–52

Kano T, Lingomo B, Idani G, Hashimoto C (1996) The challenge of Wamba. In: Cavalieri P (ed) The great spe project, Ecta and Animali 96/8, Milano, pp 68–74

Kortlandt A (1995) A survey of the geographical range, habitats and conservation of the pygmy chimpanzee (*Pan paniscus*): An ecological perspective. Primate Conservation 16:21–36.

van Krunkelsven E (2001) Density estimation of bonobos (*Pan paniscus*) in Salonga National Park, Congo. Biological Conservation 99:387–391 van Krunkelsven E, Inogwabini B, Draulans D (2000) A survey of bonobos and other large mammals in the Salonga National Park, Democratic Republic of Congo. Oryx 34:180–187

Kuroda S (1979) Grouping of pygmy chimpanzees. Primates 20:161–183

Lahm S, Kombila M, Swanepoel M, Barnes R (2006) Morbidity and mortality of wild animals in relation to outbreaks of Ebola haemorrhagic fever in Gabon, 1994—2003, Transactions of the Royal Society of Tropical Medicine and Hygiene, doi:10.1016/j.trstmh.2006.07.002

Marchesi P, Marchesi N, Fruth B, Boesch C (1995) Census and distribution of chimpanzees in Côte D'Ivoire. Primates 36: 91–607

Mitchell A (2005) The ESRI guide to GIS analysis, vol 2: Spatial measurements and statistics, ESRI Redlands, California

Mohneke M (2004) Ecological approach and comparative methodology of density estimates of the bonobo (Pan paniscus) population of Lui Kotal, in the Salonga National Park, Democratic Republic of Congo. Diploma thesis, Rheinische Friedrich Wilhelms University, Bonn

Mohneke M, Fruth B (2008) Bonobo (Pan paniscus) density estimation in the SW-Salonga National Park, DRC: common methodology revisited. In: Furuichi T, Thompson J (eds) The bonobos: ecology, behavior, and conservation. Springer, New York, pp 151–166

Myers Thompson JA (1997) The history, taxonomy and ecology of the bonobo (Pan paniscus) with a first description of a wild population living in a forest/savanna mosaic habitat. Ph-D dissertation, University of Oxford, England. 358 pgs

Oates JF, Whiteside GH, Davies AG, Waterman PG, Green SM, Dasilva GL, Mole S (1990) Determinants of variation in tropical forest primate biomass: New evidence from West Africa. Ecology 71:328–343

Plumptre AJ, Reynolds V (1996) Censusing chimpanzees in the Budongo Forest, Uganda. Int J Primatol 17:85–99

Plumptre AJ, Reynolds V (1997) Nesting behavior of chimpanzees: implications for censuses. Int J Primatol 18:475–485

Plumptre AJ (2000) Monitoring mammal populations with line transect techniques in African forests. J Applied Ecol, 37:356–368

Plumptre AJ, Cox D, Mugume S (2003) The status of chimpanzees in Uganda. Albertine Rift Technical Report Series No. 2, Wildlife Conservation Society, Kampala

Reinartz GE, Inogwabini BI, Ngamankosi M, Wema LW (2006) Effects of forest type and human presence on bonobo (Pan paniscus) density in the Salonga National Park. Int J Primatol 27:603–634

Reinartz, GE, Guislain, P, Mboyo, B, Isomana, E, Inogwabini, BI, Ndouzo B, Mafuta N, WemaWL (2008) Ecological factors influencing bonobo density and distribution in the Salonga National Park: applications for population assessment. In: Furuichi T, Thompson J (eds) The bonobos: ecology, behavior, and conservation. Springer, New York, pp 167–188

Rosenberg MS (2000) The bearing correlogram: a new method of analyzing directional spatial autocorrelation. Geographical Analysis 32:267–278

Sanderson EW, Malanding J, Levy MA, Redford KH, Wannebo AV, Woolmer G (2002) The human footprint and the last of the wild. BioScience 52: 891–904

Siegert F (2003) Assessment of land cover and land use in the Salonga National Park based on satellite imagery. Remote Sensing Solutions GmbH Spatial data applications, GTZ, Kinshasa 16 pages unpublished report

Southwell C, Weaver K (1993) Evaluation of analytical procedures for density estimation from line-transect data: data grouping, data truncation and the unit of analysis. Wildlife Research 20: 433–444

Susman RL, Badrian NL, Badrian A, Handler NT (1981) Pygmy chimpanzees in peril. Oryx 16: 179–183

Thomas L, Laake JL, Strindberg S, Marques F, Borchers DL, Buckland ST, Anderson DR, Burnham KP, Hedley SL, and Pollard JH (2001a) Distance 4.0, Beta 1. Research Unit for Wildlife Population Assessment, University of St. Andrews, UK, http://www.ruwpa.st-and. ac.uk/distance/

Thomas L, Beyers R, Hart J, Buckland ST (2001b) Recommendations for a survey design for the Central African Forest Region. MIKE Central African Pilot Project, Technical Report N° 1, CITES-MIKE

Thompson J Myers, Lubuta, N. Kabanda RB (2008) Traditional land-use practices for bonobo conservation. In: Furuichi T, Thompson J (eds) The bonobos revisited: ecology, behavior, and conservation. Springer, New York, pp 227–244

Thompson-Handler N, Malenky RK, Reinartz GE (1995) Action plan for *Pan paniscus*: Report on free-ranging populations and proposals for their preservation. Zoological Society of Milwaukee County, Milwaukee

Tutin CEG, Fernandez M (1984) Nationwide census of gorilla (*Gorilla g. gorilla*) and chimpanzee (*Pan t. troglodytes*) populations in Gabon. American Journal of Primatology 6:313–336

Urban DL (2002) Tactical monitoring of landscapes. In: Liu J, Taylor WW (eds) Integrating landscape ecology into natural resource management. Cambridge University Press, pp 294–311

# Part III
# Conservation Study Section

# Foreword to Conservation Study Section

Cosma Wilungula Balongelwa[1]

We live in a time of swirling environmental and societal changes, and the protected areas' network, as well as all other national institutions, is being transformed by the flow of modern events. The Democratic Republic of Congo is struggling to emerge from a decade of armed conflicts and from three decades of mismanagement that did not spare the ability to manage the country's spectacular national parks and led to the disastrous degradation of the state's machinery. The consequences on the social, cultural, economical and even physical structures of the country have been catastrophic. At a time when hope for a long term peace in the DR Congo is coming into sight following the devastating civil strife, we are forced to note that the scale of wildlife loss and habitat destruction represents an imminent looming threat for the preservation of the biodiversity of Salonga National Park's (SNP) and the complex Lake Tumba-Lake Mai Ndombe's (LTLM) hinterland.

The publication of this book in the wake of the International Conference on Sustainable Management of Forests in the DR Congo, held at the *Palais d'Egmont* in Brussels 26–27 February 2007, comes at the perfect time. The large and growing body of evidence has clearly established that the sheer size of this ecosystem block of humid forest means that it plays a crucial role in climate regulation of the whole central Africa. Protected areas are recognized for their part as a key tool to counter the loss of the world's biological diversity and as 'laboratories' for scientific and cultural work. To quote Her Majesty, Queen Noor, "These priceless places – national parks, wilderness preserves, community managed areas – together serve as the green lungs of our planet." The Salonga National Park and Lake Tumba-Lake Mai Ndombe's hinterland are not exceptions to the rule.

Of the both protected and unprotected areas alike, situated in the Congo Basin, SNP and LTLM hinterland are slowly coming through an extremely long and difficult period of social and political unrest. In these areas, the DR Congo shelters part of the world's heritage for humanity, including the bonobo (*Pan paniscus*), the prominent endemic species to the DR Congo and man's closest relative, sharing 98.8% DNA with humans. Biologically and behaviourally, the bonobo is our closest living relative in the animal world. The bonobo is found in Salonga National Park.

---

[1] *Executive Managing Director, Congolese Institute for Nature Conservation , Commune de la Gombe, Avenue des Cliniques 13, BP 868, Kinshasa 1, Kinshasa, D.R. Congo*

SNP is the second largest tropical forest national park in the world. It has been inscribed on the List of World Heritage Sites in Danger since 1999. Despite SNP's protected status and World Heritage Site classification, its bonobo population is at risk from important threats over large parts of the park's area. It would indeed be tragic if we did not prevent their extinction, all the more so since their fate ultimately rests in the hands of the people in whose protected areas they live, the Congolese.

The major factors leading to the loss of biodiversity include poaching, the bushmeat trade, and shifting cultivation of slash and burn agriculture. Causal factors include increased access to the forest, weak institutional capacities (inadequate resources for forest management and lack of good governance), and the application of inappropriate management strategies. Due to the complexity of these factors, various approaches and strategies are being suggested to reduce biodiversity loss. However, all of these require the best available scientific data that allows the development and implementation of sound management strategies. The goal of this section of the book is to contribute to the dissemination of up-to-date and accurate information on selected topics that are important for the conservation of biological diversity, the sustainable use of its components, and the equitable sharing of its benefits.

Given this challenging atmosphere, establishing a partnership between researchers and managers has become crucial in order to ensure the success of protected areas, especially where applied interdisciplinary research holds great potential for enabling us to better understand the many complex problems confronting the Congo Basin rain forest. Specialists on the biodiversity of this region have emphasized the need to obtain additional information on habitats and their taxa before any confident conservation decisions can be made. The results from these analyses provide direction as to where targeted research is critical for filling gaps in our understanding. According to these authors, actions are most effective when they are based on informed studies that document the specific needs of protected areas' species.

A recent and welcome change in African rain forest research is the recognised need for a landscape perspective. The heterogeneity of the forest, the long-distance movements of animals, and the varieties of human impacts all force the conservation biologist to look beyond core protected areas. This raises issues of connectivity between key ecosystems such as the SNP and the LTLM hinterland, and the need for greater sensitivity to the role that human-modified habitats must play both in the overall forest landscape and in traditional land use practices for bonobo conservation.

With a multiple use approach becoming the dominant model for management, researchers are beginning to study the long-term effects of hunting, logging, and mining on wildlife populations and ecosystems. Thus, belated attention has already brought new perspectives to bear on management and caused a rethinking of such issues as the origins of species' diversity and abundance. The authors and others in the landscape are already focused on many of these issues. This book provides a link between conservation and analysis and the practice of conservation. Furthermore, it provides an overview of trends in research and conservation over the past decade, a fascinating and dynamic period of convergent change in both fields as we have moved from scattered studies within a few narrow disciplines to a broader mix of more relevant subjects, to a recent rise in multidisciplinary studies; from a limited connection between science and conservation to a more explicit partnership that links research to the management needs of specific projects. It is not an academic review, but a practitioner's perspective based on several years of personal and professional experience.

The case studies presented in this issue illustrate a range of projects and activities where park managers, NGOs, local communities and the international community have worked together with mixed successes to achieve the ever elusive goal – sustainable biodiversity in a changing and anthropocentric world. The sort of commitment, entrepreneurship, flexibility and opportunism exhibited on the ground may be a foretaste of what will be needed to maintain many other major protected areas and sites of high biodiversity value.

Different case studies have shown that it is possible for conservation efforts to succeed even under difficult conditions, while at the same time emphasising the continued threats and challenges which the bonobo population faces. These findings therefore highlight the need to strengthen conservation efforts as we look forward to a future of improved protected area management and peace building in the region; and of international support to provide protection for this unique, highly endangered species, the bonobo. A series of management recommendations drawn from these findings and some promising lines of pursuit, and continued constraints and challenges, are outlined in each contribution. The comprehensive conservation study section of this book is a very welcome addition to the literature and armamentarium of conservation practice. It fills an important niche; we too easily forget, in the swirl of theory and global strategies, that the salvaging and management of biodiversity is eventually to be won on the ground, much like a war, by dedicated researchers and managers who know how to proceed day-to-day in particular places and times, carrying with them the tools required at a time when the DR Congo is once more, not without some difficulty, on the road to peace and national reconstruction. The protection of the huge area studied remains a challenge to the Congolese national parks service (the Congolese Institute for Nature Conservation; *l'Institut Congolais pour la Conservation de la Nature* - ICCN) given the setback from a decade of political instability that has weakened ICCN's ability to control the park. Understanding how locally based people use the park and affect its wildlife will be needed to guide efforts to involve them in supporting conservation. An analysis of the impact of subsistence and economic activities, and especially hunting, is critical to this. How can ICCN more effectively work with local communities to protect the site? What investments should be made to ensure the protection of the park's important population of bonobos? These key questions still remain to be addressed.

It is our hope that this book can act as a catalyst for structured debates about biodiversity and conservation issues relevant to the region, promote collaboration among different conservation efforts, and be a springboard for systematic conservation planning across priority landscapes and at finer scales. In this regard, this section of the book's strength is that it shares with people of many different persuasions the pursuit of a common ideal towards sustaining the quality and content of global ecosystems for the benefit of mankind. I am pleased to offer my strong endorsement of the section and do trust that it will serve to chart the way ahead for further research topics and broaden our understanding of the complexity of the issues raised, while at the same time facilitate the implementation of remedial measures to reduce or halt PNS and LTLM hinterland's current biodiversity loss. In addition, I am very pleased to make it available to the scientific community and wish to applaud any plan from the authors and publishers to distribute as many free copies as possible to wildlife managers and collaborating academic institutions.

I wish to express my sincere gratitude to all who have contributed in one way or another in the presentation and production of this book whose contribution will move management planning into the new millennium. If there is a final message for the reader, it is that protected areas are the cornerstones of biodiversity and species conservation. They will remain the single most important way to ensure long-term survival for most fauna and flora species. Despite the challenges and serious man-made threats, the SNP and LTLM hinterland still represent one of the best opportunities for long-term conservation of bonobos as a single charismatic focus species across the Congo Basin. Some engagement by ICCN with local people in favour of the parks will be essential if the parks and their vulnerable fauna are to be protected within this most important, fascinating, and threatened region.

# Avant-propos à la Section d'Etude de la Conservation

Cosma Wilungula Balongelwa[1]

Nous vivons un temps emprunt de changement social et environnemental en ébullition où le réseau d'aires protégées ainsi que d'autres institutions sont sujets au flux dicté par les événements modernes. La RD Congo se débat pour émerger de la décade de conflit armé et de trois décennies de me gestion, de la dégradation de l'appareil de l'état, y compris l'incapacité de mieux gérer le réseau spectaculaire de parcs nationaux. Les conséquences de la structure sociale, culturelle, économique et physique du pays ont été catastrophiques. En ce moment où l'espoir de paix se profile à l'horizon après la guerre civile destructrice, le niveau actuel de la perte de la faune et de l'habitat est un danger imminent pour la biodiversité de l'hinterland du Parc National de la Salonga (PNS) et du complexe Lac Tumba-Lac Mai Ndombe (LTLM).

La publication de ce livre au lendemain de la Conférence internationale sur la gestion durable de la forêt en RD Congo tenue au Palais d'Egmont à Bruxelles du 26–27 février 2007 arrive à point nommé. Il est évident que de part sa taille, ce bloc humide de l'écosystème forestier joue un rôle déterminant dans la régulation du climat de toute l'Afrique centrale. Les Aires protégées sont reconnues en tant qu'outil clé visant à faire face à la perte de la biodiversité mondiale de part de leur attribut comme laboratoires pour les travaux scientifique et culturel. Sa Majesté la Reine Noor ne disait-elle pas 'Ces places sans prix – parcs nationaux-réserves naturelles- réserves communautaires-ensemble agissent comme les poumons verts de la planète'. L'hinterland du PNS et le complexe LTLM ne sont guère une exception à cette assertion.

A l'image des aires protégées et non protégées situées dans le bassin du Congo, l'hinterland du PNS et du complexe LTLM se rétablissent peu a peu. Abritant une part du patrimoine de l'humanité, y compris le bonobo (Pan paniscus) – le dernier étant une espèce endémique en RD Congo et le parent le plus proche de l'Homme avec qui il partage 98.8% de matériel génétique. Le PNS est le second parc national forestier dans le monde, il a été inscrit sur la liste du Patrimoine mondial en danger

---

[1]*Administrateur Délégué Général de l'Institut Congolais pour la Conservation de la Nature, Commune de la Gombe, Avenue des Cliniques 13, BP 868, Kinshasa 1, Kinshasa, Republic Democratique du Congo*

en 1999. En dépit de son statut de protection et de site du patrimoine mondial, ses populations de bonobos sont en risque au regard d'importantes menaces sur une grande portion du parc. Ce serait donc tragique si nous n'arrêtons pas la vague d'extinction en cours d'autant plus que leur destin ultime repose entre les mains du peuple à qui appartiennent ces aires protégées.

Les facteurs majeurs conduisant à la perte de la biodiversité comprennent le braconnage, le commerce de la viande sauvage, l'agriculture itinérante sur brûlis. Les facteurs occasionnels incluent l'accès accru à la forêt, faible capacité institutionnelle (gestion inadéquate de ressources forestières, manque de bonne gouvernance) et gestion inappropriée des stratégies appliquées. Etant donné la complexité de ces facteurs, plusieurs approches et stratégies sont entrain d'être suggérées pour réduire la perte de la biodiversité. Elles nécessitent cependant une base scientifique d'information qui permet le développement et l'implantation des stratégies appropriées. Le but de ce livre vise à contribuer à la dissémination d'une information précise et mise à jour en tenant compte des thèmes jugés importants pour la conservation de la biodiversité, l'utilisation durable de ses composantes et le partage équitable de ses bénéfices.

Dans ce contexte de défi, établir un partenariat entre les chercheurs et les gestionnaires est devenu crucial pour garantir le succès des APs, spécialement là où la recherche interdisciplinaire est appliquée à un grand potentiel pouvant permettre de mieux comprendre plusieurs problèmes difficiles auxquels se trouve confronté le bassin de la forêt humide du Congo. Les spécialistes de la biodiversité de la région ont souligné la nécessité de l'information additionnelle sur les habitats et leurs taxons avant qu'une décision confidente sur la conservation soit prise. Les résultats de ces analyses fixeront la direction concernant le domaine de recherche ciblé afin de boucher les lacunes dans notre connaissance actuelle.

Un récent et approprié changement dans le domaine de recherche forestière se traduit à travers le besoin de l'approche Paysage. L'hétérogénéité de la forêt, les longues distances effectuées par les animaux et la variété d'impacts humains sont autant des facteurs qui forcent le biologiste de la conservation à explorer au-delà des aires protégées. Ceci fait resurgir l'approche de connectivité entre les écosystèmes clés à l'image de l'hinterland du PNS et de LTLM, ainsi que le besoin de grande sensitivité vis-à-vis du rôle que les habitats modifiés par l'homme peuvent jouer aussi bien dans le paysage forestier que dans les pratiques d'utilisation de terre pour la conservation du bonobo.

Avec l'approche d'usage multiple devenant le modèle dominant pour la gestion, les chercheurs s'adonnent de plus en plus à l'étude des effets de la chasse, de l'exploitation du bois et minière sur les populations fauniques et les écosystèmes. Ainsi, une attention tardive a déjà produit de nouvelles perspectives en termes de gestion en remettant sur la sellette l'origine de la diversité des espèces et leur abondance. Les auteurs et d'autres dans le paysage focalisent déjà sur bon nombre de ces thèmes, et ce livre fournit un lien entre la conservation, l'analyse et la pratique de la conservation. Bien plus, il donne un aperçu des tendances en recherche et conservation pendant une décennie fascinante et dynamique de changement dans les deux domaines: allant d'études éparpillées avec un nombre à la fois limité et diversifié de disciplines jusqu'à la récente vague d'études multidisciplinaires: allant d'un lien limité entre la science et la conservation jusqu'au partenariat liant la recherche aux besoin de gestion. De projets spécifiques. Il ne s'agit guère d'une révision académique, mais bien d'une perspective pratique fondée sur plusieurs années d'expérience individuelle et professionnelle.

Les études de cas présentées dans cette édition illustrent une série de projets et activités où les gestionnaires des AP, les ONG, les communautés locales et la communauté internationale ont travaillé ensemble pour réaliser l'insaisissable but, à savoir biodiversité durable dans un monde en pleine mutation et anthropocentrique. Le modèle d'engagement, d'entreprenariat, flexibilité et opportunisme montré sur le terrain peut être un avant gout de ce qui sera nécessaire pour maintenir plusieurs autres AP et sites de grande valeur en termes de biodiversité.

Différentes études de cas montrent qu'il est possible pour les efforts de conservation de réussir même dans les conditions difficiles tout en épinglant les menaces persistantes et les défis auxquels la population de bonobos fait face. Les résultats de ces études mettent en exergue le besoin de renforcer les efforts de conservation au moment où les yeux sont tournés vers la gestion améliorée des AP, le retour de la paix dans la région et le support international visant à assurer la protection de cette unique espèce hautcment menacée qu'est le bonobo. Une série de recommandations sur la gestion tirée des résultats des études couplées des orientations de poursuite, contraintes et défis sont mis en relief dans chaque contribution. La section complète sur la conservation de ce livre est une contribution à la littérature et l'arsenal de pratiques conversationnistes. Elle comble la brèche; nous perdons souvent de vue dans le tourbillon de la théorie et des stratégies globales que la sauvegarde et la gestion de la biodiversité se feront éventuellement sur terrain et ce, à l'image de la guerre, par les chercheurs et gestionnaires dédiés qui savent s'y prendre au jour le jour dans le temps et l'espace, et dotés des outils appropriés au moment où la RD Congo, non sans difficulté, se trouve sur le chemin de la paix et de la reconstruction nationale. La protection de la vaste étendue d'étude demeure un défi vis-à-vis de l'ICCN au regard du revers imposé par la décade de l'instabilité politique qui a fragilisé l'ICCN en terme du contrôle effectif du parc. Comprendre comment la population locale utilise le parc et son impact sur la faune sauvage mérite toute l'attention à dessein de l'impliquer dans les efforts pour la conservation. Une analyse de l'impact des activités économique et de subsistance, notamment la chasse s'impose. Comment l'ICCN peut-il travailler en étroite collaboration avec la communauté locale pour protéger le site ? Quels investissements mérite-t-on d'être faits pour assurer la protection d'importantes populations de bonobos? Ces questions clé restent entières.

Notre espoir est de voir ce livre servir de catalyseur des débats structurés autour des questions de la biodiversité te conservation dans la région, promouvoir la collaboration de différents efforts de conservation et être le tremplin pour la planification systématique de la conservation à travers les paysages prioritaires et à l'échelle réduite. Prise sous ce point de vue, la force de cette section du livre réside dans le fait qu'elle partage avec la population de différentes confessions un idéal commun vers la durabilité de la qualité et du contenu des écosystèmes globaux pour le bénéfice de l'Homme. Je suis heureux d'endosser cette section et reste convaincu que les contributions serviront à paver le chemin vers d'autres d'autres travaux de recherche tout en ouvrant les horizons de notre compréhension sur la complexité des sujets abordés et des propositions de mesures visant à réduire ou arrêter la perte de la biodiversité dans le complexe PNS et LTLM. Bien plus, je suis heureux de la mettre à la portée de la communauté scientifique et applaudi toute initiative des auteurs et éditeurs privilégiant la mise à disposition de quelques copies libres au bénéfice des gestionnaires de la faune et institutions académiques.

Enfin, je souhaite exprimer ma sincère gratitude à tous ceux qui ont contribué, d'une façon ou d'une autre, dans la production de ce livre et dont les apports porteront la gestion planifiée dans nouveau millenium. S'il y a un dernier message à l'attention du lecteur, il porte sur le fait que les aires protégées sont la pierre angulaire de conservation de la biodiversité et des espèces et resteront la seule voie à travers laquelle l'on pourra assurer la survie à long terme de la plupart des espèces de la faune et la flore. En dépit des défis et des menaces anthropiques, l'hinterland PNS et LTLM représente encore une des meilleures opportunités pour la conservation à long terme de bonobo en tant qu'unique espèce charismatique dans le basin du Congo. Quelques engagements de la part de l'ICCN avec les populations locales en faveur du parc seront essentiels si le parc et sa vulnérable faune doivent être protégés dans une importante et fascinante région en danger.

# Traditional Land-use Practices for Bonobo Conservation

Jo Myers Thompson[1], Lubuta Mbokoso Nestor[2],
and Richard Bovundja Kabanda[3]

## Introduction

*Pan paniscus* is among Africa's most endangered primates (Lacambra et al. 2005,
IUCN 2004. Butynski et al. 2000, Hilton-Taylor 2000, Oates 1996). They are
endemic to the Democratic Republic of Congo (Lacambra et al. 2005, IUCN 2004,
Butynski et al. 2000, Hilton-Taylor, 2000, Oates 1996), the only country where
they are found. The bonobo's survival is dependent on the human condition in a
nation ravaged by long periods of economic devastation, unmanaged exploitation
of natural resources, and civil insecurity punctuated by spasms of violence (Institute
Congolais pour la Conservation de la Nature 2006, Lacambra et al. 2005, Miles
et al. 2005, Thompson 2003, Thompson et al., 2003b Ilambu 2003, Draulans and Van
Krunkelsven 2002, Dupain and Van Elsacker 2001, Van Krunkelsven 2001) at the
heart of bonobo range (Thompson et al. 2003a). The rural Congolese people have a
rich heritage of living alongside bonobos. The range of bonobos is commensal with
the Bantu culture, which may account for particular uniformity in human traditions
across the bonobos range. Overlooking traditional customs that regulated the villagers' communal relationship with natural resources has compounded the inability of
central government to reach adequate protection levels.

Bonobos have a distinctly patchy distribution throughout their range (Lacambra
et al. 2005, Ilambu et al., 2005, Inogwabini and Ilambu 2005, Ilambu and Grossmann
2004, Draulans and Van Krunkelsven 2002, Dupain and Van Elsacker, 2001), which
is due, in part, to vegetation patterns, and more importantly, their distribution has been
postulated to be due to the negative impact of human presence (Reinartz et al. 2006).
Results of the 2003/2004 ICCN/MIKE/WCS (Blake 2005, Ilambu, et al. 2005,
Ilambu and Grossmann 2004) large mammal inventory of bonobo range occupation
and population density within the Salonga National Park (SNP / Parc National de la

[1] *Lukuru Project, D.R. Congo, c/o Lukuru Wildlife Research Foundation, 129 Pinckney Street,
PO Box 875, Circleville, Ohio, 43113 United States*

[2] *Congolese Institute for Nature Conservation, Lukuru Project Focal Point, D.R. Congo*

[3] *Congolese Institute for Nature Conservation, Anga Headquarters Post, D.R. Congo*

T. Furuichi and J. Thompson (eds.), *The Bonobos: Behavior, Ecology, and Conservation*
© Springer 2008

Salonga) indicated that the territory of the Iyaelima people is one of the two areas containing the highest relative abundance of bonobos (Ilambu et al. 2005, Ilambu and Grossmann 2004, Institute Zairois pour la Conservtation de la Nature 1980). During the initial phase of the survey, the success of the Iyaelima traditional land use practices and management of their fauna within 10 km around the villages was evident by the abundant amount of animal signs and, in particular, the relatively high rate of bonobo signs encountered across the area associated with the Iyaelima, as recorded during a two week period in April - May 2004 (Simeon Dino S'hwa, WCS team leader, written communication August 11th, 2005). A more fine-grained approach in the second phase of surveying across the whole of the park indicated that within the 2500km$^2$ sector of forest occupied by the Iyaelima, there may be up to five times more bonobos than in other sectors of the park (Grossmann et al. 2008, Gjerstad 2006).

During the process to designate a protected area within the large expanse of forest that appeared to be unoccupied on colonial maps south of the Congo River, the park evolved over more than a decade and through several versions. The impetus for creating it was originally envisioned in 1956 to preserve a large block of forest from timber harvest (Verschuren 2001) in the area that is now known as the North Block of SNP (Jacques Verschuren, personal communication during an interview in his home on 23 July 2002, Brussels, Belgique): originally named Tshuapa National Park (TNP/Parc National de la Tshuapa). Subsequently, TNP was envisioned as a place to roundup and corral the nation's forest elephants (*Loxodonta africana cyclotis*) while eliminating them elsewhere (Jacques Verschuren, personal communication 2002; Verschuren 2001). Later concepts of the national park expanded from one forest block to two discreet forest blocks (the North Block and the South Block) and changed the name to Salonga National Park (SNP), but did not incorporate the Iyaelima territory inside its boundaries (Fig.11.1). Finally, in a conscious attempt to designate a protected area greater in surface area than the Belgian state, President Mobutu expanded the park boundaries to the present description and included a Zone of Occupation for the Iyaelima (Fig. 11.2; Jacques Verschuren, personal communication, 2002; UNESCO 1988). The final version provided a corner of the park south of the Lokolo River zoned for human use.

Thus, Salonga National Park (SNP) was formally inscribed in 1970 (Institute Congolais pour la Conservation de la Nature 2006, Thompson 2003, Thompson, et al. 2003a, van Krunkelsven et al. 2000, Malenky et al. 1989, D'Huart 1988, IUCN/UNEP 1987, Susman et al. 1981, Institute Zairois pour la Conservation de la Nature 1980) as two discrete forest blocks (North Block and South Block) bisected by an occupied corridor where a national road accessed the forest and multinational palm oil plantations. One of three federally recognized protected areas for the bonobo (including the Luo Scientific Reserve and the Lomako-Yokokala Faunal Reserve), SNP is the only one categorized as a National Park. It is the world's second largest tropical rain forest national park and the largest in Africa (Lacambra et al. 2005). Following the IUCN Protected Areas Category System, the management type designation of National Park is a natural area designated a) to protect the ecological integrity of the ecosystem; b) to exclude exploitation or occupation that might be detrimental to this protection; and c) to provide a foundation for spiritual, scientific,

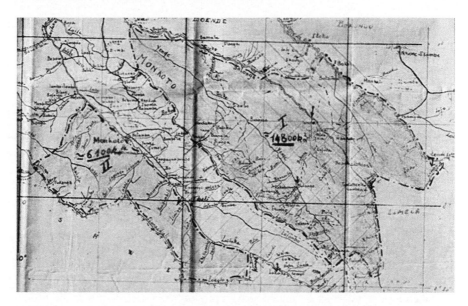

**Fig. 11.1** An early version of Tshuapa National Park (Parc National de la Tshuapa), which later became Salonga National Park (Parc National de la Salonga); documents provided by Monsieur Mokwa Vankang Izmtsho, Direction Generale, Institut Zairois Pour la Conservation de la Nature. Note that the Iyaelima traditional land falls outside the park boundaries.

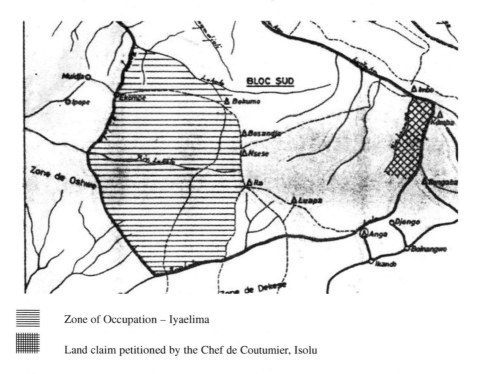

≡ Zone of Occupation – Iyaelima

▦ Land claim petitioned by the Chef de Coutumier, Isolu

**Fig. 11.2** The boundaries of the South Block of Salonga National Park as illustrated in the World Heritage Nomination Form submitted to UNESCO, March 1988.

educational, recreational and visitor opportunities that are environmentally and culturally compatible (IUCN/UNEP 1987).

Until the early 1970s, preservation of protected areas generally followed the North American model of uninhabited national parks, adhering to the notion of pristine wilderness, where the ecosystem is free of disturbance from all human beings. When SNP was created, the inhabitants were forcibly removed from the park with the exception of the Iyaelima, who fought expulsion from their ancestral land (Monsieur Mokwa Vankang Izmtsho, retired from Direction Generale, Institut Zairois Pour la Conservation de la Nature, personal communication during an interview in his home 2002, Kinshasa, DRCongo). As our approach to management of protected areas progressed, we moved away from the uninhabited wilderness model. For example, the Okapi Faunal Reserve (Reserve de Faune à Okapis) in the DRCongo was inscribed in 1996 as a human-inhabited multi-use conservation area (Peterson 2000). However, integrated conservation and development projects (ICDPs), or people-oriented approaches, have since been considered by some as not having achieved long-term management goals quickly enough. Thus, the new landscape management strategy is now moving us back towards stricter enforcement or an authoritarian protection of protected area boundaries. Protected areas represent a central strategy in biodiversity conservation and are the anchor of the new landscape paradigm. National parks are the central component of this conservation strategy built around the principal management activities of law enforcement and boundary demarcation.

The main direct threats to the survival of bonobos in SNP include chronic commercial poaching for bushmeat by communities bordering the park, poor understanding of the boundaries of the park, and pervasive large-scale commercial poaching frequently endorsed by military authorities neighboring the park (Institute Congolais pour la Conservation de la Nature 2006, Thompson et al. 2003a).

According to the Congolese Institute for Nature Conservation (Institute Congolaise pour la Conservation de la Nature 2000), plantation owners west of SNP reported increased bushmeat traffic along the rivers bordering the southern block park boundaries. Due to the vastness of the park, it is largely inaccessible except for the extensive network of river routes (Thompson 2003, Thompson et al. 2003a, Malenky et al. 1989, Susman et al. 1981). The Lokolo, Lokoro, and Lula Rivers provide the primary access routes into and out of this region of the park in the southern block. Inside the SNP, the Lokoro River flows through the center of the Iyaelima territory and the Lokolo River is in the northern part of the Iyaelima territory. Thus, the Iyaelima are the first line of defense at these access routes inside the park. Before our study, knowledge and understanding of the situation associated with the park territory inhabited by the Iyaelima was nonexistent.

## Methodology

We used a multi-method, qualitative research approach to explore the indigenous conservation practices of the Iyaelima and to identify factors that may be adapted for use in management planning; we:

- researched archival documents in Brussels, Belgium and Kinshasa, DRCongo;
- interviewed the surviving original committee members involved in the earliest planning, design, and preparation for creating the park;
- inventoried all sites of human habitation inside the study area;
- conducted a census of the human population within the study area;
- documented the socioeconomic demographics across the population;
- led nine focus group meetings for conservation education, sensitization, lobbying of authorities, and building community relations;
- met with the local authorities and group leaders for the purpose of ensuring ethical standards, including prior informed consent (PIC);
- identified and geo-referenced bushmeat trade routes and the network of human access routes through this section of the park; and
- documented the social organization, both ancestral and political/administrative, of the population.

We developed a questionnaire of semi-structured and partly open-ended questions to assess several key components, including:

1) respondent profiles to understand the composition of the villages and the population of the group;
2) socioeconomic data and domestic land use practices to determine how the people live;
3) respondents awareness about the park and perceptions of it; and
4) recognized practices of land use and wildlife conservation.

We followed a time limit of ≤ 2 hours per household interview, as agreed in the opening meeting with local authorities, during discussions of participant consent. People get tired and the time required to participate in this type of study intrudes on necessary daily survival activities. So following the suggestions of the group leaders, we asked that each participant allocate the prescribed amount of time to the household-level data collection aspect of the study. Focus group meetings were limited to ≤ 4 hours. We conducted one focus group in each village-proper, with the exception of Luapa where two focus groups were held.

Fundamental to the study, we consulted with local authorities to ensure that we took into account the rights of the local community and guarded against appropriation of their intellectual property e.g., publishing their group origins. The semi-structured interviews were administered to the household head, always with input by family members. Presence during the interviews was limited to household members. The interview team was comprised of the principal investigator, one representative of the park's department headquarters office, and two local people, one of whom was the principal officer of the park and the other who was the Chief of the Edjiki Group, Iyaelima.

## Study Site

The study site in the southern region of the South Block of SNP corresponds to the Kasai Occidental Province, Dekese Zone administrative territory, comprising 20% of the park (Fig. 11.3). The region consists of dry, upland forest habitat advanced

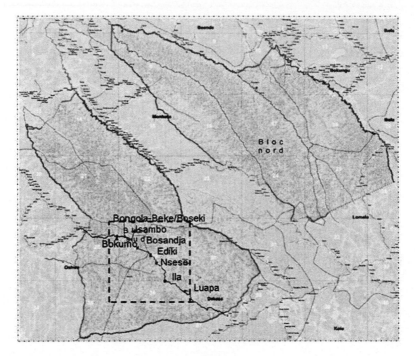

**Fig. 11.3** Study site in the Salonga National Park that includes the eight Iyaelima villages.

in elevation out of the 300m low terrain north and west of the study site, to gradients reaching plateaus of 750m above sea level at the southern periphery of the topographic Congo Basin (Thompson et al. 2003a, Myers Thompson 1997). The study site is the area identified by the Iyaelima as their ancestral land, ca. 7,000 km². Their territory encompasses a constellation of eight villages entirely within the park; from the village of Bokumo, moving east through the Bongola-Beke/Boseki, Isambo, Bosandja, Ediki, Nsese, Ila, and Luapa, with a distance of ca. 95 km between the two outer villages (Fig. 11.3). Their most eastern village, Luapa, is 56 km from the nearest neighboring village not related to the Iyaelima, and their most western village, Bokumo, is 25 km from the nearest neighboring village not related to the Iyaelima. Elevation increases as you move from west to east across the study site (Myers Thompson 1997).

## Findings: The Iyaelima

The Iyaelima are the original inhabitants of the land in the study area. The Edjiki group of the Iyaelima (the proper name spelling of the group is Edjiki and the name of the village is Ediki) people has occupied the land since the Bantu first migrated

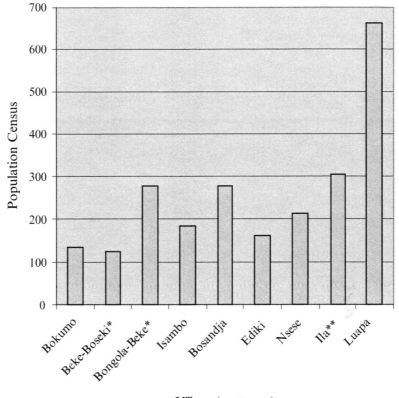

Village (west-east)

**Fig. 11.4** Population Graph by Village. * Bongola and Boseki are quartiers (paired villages) separated by a distance of approximately 1 kilometer. They share the name Beke after the river bisecting the two precincts. ** Ila-1 and Ila-2 are quartiers conjoined by a forest tract of 300 meters, here combined to make the village Ila. They represent themselves as one village.

into the region, before written history. They identify themselves as being character-ized by the region. The population totals 2,339 individuals, of which 688 are adult men and 713 are adult women (Table 11.1). They are unevenly distributed across the territory (Fig. 11.4) with the greatest number of people (28.2% of the total) living in Luapa village, which is the most eastern village and also the most remote. The Grand Chef d'Iyaelima lives in Bongola-Beke village, where his paternal ancestors are buried, and the Chef de Groupement Edjiki lives in Beke-Boseki village, which is more accessible to the outside.

Until recent decades, the Iyaelima retained traditional practices that were abandoned by the other groups under Belgian rule. Early on they had a pervasive reputation for vicious acts of intertribal warfare, retribution, menace, and defense. The Iyaelima perpetuated their image as warriors. By neighboring groups, the Iyaelima are now considered to be primitive, hostile, and of no value because they have made no progress.

**Table 11.1** Census Table – Population Iyaelima, Groupement EDJIKI

| Sites of Habitation | Population | | | | Structures | | | | Schools | | | | Pisters |
|---|---|---|---|---|---|---|---|---|---|---|---|---|---|
| | | | | | | | | | Primarie | | Secondaire | | |
| | Men | Women | Children | Total | Bureau | Dispensaire | Eglise | Houses | Students | Teachers | Students | Prof. | |
| P.P. Iyamba | 06 | 07 | 22 | 35 | – | – | – | 04 | – | – | – | – | 02 |
| P.P. Luapa | 07 | 07 | 28 | 42 | 01 | – | – | 04 | – | – | – | – | 03 |
| Luapa | 212 | 340 | 110 | 662 | – | 01 | 05 | 94 | 130 | 07 | 60 | 10 | – |
| Ila 1 * | 38 | 40 | 65 | 143 | – | – | 01 | 18 | – | – | – | – | – |
| P.P. Ila | 04 | 04 | 12 | 20 | – | – | – | 04 | – | – | – | – | 02 |
| Ila 2 * | 62 | 45 | 55 | 162 | 01 | – | 02 | 32 | 30 | 01 | – | – | – |
| Nsese | 73 | 60 | 81 | 214 | – | – | 05 | 45 | – | – | – | – | – |
| Ediki | 53 | 33 | 75 | 161 | – | – | 04 | 31 | – | – | – | – | – |
| P.P. Bosandja | 05 | 02 | 01 | 08 | – | – | – | 02 | – | – | – | – | 02 |
| Bosandja | 59 | 58 | 160 | 277 | – | – | 04 | 40 | 101 | 06 | – | – | – |
| Isambo | 49 | 42 | 93 | 184 | – | – | 04 | 34 | – | – | – | – | – |
| Bongola ** | 61 | 46 | 170 | 277 | 01 | 01 | 04 | 42 | 157 | 06 | 80 | 07 | – |
| Boseki ** | 35 | 22 | 68 | 125 | – | – | 03 | 16 | – | – | – | – | – |
| Bokumo | 46 | 27 | 61 | 134 | – | – | 03 | 30 | – | – | – | – | – |
| TOTAL | 710 | 733 | 1001 | 2444 | 03 | 02 | 35 | 396 | 418 | 20 | 140 | 17 | 09 |

Shading represents Patrol Posts inhabited by park personnel and their families.

Pisters = local trackers working for the patrol post.

P.P. = ICCN patrol post.

*Ila-1 and Ila-2 are quartiers combined to make the village Ila.

**Bongola and Boseki are quartiers (neighborhoods) combined to make the village Beke (named after the river bisecting the two precincts).

**Fig. 11.5** Aerial image of Luapa village showing shifting cultivation with very small fields of 0.5–2 ha in size; Aerial survey conducted in February 2002 (Prof. Dr. Florian Siegert, RSS - Remote Sensing Solutions GmbH).

Aerial photography (Fig. 11.5) illustrates that their impact on the surrounding forest and their land consumption practices are limited. The pattern shows shifting cultivation with very small fields of 0.5–2 ha (Florian Siegert, personal communication, 2005). In 1956–1970, the government changed the boundaries of the park numerous times, which led to the earliest misconceptions and miscommunication of park parameters that resulted in animosity and conflict between the resident Iyaelima and the park personnel. The original conservator of the sector (Conservateur Tatela) communicated that the road bisecting the Iyaelima territory was the park limit, with land southwest of the road falling outside the park limits. Today, park law enforcement is applied to the area in Fig. 11.3 without exemption for occupation, but this has not been clearly communicated to the people. Their perception is still that they are entitled to hunt legally and plant their agriculture fields southwest of the road, on land that they consider to be outside the park. This inconsistency in past and present communication about park boundaries and the application of punitive measures and law enforcement has led to confusion, disputes, and infractions that resulted in the incarceration of residents.

Resentment of the Iyaelima towards ICCN seethes under the surface of their interactions and periodically erupts. Reactions to the park restrictions against human occupation or activity, including subsistence hunting and selective timber cutting, have been manifested by an array of behaviors (Kannyo 2006) ranging from local expressions of anti-park sentiments to intentional burning of structures (IUCN 2004). It is illegal for the Iyaelima to transport meat to markets outside the park (to the centers at Oshwe, Dekese, Ilebo, Lokolama, and Mweka) and merchants are restricted from coming into the park. Transport of bushmeat is an infraction punishable by confiscation of the meat. Weapons are forbidden without ICCN authorization.

During the conflict of 1996–2002, the Iyaelima territory was occupied or traversed by various militant groups of well-armed rebel insurgents or government troops. These outside groups exploited the Iyaelima in order to extract wildlife from their land. The military groups encroached onto the Iyaelima territory and constructed camps for wildlife poaching. The well-organized military groups depleted the wildlife and other natural resources considered by the Iyaelima to be on their traditional land. The military groups forcibly coerced the people into serving as their labor to transport meat out of the park. Without enforcement authority, the Iyaelima responded by abandoning villages that had been the reservoir for human labor. For example, Iyamba village beyond the most eastern village of the study site was abandoned in 2001 because rebels used the people as forced labor to transport poached meat. The movement of these people contracted the occupied zone of the Iyaelima. Iyamba is now a Patrol Post occupied by ICCN.

During household and focus group interviews, the Iyaelima identified the principal routes used for access to their section of the park. They are used not only by the Iyaelima, but also by people from Ilebo and Oshwe (the Bungimba military post) to poach within the park. The principal access routes generally intersect at the village of Losalanga, outside the park, but traverse through the Iyaelima territory. The Iyaelima do not have the law enforcement authority to take action against outsiders who illegally poach in the park, but they are in a critical position to monitor and know about what is happening.

## Socioeconomic Feature

As one facet of our methodology, we administered the socio-economic questionnaire to a sample of 144 out of a possible 382 (37.7%) households across the population (Table 11.2). The principal activities of the households are, in rank order: 1) agriculture; 2) hunting; and 3) temporary work, including village occupations such as barber, dressmaker, or police, and work outside the park. The Iyaelima are a subsistence community dependent on crops and hunting for food security. Most interviewees reported that they cultivated cassava, bananas, sugar cane, rice, tobacco, papaya, peanuts, corn, and yams. Agriculture is generally conducted by individual households, but each village also has a community garden as a fall-back resource when individual garden production is not sufficient, either because of natural disaster or crop raiding by wildlife. When asked about factors that threaten a suitable harvest, most farmers reported that cane rats, sitatunga, red river hogs, black mangabeys, elephants, and sometimes red-tailed monkeys or soil-borne insect infestations caused damage in their gardens. In defense of their property, the people will either; kill the interloper, use fire near the fields, or dig pits around the gardens. Attacks of cassava mosaic disease (CMD) are widespread, and the

**Table 11.2**  Representation of the research sample

| Sites of Habitation | # of Households | # of Interviewed | Percent of Households | Percent of Sample |
|---|---|---|---|---|
| Luapa | 94 | 35 | 37.2 | 24.3 |
| Ila* | 50 | 19 | 38.0 | 13.2 |
| Nsese | 45 | 16 | 35.6 | 11.1 |
| Ediki | 31 | 17 | 54.8 | 11.8 |
| Bosandja | 40 | 16 | 40.0 | 11.1 |
| Isambo | 34 | 9 | 26.5 | 6.4 |
| Bongola ** | 58 | 20 | 34.5 | 12.9 |
| Beke ** | 30 | 12 | 40.0 | 8.3 |
| TOTAL | **382** | **144** | **37.7** | **100** |

*Ila-1 and Ila-2 are quartiers combined to make the village Ila.

**Bongola and Boseki are quartiers (neighborhoods) combined to make the village Beke (named after the river bisecting the two precincts).

Iyaelima do not have access to CMD resistant strains. The Iyaelima raise livestock, but are limited to goats, chickens, and ducks. The domestic animal populations are small and only support consumption on special occasions; chicken is prepared for guests and goat is prepared for ceremonies.

There is a notable variation in the number of dogs between villages. Luapa has the most (+300), with one individual owning as many as 15 dogs, and other villages having <100. Dogs are essential in the hunt. During data analysis, we used this as an indicator of who is doing the most hunting. Hunting is performed either alone or in the community group with (in rank order of preference): 1) black-fronted duiker (unanimously the preferred meat); 2) blue duiker; 3) bay duiker; 4) red river hog; 5) yellow-backed duiker; and 6) generic monkeys, as the species generally targeted. Iyaelima hunting technology is limited to dogs, spears, and bows and arrows for individual hunts, and nets are added during the village hunts. In the mid-1990s, the government confiscated all firearms from the civilian population, leaving only traditional technology for hunting. Village hunts are called once every 1–2 weeks during the dry season. Hunting for domestic consumption is the principal occupation of the men during the dry season, and fishing is their focus during the wet season. Hunting is only conducted during daylight hours. Women are generally responsible for harvesting forest products, including mushrooms, fruits, caterpillars, and honey.

The entire population (men and women) practice taboos against eating bongo, pygmy cape buffalo, bonobo, giant pangolin, congo peafowl, black-headed heron, tree hyrax, Egyptian mongoose, and African grey parrots, all of which are protected by the ancestors. When asked what methods of conservation the Iyaelima use today, they only identified one behavior, that of avoiding the use of toxic plants in the rivers. However, their chief may set a period of time when the people cannot use a part of the forest. This is a traditional practice that they do not define as conservation because their traditional practices have not been considered in their modern understanding of conservation.

They believe that the conscious rotational use of certain forest areas, either seasonally or annually, allows the animals to return, thereby eliminating the need for the hunters to go farther into the forest. In addition, they space the locations of their hunting camps to permit corridors from which animal populations can draw to replace those lost during the hunting season.

Within their land use practices they identify sites in nature that are inhabited by the spirits of their ancestors, the spirits of the water, and the spirits of the forest. For example, 20 km from Luapa, the Iyaelima recognize a magic lake, Lac Nkantotsha (Fig. 11.6). It and the corresponding forest block are not occupied or used for hunting and serve as a reservoir for wildlife populations. The Iyaelima believe that the ancestral spirit that once lived in the lake protects the villagers by killing sorcerers or those that have hurt others in the group. In areas where the Iyaelima feel they have had to limit or eliminate their use of their ancestral lands, they believe that the ancestors have not been fed since the creation of the park. The Iyaelima use this to explain their perceived increase in mortality and a general decline in their population. Thus, the park is a source of their decline.

As a protected area, the Iyaelima perceive the park as the creation of a national society. Their definitions of the park included:

- a place where animals are protected from hunting,
- a place where people cannot go,
- a place that the Christian God created.

Informants identified the park's value or importance as (in rank order): 1) only for ICCN use; 2) for the economy of the state; 3) to save wildlife; 4) for future generations; or 5) of no value. The predominant attitude about the park is that it is like a prison where the resident people are suffering and the young people are leaving because there is no development. The Iyaelima feel that they are declining in part

**Fig. 11.6** Sacred Sites: Lake Nkantotsha is a "spirit lake"; Aerial survey conducted in February 2002 (Prof. Dr. Florian Siegert, RSS - Remote Sensing Solutions GmbH).

from attrition. When asked about the laws and regulations of the park, universally they did not know that live animal trade or ownership of wild animals was illegal. There has been no effort to inform or educate the Iyaelima about the legalities of the park. The Iyaelima are distrustful towards the ICCN personnel and direct animosity towards the park guards. Their interaction with ICCN has been related to law enforcement issues. Personnel from the Iyaelima represent 2% of the park staff. Only three Iyaelima are employed by the park, two of whom are working outside Iyaelima territory.

## Human Coexisence with Bonobos

With specific reference to bonobo conservation, the Iyaelima recognize their human-like characteristics. They distinguish bonobos from all other wildlife in a number of ways, including that they are as intelligent as humans, can call their neighbors to come, can communicate with each other, build houses deep in the forest, walk upright (bipedal) like humans, have arms similar to humans and equally strong in a fight, have no tail, come to their gardens to eat sugar cane, bananas, and maize but are not considered to be crop raiders, and can dance like humans. It is noteworthy that they perceived the concept of bonobo lore - promoting stories from other groups about traditional beliefs - as having no value as a conservation tool. They viewed education campaigns to promote bonobos as our closest living relative as offensive and degrading or derogatory, especially because familial ties define all aspects of their lives. The Iyaelima report feeling no kinship with bonobos. Their folk tales usually relate fables about elephants, leopards, and duikers.

When the Iyaelima meet bonobos in the forest, they pass them quickly or detour to avoid them because they believe that, like humans, bonobos can beat you up or will kill you. As a group of people with a tradition of warfare (human versus human), intergroup conflict with neighbors has always been a feature of the Iyaelima life. Their belief that bonobos will fight them follows the historic interpretation of the Iyaelima as people in conflict with other groups. Because of concern about creating a conflict with their bonobo neighbors, bonobos are the only taboo species that the Iyaelima will not kill even when encountered. Conflict with the bonobos might result in the Iyaelima abandoning a village.

In response to questions about bonobo lore and parables from their ancestors, the predominant story related an incident of conflict. We were frequently told the parable,

> "When a hunter was calling for another animal, thinking it was an injured duiker, a bonobo came in response. The bonobo saw the hunter. The bonobo left the site and returned quickly to his group to tell them of the hunter's trick. Reinforced with his group, the bonobo and his fellow contingent went back to where the hunter was camped. The bonobo group had been brought to fight the man" (name withheld by prior consent agreement).

Amongst the Iyaelima, there are numerous tales of bonobos physically fighting with men and even wrestling them to defeat. One village leader related a personal story about his recently deceased ancestor who had broken his finger in a physical fight with a bonobo. The Iyaelima believe that bonobos routinely capture black-fronted duikers, the principal target species of hunters, carry them up into the trees, and break their legs. When the duiker cries, the bonobo is happy and laughs. When asked about bonobos dying, 100% of the interviewees declared that they had never seen a dead bonobo in the forest. The Iyaelima consistently reported that bonobos bury their dead.

When asked about the presence or absence of bonobos, the response varied across villages, but was consistent within. Some of the respondents reported that bonobos do not approach within 10 km of the village or are entirely absent near some villages, specifically the most western villages of the Iyaelima territory. The focus group interviewed at Bokumo, the western village limit of the Iyaelima occupation, reported that bonobos are very far from them and are almost never encountered. However, the people of Luapa and the guards at the patrol post of Iyamba reported that bonobos use land up to the edge of the villages. The focus group from Ila reported that bonobos come into their village especially when they have laundry drying that includes red cloth. The people cannot explain why the bonobos are drawn in by the red color. The bonobos' seeming attraction to red cloth is integrated with the Iyaelima belief that red is a sacred color. Where bonobos were reported to be encountered in proximity or up to the village limit, all indicated that bonobos are more abundant at those locations in the rainy season.

When asked about proximity to bonobos or where they know to avoid encountering bonobos, the focus group from Bosandja identified the forest along the Hale River as a bonobo village. Generally, the presence of bonobos was associated with the most remote villages and not correlated with the population size of the village or hunting pressures. For example, bonobos are reported in the proximity of Luapa where indicators denoted that hunters were most successful and abundant, and the census resulted in the highest number of people.

## Discussion and Policy Implication Recommendation

Like other sites outside the park [Wamba (Hashimoto and Furuichi 2001), Yasa (Myers Thompson 1997), Lilungu (Sabater Pi and Vea 1994)], bonobos live in high concentration near the Iyaelima villages corresponding to an area with relatively high human population. In the SNP, human settlements (specifically Luapa and Ila) with the most people and corresponding access routes and hunting areas, are the locations matching the highest relative abundance of bonobos and where bonobos are reported in close proximity to human activities. Elsewhere inside the park, ICCN was instructed to convert a large poaching camp into a research and patrol station due specifically to the presence of a notable bonobo population in residence (Reinartz 2003). At Wamba, the close proximity of bonobos to a large human population was so favorable that the Luo Scientific

Reserve (Reserve Scientifique du Luo) was established in March 1990 (Kano et al. 1996). Clearly, we can no longer universally identify human presence as a limiting factor that negatively impacts bonobo distribution and abundance. We must be cautious about any statement that identifies human presence as a major threat to bonobo conservation and maintenance of the park's ecological integrity. The factors affecting bonobo distribution and abundance are much more complex than human presence and vegetation.

In agreement with the IUCN Protected Areas Category System, integration of the Iyaelima into the management of SNP will permit occupation that might be beneficial for protection and will provide an opportunity for traditional land use that is culturally compatible. Bruner et al. (2001), found that the number of people living inside a park did not correlate significantly with success and effectiveness of protection. In the Iyaelima region of SNP, it is resident human presence that offers the best option for bonobo conservation. Here a paired association exists between human presence and rate of human activity and bonobo density and abundance. Some examples of indigenous practices of the Iyaelima that might be integrated into planned management for the conservation of bonobos include:

- local community decisions about what animals to protect/taboo;
- seasonal area rotation of hunting and fishing areas;
- local taboos against hunting and eating of select species; and
- knowledge about bonobo presence and population ecology for increased value from tourism or for focus of scientific research.

Effective monitoring of factors likely to impact bonobos, such as organized commercial poachers, can best be performed by the people in residence who feel a sense of ownership of their traditional land.

Research shows that park effectiveness correlates most strongly with density of guards (Bruner et al. 2001). Bruner, et al. (2001) calculated a benchmark that three guards per $100 \, km^2$ is sufficient staffing to effectively protect a park. Enforcement capacity – a composite of training, equipment, and salary – did not correlate with effectiveness, suggesting that it is less important than the presence of guards. Salonga National Park is $36,000 \, km^2$ with a staff of 137 total guards (Institute Congolais pour la Conservation de la Nature 2006). SNP needs 1,080 guards to reach levels of effectiveness. Recruitment efforts are currently underway. In these efforts, it would be advisable for SNP to target Iyaelima recruits to post throughout the park. Ethnicity is an important factor in cooperation between park enforcers and residents. Affirmative action hiring – selecting recruits from the Iyaelima – would improve the capacity of ICCN. Partnering local people as park employees would foster a more compatible relationship and joint focus on conservation. During periods of national or regional conflict, the Iyaelima stayed to protect their ancestral land. Their presence in the site and knowledge of the region offers the critical opportunity to achieve a meaningful presence and successful protective action across this region of the park. In order to protect and to monitor the region, ICCN must ensure non-enforcement contact between ICCN and the Iyaelima.

Cooperation with and understanding of the local people and community-based conservation efforts are required to bolster the national capacity for protecting the park and its bonobos. By considering and utilizing conservation knowledge and traditional practices that regulate the Iyaelima's relationship with natural resources, central government may augment their strategy to reach adequate protection levels. In the area of our study, interethnic group intolerance is prevalent; by incorporating the Iyaelima into the management structure, conservation efforts can be strengthened and expanded.

The local people as a community have rules and laws that regulate the systems for producing and exploiting natural resources that encourage or restrict exploitation. Only by understanding and incorporating these local structures can we encourage the conservation partnership of people. The principal threats to the survival of bonobos in SNP include commercial poaching by communities bordering the park, and organized bands of poachers endorsed by military authorities coming from outside the park. Hiring Iyaelima to build local ICCN capacity and to improve monitoring for regional protection of their ancestral land would benefit conservation by integrating them into the management structure rather than of keeping them outside of it.

## Conclusion

Recognizing and engaging the local people is essential to assure the successful management of this site and conservation of the bonobo. Integrating traditional local land use practices may have implications for a patchwork of conservation approaches weaved together for a range-wide conservation strategy. We advocate the dissemination of local knowledge as a way to provide for bonobo conservation.

It is not necessarily the presence of humans that generates a limiting effect on bonobo presences and density. High concentrations of bonobos occur in association with resident human populations. In the case of the Iyaelima, the resident people are the first line of defense to protect bonobos and to maintain the integrity of the national park. It is now up to the management committee to create a mechanism in which the Iyaelima can maintain their traditional lifestyles without demographic expansion or increase, can continue their land use practices with traditional technology, and can participate in some level of control over the management of the area. This study highlights the misconception that human presence is equivalent to human threat. In fact, concerning the Iyaelima, cooperation with and understanding of the local people are required to bolster the national capacity for protecting this park and the bonobos.

**Acknowledgements** Other expedition members who participated in this research include Monsieur Bosomba Bokuli, Chef de Groupement Edjiki and Monsieur Mvula Lomba Batshina-Tshina Joseph. This work was completed in cooperation with Grand Chef Monsieur Longanga Isako II Camille, the people of the Iyaelima, and the Congolese Institute for Nature Conservation (Institut Congolais pour la Conservation de la Nature or ICCN). Funding was provided by National Geographic Society Expeditions Council, the Columbus Zoological Park Association

Conservation and Collection Management Committee fund, the WildiZe Foundation, the Lukuru Wildlife Research Foundation, and Rachael Lee Myers. We want to recognize cooperation with Robert Mwinyihali, WCS Lokofa Block socioeconomic inventories coordinator, and Alejandra Colom, WWF Consultant socioeconomic studies in the Salonga-Lukenie-Sankuru landscape. Aerial imagery was provided by Dr. Florian Siegert 2002, Remote Sensing Solutions GmbH.

# References

Blake S (2005) Long term system for monitoring the illegal killing of elephants (MIKE). Central African Forests: final report on population surveys (2003 2004). An unpublished report by the Wildlife Conservation Society

Bruner AG, Gullison RE, Rice RE, da Fonseca GAB (2001) Effectiveness of parks in Protecting Tropical Biodiversity. Science 291(5501): 125–128

Butynski T, Members of the Primate Specialist Group (2000) *Pan paniscus*. In: IUCN (ed) 2006 IUCN red list of threatened species. *http://www.iucnredlist.org*, downloaded on 25 May 2007

D'Huart JP (1988) Parc National de la Salonga (Equateur, Zaire): conservation et gestion, developpement des collectives locales. Unpublished Report, IUCN, Gland, Switzerland

Draulans D, Van Krunkelsven E (2002) The impact of war on forest areas in the Democratic Republic of Congo. Oryx. 36(1):35–40

Dupain J, Van Elsacker L (2001) The status of the bonobo (*Pan paniscus*) in the Democratic Republic of Congo. In: Galdikas BMF, Briggs NE, Sheeran LK, Shapriro GL, Goodal J (eds) All apes great and small, Vol. 1: African Apes. Kluwer Academic/Plenum, New York

Gjerstad K (2006) News in detail. WCS electronic newsletter for March-June. Posted 26 June 2006

Grossmann F, Hart JA, Vosper A, Ilambu O (2008) Range occupation and population estimates of bonobo in the Salonga National Park: application to large scale surveys of bonobos in the Democratic Republic of Congo. In: Furuichi T, Thompson J (eds) The bonobos: behavior, ecology, and conservation. Springer Inc., New York, pp 189–216

Hashimoto C, Furuichi T (2001) Current situation of bonobos in the Luo Reserve, Equateur, Democratic Republic of Congo. In: Galdikas BMF, Briggs NE, Sheeran LK, Shapriro GL, Goodal J (eds) All apes great and small, Vol. 1: African Apes. Kluwer Academic/Plenum, New York, pp 83–93

Hilton-Taylor C (2000) 2000 IUCN red list of threatened species. IUCN, Cambridge

Ilambu O (2003) Status assessment of bonobos: a collaborative census in Salonga National Park. In: Thompson J, Hohmann G, Furuichi T (eds) Bonobo workshop: behaviour, ecology and conservation of wild bonobos. Unpublished report of Bonobo Workshop, Inuyama, Japan, p 13

Ilambu O, Grossmann F, Mbenzo P (2005) Monitoring of the illegal killing of elephants in Central African Forests: elephant and ape population surveys and human impact in the Salonga National Park, Democratic Republic of Congo. Unpublished report to CITES MIKE and Government of the Democratic Republic of Congo

Inogwabini BI, Ilambu O (2005) A landscape-wide distribution of *Pan paniscus* in the Salonga National Park, Democratic Republic of Congo. Endangered Species Update, 22:116–123

Institut Congolais pour la Conservation de la Nature (2000) The state of conservation report, 20 April

Institut Congolais pour la Conservation de la Nature (2006) Système de gestion d'information pour les aires protégées. Parc National de la Salonga Factsheet No3. Salonga National Park Factsheet

Institut Zairois pour la Conservation de la Nature (1980) Rapport annuel, pubilé avec le concours du projet IZCN-PNUD/FAO ZAI/80–002

IUCN (2004) IUCN red list of threatened species, *www.iucnredlist.org*, accessed 07 January 2006

IUCN/UNEP (1987) Parc national de la Salonga. In The IUCN Directory of Afrotropical Protected Areas. IUCN, Gland Switzerland and Cambridge, pp 918–919

Kannyo E (2006) Liberal education and global citizenship: the arts of democracy 2002–2005, globalization and wildlife conservation in Africa, module II. Conservation and Human-Animal Conflict, Rochester Institute of Technology, College of Liberal Arts *http://www.rit.edu/ ~global/glob-kannyo-conserv-conflicts.html*, accessed 16 January 2006.

Kano T, Bongoli L, Idani G, Hashimoto C (1996) The challenge of Wamba. In: Cavalier P (ed) Great ape project. ecta & animali 96/8, Milano, pp 68–74

Lacambra C, Thompson J, Furuichi T, Vervaecke H, Stevens J (2005) Bonobo (*Pan paniscus*). In Caldecott J, Miles L (eds) World atlas of great apes and their conservation. UNEP-WCMC, Cambridge, pp 83–96

Malenky RK, Thompson-Handler N, Susman R (1989) Conservation status of *Pan paniscus*. In: Heltne PG, Marquardt LA (eds) Understanding chimpanzees. Harvard University Press, Cambridge, pp 362–368

Miles L, Caldecott J, Nellemann C (2005) Challenges to great ape survival. In: Caldecott J, Miles L (eds) World atlas of great apes and their conservation. UNEP-WCMC, Cambridge, pp 217–241

Myers Thompson JA (1997) The history, taxonomy and ecology of the bonobo (*Pan paniscus*) with a first description of a wild population living in a forest/savanna mosaic habitat. Ph-D dissertation, University of Oxford, England. 358 pgs

Oates JF (1996) African Primates: status survey and conservation action plan (revised edition). IUCN/SSC Primate Specialist Group, IUCN, Glland, Switzerland

Peterson R B (2000) Conversations in the rainforest: culture, values, and the environment in Central Africa. HarperCollins

Reinartz G (2003) Conserving *Pan paniscus* in the Salonga National Park, Democratic Republic of Congo. Pan Africa News 10:23–25

Reinartz G, Inogwabini BI, Ngamankosi M, Lisalama W (2006) Effects of forest type and human presence on bonobo (*Pan paniscus*) density in the Salonga National Park. Int J Primatol 27:603–634

Sabater Pi J, Vea JJ (1994) Comparative inventory of foods consumed by wild pygmy chimpanzees (*Pan paniscus*: Mammalia) in the Lilungu-Lokote region of the Republic of Zaire. J Afr Zool 108:381–396

Susman RL, Badrian N, Badrian A, Handler NT (1981) Pygmy chimpanzees in peril. Oryx 16:179–184

Thompson J (2003) Conservation Session: discussion. In Thompson J, Hohmann G, Furuichi T (eds) Bonobo workshop: behaviour, ecology and conservation of wild bonobos. Unpublished report of Bonobo Workshopo, Inuyama, Japan, pp 21–32

Thompson J, Masunda T, Willy ID (2003a) Field research at Parc National de la Salonga, Democratic Republic of Congo. Pan Africa News 10:27–29

Thompson J, Hohmann G, Furuichi T(eds) (2003b) Bonobo workshop: behaviour, ecology and conservation of wild bonobos. Unpublished report of Bonobo Workshop, Inuyama, Japan

UNESCO (1988) World heritage nomination form. Unpublished report submitted by IUCN

Van Krunkelsven E (2001) Density estimation of bonobos (*Pan paniscus*) in Salonga National Park, Congo. Biological Conservation 99:387–391

Van Krunkelsven E, Billa-Isia I, Draulans D (2000) A survey of bonobos and other large mammals in the Salonga National Park, Democratic Republic of Congo. Oryx 34:180–187

Verschuren J (2001) Ma vie sauver la nature, editions de la dyle. Sint-Martens-latem, Belgique

# Human Hunting and its Impact on Bonobos in the Salonga National Park, Democratic Republic of Congo

John A. Hart[1], Falk Grossmann[2], Ashley Vosper[1], and Jose Ilanga[3]

## Introduction

Hunting is one of the most important threats to many great ape populations, including bonobos, in central Africa (Kano and Asato 1994, Bowen-Jones and Pendry 1999, Rose 1998, Susman et al. 1981) and it could be one of the determinants of apparent gaps in their historical range (Kingdon 1997, Kortlandt 1995 Kano1984). Butynski (2001) and Dupain et al. (2001) have attributed recent reductions in the bonobo's range over the last two or three decades to increased hunting pressure. Killing even small numbers of bonobos can have significant and long-term negative impacts on local populations, because of their long maturation, slow reproduction and cohesive social communities.

Subsistence hunting is not a new phenomenon in the bonobo's range. Traditionally, local communities near Wamba refrained from hunting bonobos for religious reasons (Kano et al. 1996, Tashiro 1995). Traditional taboos against killing bonobos prevail among the Iyaelima people who live within the Salonga National Park (Thompson ct al. 2008). Other observers report that bonobos may occasionally be taken by hunters, but are not a targeted bushmeat species (Thompson-Handler et al. 1995). Dupain et al. (2000), Thompson-Handler et al. (1995), Draulans and Van Krunklesven (2002), and Idani ct al. (2008) report that commercial hunting is increasing in the bonobo's range, implying that thcy arc at risk. Local and ethnic differences in hunting traditions are likely to have a variable impact on bonobos. Changes in hunting patterns could possibly bring new risks.

[1] Lukuru Project, Tshuapa-Lomami-Lualaba Landscape, c/o 1235 Ave. Poids Lourds, Kinshasa, D.R. Congo

[2] Wildlife Conservation Society, Africa Program, Tanzania

[3] Ministry of Environment, Nature Conservation, Water and Forests, D.R. Congo

T. Furuichi and J. Thompson (eds.), *The Bonobos: Behavior, Ecology, and Conservation*    245
© Springer 2008

Relationships between bonobos and human populations in the areas they both occupy remain ambiguous in the Salonga landscape and elsewhere (Reinartz et al., 2006). Grossmann et al. (2008) show that there is no consistent relationship between occurrence of bonobos and proximity to human settlement, or human access routes in the park and its vicinity. Some of the largest concentrations of bonobos are in the immediate vicinity of settlements. Bonobos and humans also co-exist closely at other sites in their range, e.g., Wamba, Yasa and Lilungu, (Thompson et al. 2008, Kano et al. 1996). Knowledge of the ecological, economic and cultural basis of these associations and their stability are important as human occupation, hunting, and other extractive activities increase within the bonobo's range, altering longstanding relations between bonobos and people.

Unregulated hunting remains widespread in the bonobo's range. Hunting is unlikely to be replaced by alternative subsistence and employment opportunities in the near future as long as wildlife populations remain available and alternative means to generate income remain beyond the reach of most rural Congolese. Hunting traditions are changing rapidly in many areas with the arrival of new hunting methods, the growth of commercial bushmeat trade, and the depletion of targeted wildlife populations and vulnerable species. An understanding of how hunting affects bonobos is needed to guide efforts to control and manage hunting in areas where this is possible.

Salonga National Park, covering ca. $33,346 km^2$, is one of the least disturbed areas within the bonobo's range and contains an important population of bonobos estimated at ca. 15,000 individuals (range 7,100 – 20,400) (Grossmann et al. 2008). The park covers ca. 10% of the bonobo's range and is centrally located within that area. It should be one of the most important areas for conservation of bonobos. However, despite its remoteness from major settlements, large size, and status as a UNESCO World Heritage Site, the Salonga National Park is under growing threat from uncontrolled illegal hunting (Draulans and Van Krunkelsven 2002). Hunting has already largely reduced the park's once important elephant population (Blake 2005). An important question is just how safe are the park's bonobos and what can be done to ensure their protection?

Protection of the Salonga's huge area remains a major challenge for the national park service, the Institut Congolais pour la Conservation de la Nature (ICCN). Approximately 158 guards are based in the park at 21 stations and patrol posts, an average of one guard per $211 km^2$ (Omari 2006). A decade of political instability has weakened ICCN's ability to control the park. Guards cannot effectively patrol all areas. Some engagement by ICCN with local people in favor of the park will be essential if the park and its vulnerable fauna are to be protected. Information on how local people use the park and affect its wildlife is required to guide efforts to involve them in supporting its conservation. An analysis of the impact of subsistence and other extractive activities, especially hunting, is needed. What strategies and activities will best ensure the protection of the park's important population of bonobos? What lessons can be applied to the protection of bonobos in other areas of their range?

## Objectives

We provide information to inform the public of the issues outlined above and develop guidelines for the control and management of the impact of hunting on bonobos. This information is based on a 3-year program of faunal and human activity inventories in the park and associated interviews and hunter surveys in selected villages in the park's immediate periphery. Our goal was to determine where and to what extent hunting has been and is likely to be a dominant factor in the distribution and abundance of bonobos, and to identify specific hunting practices and economic and ecological contexts that are likely to pose a significant threat. Specific objectives for this chapter include:

- describing patterns of human hunting in the Salonga National Park and selected areas of its immediate buffer zone;
- comparing landscape scale trends in the occurrence of monkeys, ungulates and elephants, the primary hunted species in the landscape, with bonobos;
- providing an assessment of the spatial distribution and intensity of hunting, and its economic and social correlates in representative blocks within the Salonga landscape;
- evaluating the current and future risk of hunting to bonobos in the park and surrounding areas; and
- developing recommendations for the conservation of bonobos and improving control of hunting in the Salonga National Park.

## Methods

We collected data from 2003 – 2006 from 3 primary sources: 1) field surveys to determine the distribution and abundance of selected fauna including bonobos, and the relative frequency of hunting, fishing and other extractive activities within the park and portions of its buffer zone; 2) analysis of satellite imagery and interviews with local people to map past settlements within the park and establish how former settlers and their descendents continue to use the park and affect bonobos; and 3) surveys of hunters to determine hunting practices, trends in commercial bushmeat trade, and their affect on bonobos in selected settlements in the vicinity of the park.

### *Surveys of Fauna and Human Activities in the Park*

We conducted surveys of large mammals, including bonobos, elephants, ungulates and monkeys (guenons, colobus and mangebeys), and human activities, including hunting, fishing and passage (paths and machete cuts), at 2 spatial scales via a multiphase, nested survey design (Grossmann et al. 2008). In Phase I surveys we sampled most of the northern and southern sectors of the park, and a block of over 2000 km$^2$ of the corridor separating the two. We collected data on GPS-referenced and compass oriented reconnaissance walks (termed recces) that were systematically allocated on a spatial grid of ca. 10 × 10 km.

In Phase II we surveyed 3 inventory blocks – Lokofa, Iyaelima and Lomela – each with an area of 2000 – 2750 km$^2$. We identified the blocks during Phase I surveys as being representative of the range of bonobo occurrence, human settlement and hunting patterns within the park. We made Phase II observations from both recces and line transects which were allocated systematically at a spatial grain of ca. 5 × 5 km. Line transects were 1.4 km in length and measured on the ground with GPS and topofil. We measured perpendicular distances from line of travel to the observations on line transects and documented their location with GPS. We used DISTANCE software (Thomas et al. 2001) to analyze results.

We collected field data on indicators of large mammal occurrence, including direct encounters with animals (seen, heard or both), observations of dung and feeding signs, and for bonobos, nests. We identified dung and feeding signs to species or to a broader taxonomic group when specific identification was not possible. We recorded age class (fresh, recent, old, and disappearing) for dung and nests. Further information on nest count methods and field team deployment is found in Grossmann et al. (2008).

Field indicators of human hunting included direct encounters with hunters, observations of snares (classed as active, or inactive, and by the size of the sapling anchor), spent cartridges, gunshots and hunting camps. We recorded hunting camp activity (occupied, recently abandoned, and long-abandoned), the number of shelters and beds, and the presence and size of meat drying racks. Field teams also recorded fishing camps, other fishing signs (dammed streams, fish traps), trail crossings, machete cuts, and evidence of all other extractive activities. We photographed most of the illegal hunting and fishing camps we encountered in the park.

We integrated the field indicators recorded on the Phase I surveys for each of the faunal groups– ungulates, monkeys, elephants and bonobos– and the indicators of hunting into composite indices of relative occurrence for each of the 233, 10 × 10 km analytical quadrats that had at least 5 km of recce coverage. We calculated indices for each quadrat by summing the indicator encounter rates in the cell (observations/km) weighted by an integrated score based on each indicator's probability of detection, the certainty of its identity, possible time lapse between the detection of the indicator and the actual occurrence of the animal or hunting activity, and the production and decay rates of the indicator. The criteria for the scoring are presented in detail in this volume (Grossmann et al. 2008). The integrated weighting scores for the field indicators are in Table 12.1.

The composite indices have a log normal distribution. We transformed raw index values to base 10 logarithms and classed these on an ordinal scale as low, average, and high. The mean +/− one standard deviation of the log-transformed value is average. We classed the grid cells with the highest 12 index values as very high.

**Table 12.1** Weighting scores for observed field indictors used to develop composite indices of faunal occurrence and hunting intensity. See Table 12.2 in Grossmann et al. (2008) for criteria definitions.

A) Faunal indicators:

| | Criteria | | | | | |
|---|---|---|---|---|---|---|
| Indicator | Certainty of identity | Probability of detection | Time lapse | Production rate | Decay rate | Total score |
| Feeding sign / km | 0 | 1 | 1 | 0 | 0 | 2 |
| nests / km | 1 | 2 | 0 | 1 | 1 | 5 |
| Dung / km | 1 | 1 | 1 | 1 | 1 | 5 |
| Fauna seen or heard | 2 | 0 | 2 | 0 | 0 | 4 |

B) Hunting indicators:

| | Criteria | | | | | |
|---|---|---|---|---|---|---|
| Indicator | Certainty of identity | Probability of detection | Time lapse | Production rate | Decay rate | Total score |
| Camps / km | 1 | 2 | 0 | 1 | 2 | 6 |
| Snares / km | 2 | 2 | 1 | 2 | 1 | 8 |
| Hunters encountered | 2 | 0 | 2 | 0 | 0 | 4 |

## Imagery and Field-based Mapping of Former Settlements in the Park

We used satellite imagery to locate and map existing and former village settlements within the park, which have a distinct visual signature of regenerating vegetation. We investigated the history of most of the former settlement sites in the Phase II Lokofa Block via field visits and interviews with local guides. We gathered information on the identity of former occupants, approximate date and cause of abandonment, location of the displaced population, and current use of the former settlement and its surrounding forest.

## Village-based Data Collection

We complimented field data on hunting indicators recorded on recces and transects with information on hunters and hunting practices gathered from interviews and direct observations by trained observers in villages located outside the park in the vicinity of the Phase II Lokofa Block and an immediately adjoining area covering about 2000 km$^2$, the Lokolo Block. The data includes village censuses, inventories of hunting equipment (snares and shot guns) and counts of hunting dogs. We assessed the level of involvement in the bushmeat trade of a sample of local hunters from different ethnic groups who hunted in the park. We made additional observations on the export of bushmeat at key transit points on the Luilaka and Lokolo Rivers used by hunters to gain access to the western half of the southern sector of the park. We interviewed hunters that we encountered in the park during the Phase I and Phase II field surveys to obtain information on their community of origin, current village base, and where they hunted in the park.

## Human Occupation and Hunting in the Salonga National Park

The Salonga National Park is one of the largest and most intact forest ecosystems in Central Africa (Siegert 2003). Humans occupied areas that are now included in the park at probably < 1 inhabitant/km$^2$ overall in the past. Forty percent of the park, ca. 13,300 km$^2$, is located > 15 km from the nearest human settlement. Yet despite this low level of human occupation, the park remains relatively accessible along a network of rivers, navigable by dugout canoe, that traverse both sectors of the park in an east-west direction and by several abandoned roads around parts of the park periphery that are still traversable by foot and bicycle (Grossmann et al. 2008).

Communities living in the vicinity of the park belong to several ethnic groups classed as pygmies (Iyeke) or villagers (Nkundo). Most speak related languages within the Mongo language group, widespread in Congo's central cuvette region. The people share the same basic subsistence economy based on shifting cultivation, forest gathering, fishing and hunting of small to medium sized animals (monkeys, ungulates, large rodents), with the emphasis on fishing and riverine settlement versus hunting and upland settlement varying by ethnic background.

About 215 villages are located around the periphery of the park. Most are small. Nine villages are located inside the park border. Kitawala in the northern sector (population ca. 5,000–7,000) is the largest, established in the early 1960s by members of a syncretic religious sect. Eight villages of the Iyaelima, totaling ca. 2,500 inhabitants are located in the southern sector. Residents of most of the communities living in and around the park regularly hunt and fish within the park limits. Although settlement and extractive activities within its limits are illegal, fishing and hunting in the park have persisted since the park's creation in 1970, and traditional land claims have not been fully resolved. See also Thompson et al. (2008) in this volume for a detailed case study of the Iyaelima.

Habitat modification, mainly by shifting agriculture, and hunting are the human influences of greatest concern for the conservation of bonobos. Shifting agriculture may create both favorable and unfavorable sites, depending on the extent of clearing, age of regeneration and occurrence of favored food trees. Conversely, hunting is at best a neutral factor, but more likely to have a negative impact on bonobos. The impact of hunting is likely to vary depending upon hunting methods, frequency of use of an area by hunters, and whether bonobos are targeted species.

Hunters operating in the Salonga National Park use active pursuit with bow-and-arrow or 12-guage shotguns, and snares and traps. The primary targeted species are medium-sized ungulates (duikers and pigs) and monkeys. Hunters also target larger rodents, birds, reptiles and small carnivores using pitfalls, snare lines along barriers and other specialized methods; however, these smaller species comprise only a minor portion of the bushmeat consumed and sold. Differences in hunting methods between communities are mainly in the relative importance of pursuit versus snares, in the prevalence of the use of firearms, and the degree to which dogs are used.

We classified hunters as either locally based or mobile professionals. Locally based hunters are more likely to own and use dogs and bows and arrows. Mobile professionals often reside outside the Salonga National Park area and visit temporarily to hunt meat to sell. Many mobile professionals specialize in the use of large numbers of snares. Locally based hunters hunt for subsistence but also sell varying amounts of surplus meat to itinerant meat buyers who gain access to the area along the network of rivers. Bushmeat is exported from the Salonga National Park and its vicinity to Mbandaka and Kinshasa to the west and to the mining centers of the Kasai to the east and south.

A special category of professional hunter is dedicated to elephant hunting. They are armed with military-grade weapons, are highly mobile, and use expeditionary operations including porters and local guides. Despite seriously depleted elephant populations in the park (Blake 1995) ≥ 4 elephant hunters were recorded in the park over the course of the surveys.

Table 12.2 is a summary of hunting methods recorded in the Salonga National Park, and an assessment of the risk they pose to bonobos. Appendix 1 is a list of the large mammals of the Salonga National Park, and their frequency as hunter kills based on observations and interviews.

# Results

## *Faunal Occurrence and Hunting Indices*

In Phase I (2003 – 2004), we surveyed a total of 325, $10 \times 10$ km ($100 \text{km}^2$) quadrats via 2,900 km of systematic reconnaissance walks covering ca. 82% of the park area and 2,100 km² within the corridor separating the northern and southern sectors. In Phase II surveys (2005– 2006), we conducted inventories on 186 transects (260 km in total) and 1,509 km of systematic reconnaissance walks in three blocks: Lokofa, Iyaelima and Lomela. Grossmann et al. (2008) provide further details on the survey deployment and maps of survey coverage.

Figure 12.1 is a map of occurrence of bonobos, ungulates, monkeys and elephants in $10 \times 10$ km quadrats with ≥ 5 km Phase I reconnaissance coverage. Contiguous grid cells have been grouped into larger blocks covering $900 – 4,750 \text{km}^2$ each used to estimate bonobo populations (Grossmann et al. 2008) and evaluate the impact of hunting.

**Table 12.2** Hunting and trapping methods observed in the Salonga National Park and the threat they likely pose to bonobos.

| Hunting type | Class | Subclass | Note | Threat to Bonobos |
|---|---|---|---|---|
| Active Pursuit | Firearms | 12 gauge shotgun | Includes both imported and locally manufactured weapons; locally reload cartridges primary ammunition. | Major: probably highest cause of opportunistic kills. |
| | | Military weapons (FAL, Kalashnikov) | Weapons and ammunition obtained through military or police channels. | Potentially major: limited numbers of weapons in use. |
| | Bow and arrow | Steel tipped arrow | Used for pigs; hunters often use dogs. | Moderate: may be used for terrestrial bonobos. |
| | | Poison arrows | Specialized use for primates and smaller game | Low: not likely to deliver lethal dose to large animal. |
| Trapping | Snares | Large cane cable noose | Designed to hold pigs and large antelope. | Major: death or serious injury likely. |
| | | mid-sized cane, cable or nylon noose | Designed to hold small to mid-sized ungulates | Major to moderate: Death possible serious injury probable |
| | | Small cane, nylon noose | Designed to hold rodents, birds | Low: death unlikely, injury to hands or feet possible |
| | | Barrier | Designed for small animals | Low: Visible and avoided |
| | | Arboreal | Designed for squirrels, pangolins, birds | Negligible |
| | Pitfall | Large mammal | Designed for larger ungulates. | Low: bonobos can climb out |

Large mammals occur widely in the park and buffer zone. Patterns of relative abundance vary significantly between the four taxa (pair-wise $X^2$ tests, $P \leq 0.05$). Important concentrations of bonobos occur in the southern sector of the park in the Iyaelima (B) and Southwest (F) blocks, the northern sector in the Lomela (C) and West Lomela (L) blocks, and the Corridor (H). In contrast, bonobos are absent or occur in low, widely dispersed numbers in the Lokolo (D), Lokofa (A) and South Central blocks (E) in the southern sector, and in the Northwest (I) and North Central (J) blocks in the northern sector. Grossmann et al. (2008) in this volume provide further analysis of distribution and population estimates of bonobos for the park and eastern corridor.

Small and mid-sized ungulates, including primarily duikers (*Cephalophus spp*), chevrotain (*Hyemoschus aquaticus*) and red-river hogs (*Potamochoerus porcus*) are

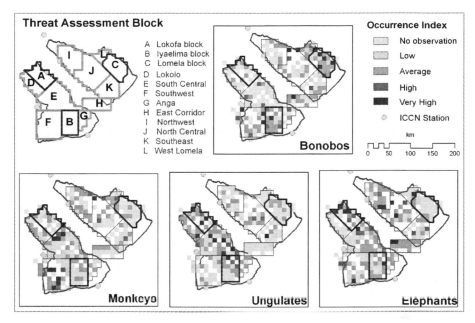

**Fig. 12.1** Occurrence of bonobos, ungulates, monkeys and elephants in the Salonga National Park, and surveyed corridor integrating weighted Phase I encounter rates of field indicators for 10 × 10 km quadrats having ≥ 5 km recce survey coverage. Contiguous quadrats are grouped into Threat Assessment Blocks to evaluate impact of hunting on bonobos and other fauna. Lokofa, Iyaelima, and Lomela Threat Assessment Blocks cover Phase II Population Inventory Blocks described in Grossmann et al (2008) (*See Color Plates*).

most abundant in the southern sector of the park in the Lokofa (A), Lokolo (D) and South Central (E) blocks and in the northern half of the Iyaelima block (B). They are less abundant in the northern sector of the park in the Northwest (I), Lomela (C) and West Lomela (L) blocks and in the southern sector in the Southwest (F) block and Corridor (H). Monkeys in contrast are abundant in the Southwest (F) block, and locally in the South Central (E) block (southern sector), but were found only in low numbers in the Lomela (C) and Southeast (K) blocks (northern sector) and in the Corridor (H).

Indicators of elephants were concentrated locally around large swampy clearings (termed *Botoka ndjoku*, or elephant baths) in the Lokofa (A), Northwest (I), South Central (E), North Central (J) and Iyaelima (B) blocks, and in the Corridor (H). In contrast, they were markedly absent in the Lomela (C) and Lokolo (D) blocks. *Botoka ndjoku* in blocks with low elephant abundance had low levels of visitation.

We recorded indicators of hunting, including 26 direct encounters with hunters, in 165 (51%) of 325, 10 × 10 km Phase I quadrats (Fig. 12.2). Snares and hunting camps were the most frequently observed indicators. We noted spent shot gun cartridges and specialized traps, such as pits falls, on just a few occasions, and rarely heard gunshots. The infrequency of spent ammunition can be accounted for by the fact that most hunters retrieve and reload spent cartridges. We recorded fishing

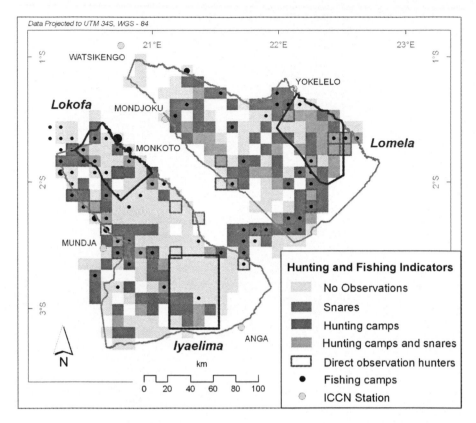

**Fig. 12.2** Hunting and fishing indicators recorded on Phase I surveys in the Salonga National Park and surveyed corridor. Phase II inventory blocks are shown in outline (*See Color Plates*).

camps in 61 survey quadrats in the park. Most were concentrated along larger water courses. Hunting camps were in 75 quadrats with a wide distribution throughout the park. Some camps served as bases for both fishing and hunting.

Figure 12.3 is the distribution of composite hunting index for the 233, $10 \times 10$ km quadrats with $\geq 5$ km of recce coverage. While hunting was widespread throughout the park, intensive hunting was concentrated in the eastern quarter of the northern sector and along the Lomela River. Large areas of the southern sector, in contrast, had low hunting indices. Hunting indices in the corridor between the two park sectors were comparable to indices in many areas within the park itself, and were notably lower than indices for large areas of the northern sector of the park.

## Former Settlements in the Park

Figure 12.4 gives the location of former settlements and associated clearings in the Salonga National Park as determined by analysis of satellite imagery, site visits and interviews. Most former settlements, locally termed *mpumba* or *eladji*, were abandoned

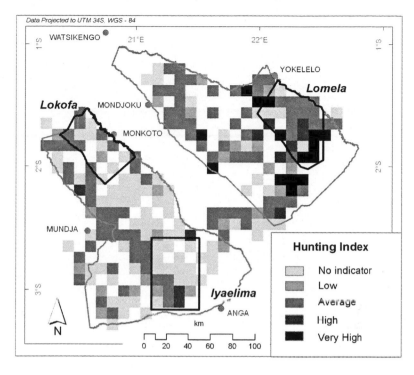

**Fig. 12.3** Composite hunting index integrating weighted Phase I encounter rates of field indicators for 10 × 10 km quadrats having ≥ 5 km recce survey coverage in the Salonga National Park and surveyed corridor. Phase II inventory blocks are shown in outline.

before the park's creation in 1970 as a result of colonial and post-colonial policies to regroup human people along roads. Although these settlements are no longer permanently occupied, the former clearings and surrounding forests are still claimed and used by descendants of the original occupants mostly for hunting and fishing. The Lokofa Block contains a relatively high proportion of the park's former settlements.

Bonobo abundance is negatively correlated with proximity to areas of former settlement (Fig. 12.5). The relative depletion of bonobos extends out from the area of regenerating secondary vegetation in the abandoned clearings and gardens, several kilometers into the surrounding undisturbed forest. Thus, the reduced rates of occupation by bonobos can not be attributed to the direct effects of habitat modification.

Descendants of former occupants return to *mpumba* in the park to hunt and fish years after the villages have been abandoned. Those interviewed stated that most of the displaced communities had no problem gaining access to land for gardens in their new settlement areas, but that access to new hunting and fishing territories remained difficult and a source of conflict between communities even presently. Thus, the *mpumba* and *eladji* within the park remain the primary access to bushmeat and fish for many displaced communities.

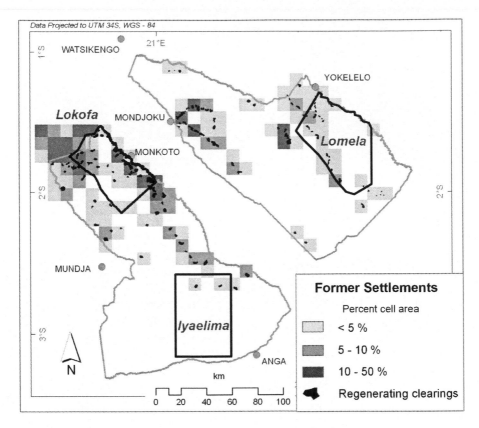

**Fig. 12.4** Former settlements (*mpumba*) within the Salonga National Park abandoned before creation of the park in 1970. The proportion of the quadrat area covered by regenerating vegetation is given for 10 × 10 km quadrats with *mpumba*. Phase II inventory blocks are shown in outline.

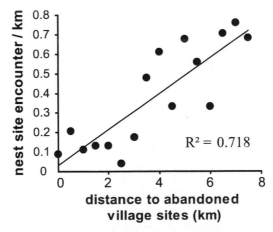

**Fig. 12.5** Bonobo abundance (nest group encounter rate) and distance to former settlements (*mpumba*) for all surveyed *mpumba* in the Salonga National Park.

This apparent limitation of fish and wildlife resources, even where human densities are low, is characteristic of areas with low nutrient soils such as the leached sands and weathered substrates found over much of the western Salonga National Park (Barnes and Lahm 1997). Similar substrate-linked constraints may determine settlement and hunting patterns in other areas and could also play a role in determining occupation by bonobos.

## Village and Hunter Surveys

Systematic surveys were conducted in 37 villages along the Wafania-Boleko road bordering the western border of the southern sector of the park. Hunters based in these villages hunt within the Salonga National Park, including 13 villages totaling 5,800 inhabitants bordering the Lokofa Block and 24 villages totaling 6,335 inhabitants bordering the adjacent Lokolo Block (Fig. 12.1). We conducted interviews in 147 households and with an additional 62 hunters, including both pygmy (Iyeke) and villager (Nkundo) ethnic groups identified in the villages or encountered within the park in the Lokofa Block during Phase II surveys.

Table 12.3 is a summary profile of the surveyed villages. Although the Lokofa and Lokolo communities have approximately the same number of inhabitants, the villages bordering the Lokolo Block have more hunters than the villages bordering the Lokofa Block. Both Lokofa and Lokolo villages had comparable equipment indices (snares per hunter and shotguns per hunter) with dogs used frequently in both areas. Lokolo villages had higher involvement in the commercial meat trade than Lokofa Block villages (7 vs. 4) and a greater presence of mobile professional hunters (12 vs. 3). All of the 20 hunters encountered by survey teams within the park in the Lokofa Block during Phase II inventories were based in just two villages, and almost all were Iyeke pygmies.

In the Lokofa villages, only one of the three professional hunters we interviewed had his own camp in the forest at the time of the survey. Two had rented their cable snares and shot guns to local hunters in exchange for a share of the meat. All of the 12 mobile professionals we interviewed in the Lokolo villages had their own hunting camps in the park or joined forces with local hunters. One elephant hunter, with links to the Congolese military, operated in the Lokolo Block during the survey period. In Mangilombe village (total population 807, including both Iyeke and Nkundo), the most active hunting village among the communities bordering the Lokolo Block, 3 of the 4 professional hunters present came from Mbandaka and were related by marriage to the traditional chief who provided them with illegal authorizations to hunt inside the park.

Ungulates and monkeys comprised the near totality of the meat recorded in transport from the park or along the road to the dugout canoe ports of Wafania (export point for the Lokfa Block) or Boleko (export point for the Lokolo Block). Other observed bushmeat species included pangolins, porcupines and possibly elephant. Survey teams recorded no bonobos in the bushmeat samples examined, and no hunters admitted to killing or selling bonobos, although several said that dead bonobos were occasionally brought into their village by hunters.

**Table 12.3** Profiles of communities with significant hunting in the Lokofa and Lokolo Blocks (southern sector, Salonga National Park) [1]

| Block | | Total inhabitants | Villages with major commercial hunting | Hunter census | | Equipment and hunting dog inventory | | | |
|---|---|---|---|---|---|---|---|---|---|
| | | | | Locally based | Mobile professional | Snares | Shotguns | Bows | Dogs |
| Lokofa 13 villages | Total census | 5,800 | 4 | 290 | 3 | 33,242 | 89 | 603 | 342 |
| | average per village | 446 | – | 22 | – | 2,557 | 7 | 46 | 26 |
| | range | 83 – 1663 | – | No data | – | 0 – 12,800 | 0 – 15 | 0 – 119 | 0 – 78 |
| Lokolo 24 villages | Total census | 6,330 | 7 | 830 | 12 | 93,650 | 277 | No data | No data |
| | average per village | 264 | – | 35 | – | 3,902 | 12 | No data | No data |
| | range | 41 – 807 | – | 4 – 200 | – | 400 – 20,000 | 4 – 36 | No data | No data |

[1]Survey period March – August, 2005. See Fig. 12.1 for Block location.

Although many locally based hunters participate in the commercial trade, most do so in a limited way to provide cash to buy basic supplies (machetes, salt, soap, petrol, cooking pots), luxury or investment items (radios, shotguns, bicycles, sewing machines), to pay school fees, or less frequently, to produce capital for family events such as marriages. In a list of 21 exchange equivalents for bushmeat developed from interviews in the villages bordering the Lokofa Block, 9 items were clothing, 4 were luxury or capital investments and 9 were subsistence supplies.

Mobile professional hunters reported that they exported their meat to urban markets in Mbandaka and rarely traded it locally. Exchange rates (Congolese franc equivalents) for bushmeat traded at the village or hunting camp at the beginning of the bushmeat chain are one fourth to one tenth the prices paid for the same item once it reaches the urban market in Mbandaka.

## Discussion

### *Impact of Hunting on Bonobos and Other Fauna*

Bonobos, ungulates, monkeys and elephants differ in their distribution and abundance in the park. No single species or taxonomic group can be used to provide a comprehensive index of the impact of hunting in the park. Each species responds differently to ecological factors and to the effects of hunting. Ungulate abundance decreased consistently from low hunting to high hunting quadrats in Phase I surveys. This expected relationship between relative faunal abundance and hunting pressure was not recorded for any other species, although for all taxa, including bonobos, the proportion of quadrats with high faunal abundance was lowest for the grids with high hunting indices (Fig. 12.6).

Table 12.4 is a summary of bonobo densities, human settlement and hunting practices in the three Phase II inventory blocks. In the Lokofa Block, bonobo nests were significantly less abundant in quadrats with the highest hunting indices (T tests, $P<0.5$). Locations with high snare encounters had fewer nests. Most nests were

**Table 12.4** Profiles of Phase II inventory blocks

| Block | Bonobo densities[1] (per km$^2$) | Human settlement | Hunting methods | Bonobo kills recorded | Commercial meat trade | Hunter attitude toward bonobos |
|---|---|---|---|---|---|---|
| Lokofa | 0.278 (0.102 – 0.395) | peripheral | Snare, shotgun, archery | None | present | Neutral / unknown |
| Iyaelima | 0.670 (0.328 – 0.803) | interior | Snare, archery, shotgun | None | absent or low | Avoid |
| Lomela | 0.865 (0.441 – 1.00) | Interior and peripheral | Snare, shotgun | Yes | high | Possibly targeted |

[1] Mean and 95% confidence limits for estimate.

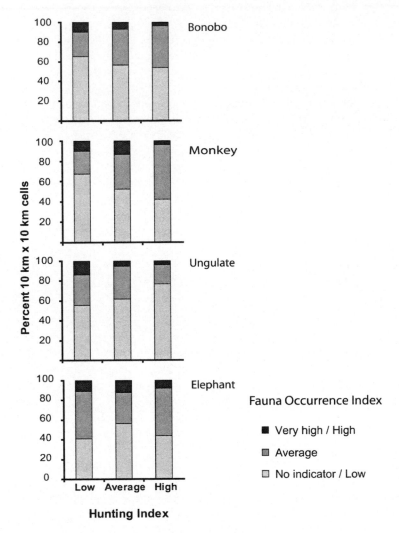

**Fig. 12.6** Faunal occurrence in relation to hunting indices for 10 × 10 km Phase I quadrats with ≥ 5 km recce coverage in the Salonga National Park and surveyed corridor.

found in quadrats with low hunting indices or where no hunting indicators were found (Fig. 12.7). In the Iyaelima Block, in contrast, bonobos were widespread and abundant in areas used by local hunters, and we detected no relationship between bonobo abundance and hunting indices. A similar situation was initially detected in the Lomela Block during the Phase I survey (2005), but by the time of the Phase II survey (ca. 1.5 years later), a number of quadrats in the area of the Kitawala village where we had recorded average to high bonobo indices during Phase I, contained few or no bonobo nests during the Phase II inventory. The Lomela Block was the only area in the Salonga National Park where survey teams recorded bonobos killed

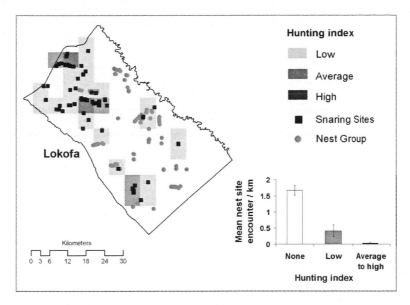

**Fig. 12.7** Hunting indices, snare localtions, and bonobo nest groups recorded on Phase II inventories in the Lokofa Block. Histogram shows mean +/− SE nest group encounters for 5 × 5 km quadrats.

by hunters. We estimated that over 20 mobile professional hunters operated in the Lomela Block during the period of the surveys. We recorded no mobile professional hunters in the Iyaelima Block, where the local Iyaelima actively discourage outsiders from hunting in their forest (Thompson et al. 2008). We recorded 3 mobile professional hunters in the villages bordering the Lokofa Block.

## Human Settlement and Bonobos

Bonobos live in close proximity to some villages in the Salonga National Park and the surrounding area. However, we found a negative relationship between bonobo abundance and proximity to former settlements in the park, most of which are used by descendants of the former inhabitants for hunting and fishing. The reduction of bonobo populations around former settlements is most likely the result of sustained hunting at the site over many years. Hunters that we interviewed stated that some *mpumba* in the Lokofa Block have been hunted consistently for over six decades. Reductions of some of the bonobo populations around *mpumba* are probably not recent and continued hunting may prevent re-colonization of depleted sites. These observations support Butynski (2001), Dupain et al. (2001) Kingdon (1997), Kortlandt (1995) and Kano (1984), who argued that past hunting pressure may have produced gaps in the bonobo's historic range, and that current hunting promotes ongoing range reduction.

Displaced communities will continue to seek access to their former *mpumba* hunting grounds as long as areas outside the park are over-exploited or until alternative sources of income – and protein – become available and accepted. The relationship between human occupation of the forest and occurrence of bonobos may be highly dynamic as suggested by the apparent decrease in bonobos in the Lomela Block from 2005 – 2006, during a major increase in hunting.

The key point in the relationship between people and bonobos is not where human settlements occur, but rather where and how people hunt. Bonobos and humans are likely to coexist, even in close spatial proximity, where hunters do not target bonobos because of cultural taboos or where hunters use methods that do not put bonobos at risk. Bonobos are vulnerable where hunters unselectively target large bodied species, or broaden their range of targets to include bonobos as preferred game species are depleted.

## Threats to Bonobos in the Salonga National Park

We identified three primary threats to bonobos occupying the Salonga National Park. These included:

1) **High hunting indices:** Intensive hunting is a threat to bonobos even when they are not targeted, since the non-selective hunting methods widely used in the park (cable snares) are likely to catch, maim or kill bonobos, as has been documented with chimpanzees (Hashimoto 1999, Reynolds et al. 1996). Bonobos are also likely to be killed opportunistically by hunters with firearms when they are encountered. Areas with high hunting levels are likely to include a higher proportion of mobile professional hunters who may be more inclined to seek and kill bonobos.

2) **Commercial bushmeat:** Market hunting leads to an intensification and spatial expansion of hunting. By controlling the prices they pay for meat, commercial buyers can manipulate locally based hunters to produce unsustainable off takes. Large bodied, social bonobos are especially at risk where commercial hunting prevails, as each individual animal provides large quantities of meat and multiple kills are possible for each encounter. The perception (true or not) is that higher populations of wildlife in the park attract commercial hunters. They claim to have ready access to areas of the park that are not patrolled by ICCN. Hunters in the park may also avoid the need to pay fees or provide tribute to traditional authorities in exchange for hunting rights. These are significant gains to hunters pursuing marginally higher profits.

3) **Absence of active protection:** The control of the Salonga National Park by the ICCN is incomplete and ineffective. Some areas of the park have never been patrolled.

Three additional factors represent indirect threats and reduced or uncertain levels of direct risk. They can potentially affect the impact of hunting on bonobos.

4) *Faunal depletions:* Reductions in populations of ungulates and monkeys that are selected by most hunters could put bonobos at risk if hunters turn to bonobos as alternative targets.

5) *Former settlements:* Bonobo populations are likely to be threatened in areas that have an extended history of hunting, in particular in former settlement areas within the park that are used as traditional hunting grounds by local communities. Given their long life spans and wide daily ranging, bonobos will be exposed to accumulating risk, even under lower hunting levels, if the areas they occupy are hunted consistently over time.

6) *Hostility of local populations:* We found that some local communities that are hostile to the park and the presence of ICCN staff threaten guards and prevent their deployment in areas of the park where they are hunting illegally. This may also facilitate expansion of direct threats such as commercial hunting.

These six threats vary spatially in their influence and they do not affect the park's bonobos equally. Table 12.5 is an evaluation of the relative importance of each of these threats in the 12 threat assessment blocks (eleven within the park and one in the eastern corridor between the two park sectors) that were delimited to develop estimates of bonobo populations given in this volume in Grossmann et al. (2008) and mapped in Fig. 12.1. For each threat, the level of risk is graded from low to high, on a three point scale (1–3). Composite threat scores are calculated for each block by multiplying the sum of the scores of the three direct threats by 2 and then calculating the average of the direct and indirect threat scores combined.

While the scoring of threats is approximate and the calculation of risk is just one of several possible computations, several trends are nevertheless evident:

1) Over 14% of the park area and over 22% of its bonobos are highly threatened. Another 25% of the bonobos are only slightly less threatened. Just 3% of the park area and < 3% of its bonobos could be classified as low risk.

2) Threats to bonobos are not distributed equally. The Lomela and West Lomela blocks, both of which have high bonobo occurrence, also have among the highest threat scores.

3) High hunting levels and commercial bushmeat trade were recorded in five of the 12 blocks, covering over half of the park's area.


## Conservation Potential

The ICCN is the sole authority legally mandated to patrol the Salonga National Park and ensure protection of its fauna. An index of the potential protection of the park area by ICCN staff can be defined as a function of the distance from park stations or patrol posts, weighted by the number of guards present at each location. The potential protection index can be compared with an index of hunting levels to provide an overall assessment of the vulnerability of the park's bonobos (Fig. 12.8).

**Table 12.5** Threat to bonobos posed by hunting and related risks in the Salonga National Park and eastern corridor [a]

| Sector | Threat Assessment Block [b] | area (km²) | Bonobo population (mean estimate) [c] | Hunting pressure | Commercial bushmeat trade | Absence active protection | Faunal depletion | Hunting history | Local Hostility | Composite score | Level of threat to bonobos [d] |
|---|---|---|---|---|---|---|---|---|---|---|---|
| South | A. Lokofa | 2300 | 611 | 2 | 2 | 1 | 1 | 3 | 1 | 2.5 | intermediate |
| | B. Iyaelima | 2400 | 1526 | 2 | 1 | 2 | 2 | 1 | 2 | 2.5 | intermediate |
| | D. Lokolo | 1550 | 415 | 2 | 3 | 3 | 2 | 3 | 3 | 4.0 | high |
| | E. South Central | 4750 | 1213 | 1 | 2 | 2 | 1 | 2 | 2 | 2.5 | intermediate |
| | F. Southwest | 4200 | 2212 | 1 | 2 | 3 | 2 | 1 | 1 | 2.7 | intermediate |
| | G. Anga | 1100 | 412 | 1 | 1 | 1 | 2 | 2 | 1 | 1.8 | low |
| | Unsampled zone | 1800 | 667 | 2 | 2 | 2 | 2 | 2 | 2 | 3.0 | intermediate |
| North | C. Lomela | 2700 | 2242 | 3 | 3 | 3 | 3 | 2 | 3 | 4.0 | high |
| | I. Northwest | 3100 | 1064 | 2 | 2 | 1 | 3 | 3 | 1 | 2.8 | intermediate |
| | L. West Lomela | 4700 | 1391 | 3 | 3 | 2 | 2 | 2 | 2 | 3.7 | intermediate |
| | J. North Central | 2500 | 1106 | 3 | 2 | 3 | 3 | 1 | 2 | 3.7 | intermediate |
| | K. Southeast | 900 | 643 | 3 | 3 | 2 | 3 | 3 | 3 | 3.8 | high |
| | Unsampled zone | 3700 | 1380 | 3 | 3 | 2 | 2 | 2 | 2 | 3.7 | intermediate |
| **Total Park** | **Totals / Mean** | **35700** | **14883** | **2.2** | **2.2** | **1.9** | **2.2** | **2.1** | **1.9** | **3.1** | **intermediate** |
| Buffer Zone | East Corridor | 2100 | 809 | 2 | 3 | 3 | 3 | 1 | 2 | 3.7 | intermediate |

[a] See text for definitions and scoring of threat criteria.

[b] See Fig. 12-1 for block locations.

[c] Bonobo population estimates developed in Grossmann et al. (2008).

[d] Level of threat to bonobos given by composite score: < 2.5, low; 2.5 – 3.7, intermediate; > 3.7, high.

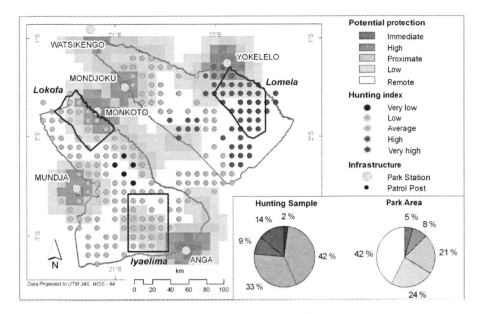

**Fig. 12.8** Getis-Ord G, hunting hot spots, and potential protection indices computed as a function of distance from established ICCN infrastructure (patrol posts and park stations) weighted by number of park guards based at the site (*See Color Plates*).

Based on this index, only about one third of the park area is classified as having proximate or higher potential protection. Over 40% of the park is remote (> 20 km and > 30 km respectively) from manned patrol posts or park stations. Elevated hunting indices, as grouped by the Getis Ord G, hot spot analysis (Mitchell 2005), are concentrated in the Lomela Block and in the eastern third of the northern sector where ICCN bases are remote and potential protection very low.

The analysis also shows that proximity to ICCN patrol bases is unlikely to explain the low levels of illegal hunting we recorded in portions of the southern sector of the park. Many of these areas are remote from ICCN bases and infrequently patrolled. They have no *mpumba*, and may not have been claimed as traditional hunting territories before the creation of the park. Professional mobile hunters simply may not yet have reached these areas, or alternatively, they may have hunted these areas and left before the surveys. In the northern sector, high levels of hunting were found even in areas of proximate potential protection. Under its current deployment, ICCN's infrastructure and staff are poorly placed to deal with some of the most significant threats to the park.

Hostility between ICCN staff and local communities hinders deployment of park guards and reduces their efficiency. The inability of ICCN and local communities to resolve seemingly simple problems ingrains antagonism against the park. Unresolved land claims and disputes over park limits are a constant distraction to the ICCN. Congo's decade of conflict and political instability left the ICCN weakened and opened some of the country's parks to land grabs and destruction that are not easily reversed (Hart 2005). The Salonga National Park, due to its large area, remoteness and lack of known mineral reserves, still remains intact compared to some other protected areas in Congo. However, more than 20% of the 215 villages surrounding the Salonga National Park, and all of the nine villages located within the park have long standing disputes with the ICCN that impede collaboration (Thompson et al. 2008).

## Recommendations to Improve Protection of Bonobos

Protected areas are one of the basic mechanisms to ensure conservation of vulnerable species. Yet legal gazettement, and even the deployment of park guards, are not sufficient to ensure the integrity of the protected area or the conservation of its fauna. Salonga National Park contains a significant population of bonobos and covers ca. 10% of their range. Despite the park's World Heritage status, illegal hunting occurs over wide areas. Almost a quarter of the park's bonobos are at high risk from illegal hunting. It is unlikely that the ICCN will be able to deal with the threats fully. Solutions are urgently required or the park risks having one of its most valuable assets seriously reduced. Failure to protect the Salonga National Park and its bonobos could compromise efforts to establish other protected areas within the bonobo's range.

We offer the following recommendations as guidelines for immediate actions:

*Recommendation 1:* Control the most dangerous hunting. High levels of hunting in areas that contain concentrations of bonobos present the most important threat. It may not be possible to eliminate high levels of hunting everywhere, but a focused approach to reduce hunting in areas that are most important for bonobos is needed.

*Recommendation 2*: Target specific hunters. Our interviews and observations in the field revealed that the bulk of the intensive hunting in any given area of the park is likely led by a small, readily identified group of hunters, most of whom are involved in the commercial bushmeat trade. These hunters and their associated buyers should be the first focus of control by ICCN. In some cases, controls on specific hunting practices might mitigate the impact of hunting on bonobos. The advantage of the focused approach is that it reduces the likelihood of misdirected punishment of hunters who, though operating illegally within the park, are less threatening to bonobos.

We recommend a combination of individualized and collective approaches for hunter education. Individualized education programs can be tailored to specific hunting territories, ethnic concerns, hunting methods and specific hunters. Individualized approaches also develop a basis for personal responsibility – and its benefits – a key element in legal recourse and for certain opportunities such as alternative sources of income or employment, including the possibility of hiring former poachers as park guards.

*Recommendation 3*: Reduce local hostility. We recommend that selected demands by local communities for access to key sites and resources within the park be evaluated, and agreements be developed where access and utilization can be controlled and will not damage park values. Arrangements with local communities could include agreements that commit them to support protection of the park in exchange for access rights. Locals could be hired to participate in the monitoring of park use, an approach already initiated by some international NGOs supporting the park. Novel approaches, like "cultural tourism," that permit controlled access and managed use of the park for cultural events, such as some types of seasonal fishing, might be possible. Not all proposed uses will be compatible with the protection of the park, and negotiations should ensure that the overall outcome is improved conservation of the site. This approach will require that the ICCN acquire capacities such as community outreach and conflict resolution. A pilot project in the Lokofa Block to delimit contested park limits with participation of local communities improved relations with the ICCN, but it is not certain if this approach can be used to resolve more difficult issues such as illegal hunting and fishing.

# Conclusions

Salonga National Park's globally significant population of bonobos are at best only partially protected and secured. Illegal hunting is widespread in the park. High levels of hunting in areas where bonobos are most abundant is the most important threats. Salonga National Park has a long history of human use focused mainly on fishing and hunting. Demands by local communities for access to the park's resources will continue to be made, and new approaches are needed to respond to these while at the same time ensuring the integrity of the park and protection of its key fauna, including bonobos. The multiphase program of field inventories in conjunction with village and hunter surveys is an efficient way to identify areas with high concentrations of bonobos, and to evaluate dangers to them. Despite the serious threats, Salonga National Park represents one of the best opportunities for long term conservation of bonobos. Support of both local communities and ICCN are required to secure the park and protect its bonobos.

**Acknowledgements**  We acknowledge the support of the Institut Congolais pour la Conservation de la Nature (ICCN) for the Salonga survey project We thank the CITES-MIKE program, Wildlife Conservation Society (WCS), USAID's Central African Regional Program for the Environment (CARPE), the Alexander Abraham Foundation and World Wide Fund for Nature (WWF) for financial support.

We are grateful to the more than one hundred individuals who participated in surveys and assisted in providing the major logistical effort to support the survey teams in remote locations. Field team leaders who deserve special mention include: Maurice Emetshu, Aime Bonyenge, Bernard Ikembelo, Simeon Dino, Pupa Mbenzo, Samy Matungila, Menard Mbende and Pele Misenga. Community surveys were led by Georges Lombombe and Jeef Ikwange. Inogwabini Bila Isia and Omari Ilambu assisted in the training of the field teams.

We thank WCS and WWF for their support in developing a series of workshops in the field in the Lokofa Block, and supporting workshops in Kinshasa on conflict resolution and management plan development that provided input for the development of some of the conclusion presented here.

We extend a special expression of appreciation to Jo Thompson for sharing her experiences with the Iyaelima with us and for providing essential support and encouragement in the preparation of this chapter.

# References

Barnes RFW, Lahm SA (1997) An ecological perspective on human densities in the central African forest. Journal of Applied Ecology 43: 245–260

Blake S (2005) Forêts d'Afrique centrale: rapport final sur les relevés démographiques d'éléphants (2003–2004). CITES –MIKE, Nairobi, pp 135

Bowen-Jones E, Pendry S (1999) The threat to primates and other mammals from the bushmeat trade in Africa, and how this threat could be diminished. Oryx 33: 233–246

Butynski TM (2001) Africa's great apes In: Beck B, Stoinski TS, Hutchins M, Maple TL, Norton B, Rowan A, Stevens EF, Arluke A (eds) Great apes and humans: the ethics of coexistence. Smithsonian Institution Press, Washington DC, pp 3–56

Draulans D, Van Krunkelsven E (2002) The impact of war on forest areas in the Democratic Republic of Congo. Oryx, 36: 35–40

Dupain JL, Van Krunkelsven E, Van Elsacker L, Verheyen RF (2000) Current status of the bonobo (*Pan pansicus*) in the proposed Lomako Reserve (Democratic Republic of Congo). Biological Conservation 94: 265–272

Dupain JL, Van Elsacker L, Verheyen RF (2001) The status of the bonobo in the new Democratic Republic of Congo. In: Galdikas B, Briggs N, Sheeran L, Shapiro G (eds) All apes great and small, Vol. I: Chimpanzees, Gorillas and Bonobos. Kluwer, New York, pp 57–74

Grossmann F, Hart JA, Vosper A, Ilambu O (2008) Range occupation and population estimates of bonobos in the Salonga National Park: application to large scale surveys of bonobos in the Democratic Republic of Congo. In: Furuichi T, Thompson J (eds) The bonobos: behavior, ecology, and conservation. Springer, New York, pp 189–216

Hart JA (2005) Wildlife Wars – The Democratic Republic of Congo's national parks and their guardians under siege. International Herald Tribune Op-Ed June 7 Paris

Hashimoto C (1999) Snare injuries of chimpanzees in the Kalinzu Forest, Uganda. Pan Africa News 6: 5

Idani G, Mwanza N, Ihobe H, Hashimoto C,Tashiro Y, Furuichi T (2008) Changes in the status of bonobos, their habitat, and the situation of humans at Wamba, in the Luo Scientific Reserve, Democratic Republic of the Congo. In: Furuichi T, Thompson J (eds) The bonobos: behavior, ecology, and conservation. Springer, New York, pp 291–302

Kano T (1984) Distribution of pygmy chimpanzees (*Pan paniscus*) in the Central Zaire Basin. Folia Primatologica 43: 36–52

Kano T, Asato R (1994) Hunting pressure on chimpanzees and gorillas in the Motaba River area, northeastern Congo. African Study Monographs15: 143–162

Kano T, Lingomo B, Idani G, Hashimoto C (1996) The challenge of Wamba. In: Cavalier P (ed) Great ape project: Ecta & Animali 96/8. Milano, pp 68–74

Kingdon J (1997) The Kingdon field guide to African mammals. Princeton University Press

Kortlandt A (1995) A survey of the geographical range, habitats and conservation of the pygmy chimpanzee (*Pan paniscus*): An ecological perspective. Primate Conservation 16: 21–36

Mitchell A (2005) The ESRI guide to GIS analysis, vol 2: Spatial measurements and statistics. ESRI Redlands, California

Omari I (2006) Review of the ICCN infrastructure of PN Salonga. WWF, Kinshasa

Reinartz GE, Inogwabini BI, Ngamankosi M, Wema LW (2006) Effects of forest type and human presence on bonobo (*Pan paniscus*) density in the Salonga National Park. International Journal of Primatology 27: 603–634

Reynolds V, Burch D, Knight J, Smith R, Walter J (1996) The nature, causes and consequences of injuries sustained by wild chimpanzees in the Budongo Forest, Uganda. Abstract, XVIth Congress of the International Primatological Society

Rose AL (1998) Growing commerce in bushmeat destroys great apes and threatens humanity. African Primates 3: 6–10

Siegert F (2003) Assessment of land cover and land use in the Salonga NP based on satellite imagery. Remote Sensing Solutions (RSS-GmbH) and GTZ, Kinshasa. 16 pages Unpublished report

Susman R, Badrian N, Badrian A, Thompson-Handler N (1981) Pygmy chimpanzees in peril. Oryx 16: 179–183

Tashiro Y (1995) Economic difficulties in Zaire and the disappearing taboo against hunting bonobos in the Wamba area. Pan Africa News 2

Thomas L, Laake JL, Strindberg S, Marques F, Borchers DL, Buckland ST, Anderson DR, Burnham KP, Hedley SL, Pollard JH (2001) Distance 4.0. Beta 1. Research Unit for Wildlife Population Assessment, University of St. Andrews, UK, *http://www.ruwpa.st-and.ac.uk /distance/*

Thompson J Myers, Lubuta MN, Kabanda RB (2008) Traditional land-use practices for bonobo conservation. In: Furuichi T, Thompson J (eds) The bonobos: behavior, ecology, and conservation. Springer, New York, pp 227–244

Thompson-Handler N, Malenky RK and Reinartz GE (1995) Action plan for *Pan paniscus*: report on free-ranging populations and proposals for their preservation. Zoological Society of Milwaukee, Wisconsin, United States

**Annex 12.1** Large Mammals of the Salonga National Park (nomenclature following Kingdon 1997)

| Taxon | French | English | Vernacular | Occurrence in hunter kills (2003 – 2006) |
|---|---|---|---|---|
| **ELEPHANT** | | | | |
| *Loxodonta africana cyclotis* | Eléphant de fôret | Forest elephant | Ndjoku | Uncommon |
| **GREAT APES** | | | | |
| *Pan paniscus* | Bonobo | Bonobo | Edja, Esumbu, Mokomboso | Rare |
| **ANTHROPOID PRIMATES** | | | | |
| *Piliocolobus tholloni* | Colobe bai | Tshuapa red colobus | Djofe, Kolongo | Common |
| *Colobus angolensis* | Colobe d'Angola | Angola pied colobus | Libuka | Uncommon |
| *Cercopithecus ascanius* | Cercopithèque ascagne | Red-tailed guenon | Nsoli | Common |
| *Cercopithecus (mona) wolfi* | Cercopithèque de wolf | Wolf's guenon | Mbeka | Common |
| *Cercopithecus neglectus* | Cercopithèque de Brazza | De brazza's guenon | Bosila | Not recorded |
| *Allenopithecus nigroviridis* | Singe de marais | Allen's swamp monkey | Ekele | Rare |
| *Lophocebus aterrimus* | Cercocebe noir | Black mangabey | Ngila | Common |
| *Cercocebus chrysogaster* | Cercocebe à ventre doré | Golden-bellied mangabey | Inku | Uncommon. No observations from park |
| **UNGULATES** | | | | |
| *Potamochoerus porcus* | Potamochère | Red river hog | Nsombo | Common |
| *Hyemoschus aquaticus* | Chevrotain aquatique | Water chevrotain | Etambe | Uncommon |
| *Tragelaphus spekei* | Sitatunga | Sitatunga | Mbuli | Common |
| *Tragelaphus euryceros* | Bongo | Bongo | Mpanga | Not recorded |
| *Cephalophus monticola* | Cephalophe bleu | Blue duiker | Mboloko | Common |
| *Cephalophus nigrifrons* | Cephalophe à front noir | Black-fronted duiker | Bombende, Nkulufa, Nginda | Common |
| *Cephalophus dorsalis* | Cephalophe bai | Bay duiker | Bofala | Common |
| *Cephalophus silvicultor* | Cephalophe à dos jaune | Yellow-backed duiker | Mbende, Lisoko | Uncommon |
| *Cephalophus callipygus* | Cephalophe de peter | Peter's duiker | Mpambi, Mbengele | Common |

(continued)

**Annex 12.1** (continued)

| Taxon | French | English | Vernacular | Occurrence in hunter kills (2003 – 2006) |
|---|---|---|---|---|
| *Syncerus caffer* | Buffle de fôret | Forest Buffalo | Mpakasa, Ngombo | Not Recorded |
| CARNIVORES | | | | |
| *Panthera pardus* | Léopard (Panthère) | Leopard | Nkoy | Uncommon |
| *Civettictis civetta* | Civette d' Afrique | Afrique Civet | Bowane, Ibobi | Rare |
| *Genetta servalina* | Genette servaline | Servaline genet | Iyeni | Common |
| *Genetta tigrina* | Genette tigrine | Blotched genet | Nsima | Common |
| *Nandinia binotata* | Nandinie | African palm civet | Mbio | Common |
| *Aonyx congica* | Loutre du Congo | Swamp otter | Iyoko | Not recorded |
| *Mellivora capensis* | Ratel | Honey badger | Esisi, Lokusa, Mpolo | Not recorded |
| *Atilax paludinosus* | Mangouste des marais | Marsh mongoose | Bolia ya mai | Not recorded |
| OTHER | | | | |
| *Galagoides demidoff* | Galago de Demidoff | Demidoff's galago | Isile | Not recorded |
| *Perodicticus potto* | Potto de Bosman | Potto | Nkatshu | Not recorded |
| *Smutsia gigantea* | Pangolin géant | Giant pangolin | Ikaka | Rare |
| *Phataginus tricuspis* | Pangolin commun | Tree pangolin | Nkalamonyo | Common |
| *Uromanis tetradactyla* | Pangolin à longue queue | Long-tailed pangolin | | Not recorded |
| *Potamogale velox* | Potamogale | Giant otter shrew | Yoongo | Rare |
| *Dendrohyrax arboreus* | Daman d'arbre | Tree hyrax | | Rare |
| *Atherurus africanus* | Athérure africain | Brush tailed porcupine | Iko | Uncommon |
| *Anomalurus derbianus* | Anomalure de Derby | Lord Derby's anomalure | Lokiyo | Uncommon |
| *Protoxerus stangeri* | Ecureuil géant | Giant squirrel | | Not recorded |

# The Bonobos of the Lake Tumba – Lake Maindombe Hinterland: Threats and Opportunities for Population Conservation

Bila-Isia Inogwabini[1], Matungila Bewa[1], Mbende Longwango[1], Mbenzo Abokome[1], and Miezi Vuvu[1]

## Introduction

### Bonobos in the Lake Tumba – Lake Maindombe Region: an Historical Chronicle

Anecdotally the term bonobo, now the popular name of *Pan paniscus*, is a misuse of the village name Bolobo (Thompson 1997, de Waal and Lanting 1997). Historically, Bolobo has served as one of the Congo's main river commercial centers, which had seen the exportation of quantities of the Congo's minerals and other natural resources. It is not a surprise that bonobos and other species may have come through that village. Despite confirmation of their presence at the vicinity of Lake Tumba in the early 1970s and 1980s (Fenart and Deblock 1973, Horn 1980, 1976, Deblock 1973), the presence of bonobos (Fig. 13.1) was debated because the evidence provided by Horn (1976) was limited to a few signs (Thompson 1997, Thompson-Handler et al. 1995). Nevertheless, overviews of the nation-wide distribution of the species (Kortlandt 1995, Thompson-Handler et al. 1995, Kano 1984, Fenart and Deblock 1973) included some localities within the Tumba – Maindombe hinterland in the list of locations where bonobos were present, albeit most of them without precise mapped distribution. The imprecision of these data, or rather the lack of it including simple distributional maps, hampered any conservation effort for bonobos in the region for many years.

The presence of bonobos in the Lake Tumba – Lake Maindombe hinterland has now become fully recognized. This started to raise pertinent questions within the conservation community only very recently when Mwanza et al. (2003) located bonobos in the forest of Botuali.

Despite this good news, the scope of their work was geographically limited, and attempts to carry out a wider survey to document the presence of bonobos across

---

[1] World Wide Fund for Nature, WWF-DRC Program, 06 Lodja Av., Gombe, Kinshasa, Democratic Republic of Congo

this region would come only with the arrival of the World Wide Fund for Nature (WWF) under the umbrella of the Central African Regional Program for Environment (CARPE). This program aims to provide a land-use plan for 11 pre-selected landscapes in order to reduce rates of biodiversity loss. To reach this overall goal, WWF initiated a research program to document new wildlife populations, habitats, threats and opportunities. This assessment was the highest priority for conservation before any action can be envisaged. In this chapter we describe the results of that effort, specifically those that pertain to bonobos in the Lake Tumba – Lake Maindombe hinterland, describing their distribution, population size, and conservation challenges and opportunities.

## *Lake Tumba - Lake Maindombe Hinterland: Biophysical Attributes and Conservation Value of the Landscape*

The Lake Tumba – Lake Maindombe hinterland (ca. 78.972 km²; Fig. 13.1) is part of the Lake Télé – Lake Tumba landscape, which constitutes the most extensive inundated forest, and therefore the most extensive freshwater habitat in Africa

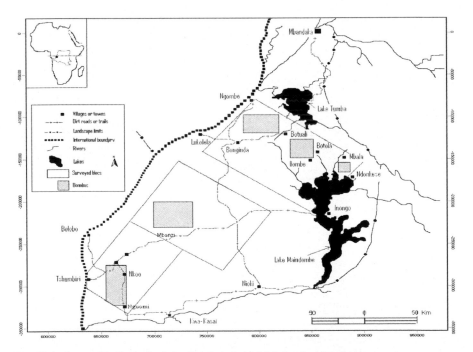

**Fig. 13.1** Survey blocks in the Lake Tumba – Lake Maindombe hinterland.

(Kamdem et al. 2006, Birdlife International 2005), and comprises some of the largest interior lakes of the western central African forests: Lake Tumba (765km$^2$), Lake Etongo, and Lake Maindombe (2400km$^2$) (Aveling et al. 2003). The landscape is divided into two blocks separated by the majestic Congo River, which is fed by major effluents such as the Lulonga River, the Ikelemba River, and the Kasai-Kwa River, which is the southern border of the landscape. The Congo River and adjacent flooding buffer zones offer an assemblage of biotopes that enable the existence of diversified aggregates of species and high biodiversity, such as fish (Brown 2005, Bailey 1986, Banister 1986) and herpetological taxa (Marlier 1958), and high ornithological diversity that makes it one of the most important bird areas in the region (Birdlife International 2005).

The region is within the equatorial zone, where it rains most of the year, punctuated by two minor dry seasons alternating in February and July (Bultot and Griffiths 1972). The mean temperature of the zone was thought to be somewhat stable at 25°C (Vancutsem et al. 2006, Bultot and Griffiths 1972). However, a more recent study carried out on a series of 34-year meteorological data indicated a decrease in temperature, with the average fluctuating around 19°C (Inogwabini et al. 2006). The decrease in temperature has been explained as being caused by permanent clouds consequent to the evaporation of the lake's surface water during prolonged dry seasons that Lake Tumba has gone through in the last 30 years, a situation linked to the overall climate change phenomenon in the last 30 years. In the northern part of the landscape, maximum rainfall oscillates around 1770 mm per year (Birdlife International 2005), while it decreases from 1500 – 1600 mm as one moves south (Inogwabini et al. 2006). The annual average relative humidity is ca. 85% (Bultot and Griffiths 1972). The altitude follows a gradient rising smoothly from the north where it does not exceed 300 – 350m, to reach 650 – 700m above sea level at the southern edges of the area, on the Batéké plateau. The northern part and the shoulders of the lakes are swamps and seasonally inundated forest dominated by communities of *Entandrophragma palustre, Coelocaryon botryoides, Raphia sese, Pandanus candelabrum, Guibourtia demeusei, Oubanguia africana, Uapaca guineensis,* and *Uapaca heudelotii* (Birdlife International 2005, Inogwabini 2005a). The southern block has forest stretching from Lake Tumba down to the edges of Lake Maindombe (the Lake Tumba – Lake Maindombe hinterland). It is predominantly mixed mature forest undulating between swampy and *terra firma* strata. Notable tree species of flooded forests remain identical to the specific composition of swampy forests in the northern block. but with clearly differentiable *terra firma* forests harboring a canopy covered (70 – 75%) by species such as the *Mellitia laurentii, Entandrophragma cylindricum, Entandrophragma angolense, Anonidium mannii, Polyalthia suaveolens, Diospyros sp.* Further south, the landscape becomes a beautiful succession of forest-savannah mosaic, exhibiting an interesting savannah re-colonization dynamic wherein *Uapaca* species pioneer the process. Within *terra firma* forests and gallery forests in the south, stands of Marantaceae, particularly *Haumania liebrechtsiana* and *Megaphrynium macrostachii,* are abundant (Inogwabini 2005a).

# Methods

In 2004, the WWF program hired and trained survey teams to explore the Lake Tumba – Lake Maindombe hinterland. The main goal of the survey was to significantly enhance the long-term conservation of biodiversity in the Lake Tumba area, particularly focusing on bonobos, bongos, elephants, hippopotami, and diurnal monkeys. Toward that end, we had two bonobo specific objectives: (1) provide an estimate of the bonobo population across the southern portion of the Lake Tumba landscape and (2) describe major habitat types quantitatively, with a corollary to determine whether bonobos exhibit preferences for specific habitats. A third conservation objective was to assess the extent of human activities within the Lake Tumba – Lake Maindombe hinterland in order to determine their impact on bonobo distribution.

To begin the survey, teams had to explore the entire area to collect baseline data. This was then used to help design a sampling plan to collect data via conventional distance sampling methods, combining reconnaissance routes and line-transects (Walsh and White 1999, Buckland et al. 1993). Some reconnaissance routes were guided by the information provided by local people. We also used satellite images of forest cover provided by the Observatoire Satellital des Forêts d'Afrique Centrale (OSFAC) and the University of Maryland (UMD) to trace some other routes, choosing relatively intact forest blocks for surveying.

As in other surveys (Reinartz et al. 2006, Hall et al. 1998, 1997), we first gathered the information on bonobo distribution within the region from the literature (Mwanza et al. 2003, Horn 1980, 1976, Fenart and Deblock 1973). We also gathered information from local communities on the bonobo distribution in the vicinities of their villages. Combining information, we then conducted the pilot field work throughout the region to collect the baseline information. Based on this preliminary information, we dissected the Lake Tumba – Lake Maindombe hinterland into three blocs by survey intensity levels: intensive survey zone (a zone with relatively more bonobo indicators), a semi-intensive survey zone (a zone with moderate presence of bonobo indicators), and low intensive zone (a zone with low bonobo indicators). The three survey blocks fall into two major habitat zones: the swampy forest around and near the two lakes, and the plateau at the southern edges of the Lake Tumba – Lake Maindombe hinterland. The first zone is mostly composed of mixed mature inundated forest (Table 13.1). The second zone is a transition zone between forest and savannah habitats (Table 13.1).

We calculated efforts, defined as the total length of transect to be sampled, based on encounter rates during the exploratory missions. We assigned two different coefficients of variation for the intensive and semi-intensive zones (CV $= 10\%$ and $20\%$ respectively) and used the equation $CV(Đ) = (b/\ L\ (n_o/l_o))^{1/2}$ (Buckland et al. 1993) wherein $CV(Đ)$ is the coefficient of variation for the estimate of density Đ, b is of the range $(1.5 - 3)$, L is the predicted sampling effort (i.e. the total length of transects), and $n_o/l_o$ is the encounter rate from the pilot study. As suggested by Buckland et al. (1993) and used by other wildlife survey designs

**Table 13.1** Population estimates of bonobos in the Lake Tumba – Lake Maindombe hinterland

| Name | Area (km²) | Swamps [1] or savan-nah [2] | Suitable habitat (km²) | Density | | | Population estimate | | |
|---|---|---|---|---|---|---|---|---|---|
| | | | | Low | Mean | High | Low | Mean | High |
| Botuali-Ilombe | 955 | 573[1] | 382 | 0.24 | 0.27 | 0.29 | 91.7 | 103.1 | 110.8 |
| Ngombe-Bonginda | 1,829 | 1098[1] | 731 | 0.24 | 0.27 | 0.29 | 175.6 | 197.5 | 212.1 |
| Mbala-Donkese | 160 | 94[1] | 64 | 0.24 | 0.27 | 0.29 | 15.4 | 17.3 | 18.6 |
| Northeast Mbanzi | 1,380,390 | 552156[2] | 828,234 | ** | ** | ** | ** | ** | ** |
| Malebo-Nguomi | 1,993 | 949[2] | 1,044 | 1.8 | 2.2 | 3.4 | 1879.5 | 2297.1 | 3550.1 |

across Central Africa, we used b = 3, which was incorporated in the above equation to yield 76 km of transect for the intensive survey zone. In semi-intensive survey zones, we combined 30 km of transect with reconnaissance trips. Only exploratory trips were used within the low intensity zone.

Following similar studies on bonobos and other ape species (Reinartz et al. 2006, Inogwabini et al. 2000, Van Krunkelsven et al. 2000, Fay 1997, Carroll 1986), essential bonobo data consists of nest sites, dung piles, direct sightings, food remains, and calls. We measured perpendicular distances to transect lines for each nest. We recorded human signs such as machete cuts, snares, open footpaths, and campsites. We took no measurement of perpendicular distance for the reconnaissance data or for human signs on transects.

Also, in agreement with previous ape studies (Reinartz et al. 2006, Inogwabini et al. 2000, Hall et al. 1998), we analyzed only data from intensive survey areas using the DISTANCE program (Buckland et al. 1993), whereas we analyzed data from semi-intensive and low intensity survey zones to provide simple encounter rates. Estimates were based on the steady-state assumption, using the Barnes and Jensen (1987) and McClanahan (1986) formula, A = D/Yη; wherein A = estimated number of a given species; D = indirect signs' density (generated by DISTANCE); Y = daily rate of sign production by each individual; and η = average lifespan of sign. We assumed that each weaned individual built one nest per night, ignoring, therefore, reports from some sites on nest sharing and revisiting sites phenomena (Sabater Pi and Vea 1990) or the existence of day nest sites (Fruth and Hohmann, 1994), albeit with physical descriptions difficult to distinguish from night and day nests. Because nest decay had not been studied in the Lake Tumba landscape, we used η = ca. 99 days, the lifespan of sign from Lomako (Reinartz et al. 2006, Van Krunkelsven et al. 2000).

We classified major habitat types into mature forest and secondary forest. We defined mature forest as old-growth forest with high and continuous canopy cover, sparse woody under-storey vegetation (White and Edwards 2000) generally referring to stands dominated by tall and robust trees (dbh ≥ 30cm and heights ≥ 20m).

We described the substrate of the mature forest in three subclasses: mixed mature swampy (permanently inundated or on hydromorphous soil with tree communities including *Raphia sese, Raphia gilletii,* and *Marantochloa congensis,* etc. (White and Edwards 2000, Evrard 1958) ); mixed mature seasonally inundated with immersed floor during the rainy seasons, but dried up at low water seasons with tree communities including *Uapaca guineensis* and *Guibourtia demeusei,* etc. (White and Edwards 2000); and mixed mature *terra firma* forests with dry floor all year round and tree communities including *Scorodophloeus zenkeri, Anonidium mannii, Entandrophragma cylindricum, Entandrophragma angolense,* and *Polyalthia suaveolens,* etc. (White and Edwards 2000, Evrard 1958).

Secondary forest was classified for human-induced coverage clearing activities such as farms, logging, villages, fire, etc. We noted two age-dependent subclasses as: young secondary (25 years old, including fast growing species such as *Anthocleista liebrechtsiana, Anthocleista vogelii, Musanga cercropoides, Macaranga laurentii,* and *Macaranga spinoza,* etc. (White and Edwards 2000, Evrard 1958) ) and old secondary forest (modified in the distant past, 30 – 100 years, with tree communities including species such as *Alstonia boonei, Canarium schweinfurthii* and *Fagara macrophylla,* etc. (White and Edwards 2000, Evrard 1958) ).

We systematically recorded each type of forest category whenever changes appeared on transects or on reconnaissance trips. We associated each record with qualitative under-storey categories (woody, open, Marantaceae, Zingiberaceae, and liana) and canopy classes (open, closed). We estimated portions of the forest by category as the simple mean calculated for each forest type summing up all forest types and dividing by the grand total of all forests (N) (Greig-Smith 1964, Cottam et al. 1953).

# Results

## *Bonobo Distribution and Population Size*

Survey teams executed 76 km of transects and traveled over 261 km of reconnaissance (excluding the travel distance to reach the sites) within the Lake Tumba – Lake Maindombe hinterland over 8 months in 2005, from April through December. We located four separate communities of bonobos within the landscape at: (1) Botuali – Botola, (2) Bonginda – Ngombe, (3) Dongese – Mbala, and (4) Malebo – Nguomi – Northeast Mbanzi. The Malebo – Nguomi – Northeast Mbanzi group (Fig. 13.1) comprised two subgroups. These two were thought to be subgroups because, despite the fact that they were separated by a distance of 50 km (Fig. 13.1), the two locations remained connected by a swap of gallery forest which could allow contact between individuals from each subgroup. Their mean densities ranged between 0.27 individuals/km$^2$ near Lake Tumba (Botuali-Botola, Bonginda-Gombe, and Mbala-Donkese) and 2.2 individuals/km$^2$ within 90% CI = (1.8 1.4-3.4] individuals/km$^2$ further south

(Malebo-Nguomi and Northeast of Mbanzi) when savannahs are excluded from the surface area measure. Densities differed greatly across the zone, varying between 0.24 individuals/km$^2$ (in the vicinities of Lake Tumba) to 3.4 individuals/km$^2$ (in the Malebo zone, Inogwabini et al. 2007, Table 13.1). Comparing population estimates across the landscape, the populations at the vicinities of Lake Tumba (Bonginda–Ngombe and Botuali-Ilombe) fall within the range of 267 – 323 individuals (Inogwabini et al. 2007, Table 13.1). The mean estimate for the Malebo zone is 2,297 individuals, within 90% CI margins (1,879.5 – 3,550 individuals, Table 13.1). Using fresh nest counts only, we calculated a mean party size of 13.6 individuals/group (SD = ± 7.7 individuals), which is higher than in any other site identified, with the exception of Wamba (Thompson 1997).

At the edges of Lake Tumba, the habitat was composed of 57% permanent swamps and seasonally inundated forests, while 23.7% of *terra firma* mixed mature forest (with open and closed under-storeys), and 2.3% forest-encapsulated savannahs. By contrast, the habitat of the Malebo zone was composed of 40% savannah and 60% of forests divided into inundated and *terra firma* forest (Table 13.1).

## *Humans and the Bonobos of the Lake Tumba -Lake Maindombe Hinterland*

Two main human activities were identified as serious threats to biodiversity between Lake Tumba and Lake Maindombe. They are habitat loss from timber extraction activities and hunting, both for subsistence and the bushmeat trade. Logging is practiced by both local communities and large scale logging companies. Hunting is done via traditional techniques such as dogs, nets, spears, and bow and arrows. Commercial hunting is essentially focused on large species such as elephants and hippopotami. Data from the three survey blocks (intensive, semi-intensive survey zone, and low intensive zones) showed that there was a tendency towards a negative relationship between bonobo signs and human activity signs. In other words, there appears to be a bias that fewer bonobo signs occur in proximity to human activity. However, the sample size was too small to allow regression analysis.

## Discussions

### *Bonobo Distribution and Abundance*

Many years after the only two nation-wide reviews of bonobo distribution were carried out (Kano 1984, Fenart and Deblock 1973), only 10 of the 38 populations identified are confirmed by field work and described by the scientific community. Therefore, information on bonobo distribution in the Lake Tumba – Lake Maindombe hinterland is highly important for the overall effort to conserve bonobos. Furthermore, apart from the Botuali-Botola bonobos (Fig. 13-1) previously identified by Mwanza et al. (2003) and Horn (1980, 1976), the bonobo communities in the zones of Bonginda-Ngombe, Mbala-Ndonkese, Malebo–Nguomi – Northeast Mbanzi (Fig. 13-1) were discovered during our survey. Our data on the three groups Bonginda-Gombe, Botuali-Botola, and Mbala-Donkese provide current confirmation of the bonobos previously known as the Lukolela population (Fenart and Deblock 1973) or the Lake Tumba population (Kano 1984, Horn 1976). The two bonobo communities confirmed at Malebo-Nguomi and Northeast of Mbanzi are a new finding, extending the known geographic range of the species.

Near Lake Tumba, bonobos of Bonginda-Gombe, Botuali-Botola, and Mbala-Donkese live in mixed mature forest, but it is differentiated from other habitats by higher proportions of inundated and/or seasonally flooded areas during the high water seasons. During this time, bonobos live on small *terra firma* forests found undulating throughout large expanses of water. The bonobo populations that we described have not been observed ranging in such wet areas. During the heavy rainy season, we found bonobo signs on a very compacted islet of *terra firma* encapsulated by a vast inundated area, which may influence the feeding and ranging behaviour of those bonobos. Moreover, local hunters reported seeing bonobos digging through mud looking for worms and crossing through shallow water. These findings are consistent with results from Wamba where Hashimoto et al. (1998) reported that bonobo home ranges consist mainly of dry forest, but also include swamp forest. Additionally, Thompson (2002) observed bonobos regularly using perennial pools in the Lukuru to harvest and consume subaquatic vegetation. Unpublished data collected from the northern sector of the Salonga National Park also indicated that bonobos used swamp forest for foraging (Inogwabini, unpublished data).

The Malebo – Nguomi – Northeast Mbanzi community lives in a forest-savannah mosaic habitat, a zone comparable only to the Lukuru area (Thompson 2002, 1997). Bonobo trails connect adjacent forest galleries, indicating that, indeed, at times bonobos venture into savannahs. This population has the highest density and largest group size of bonobos ever recorded. These results may suggest that bonobos prefer mixed gallery forest/savanna habitat to the dense rainforest they were previously believed to prefer. One reason for their success in this type of habitat could be that the asynchronous fruiting seasons between savannah and forest trees lead to the availability of fruits year round, instead of just seasonal fruits in the dense forest. Also, bonobos often walk upright (D'Août et al. 2004). One theory as to why they prefer this habitat may be that it gives them greater opportunity to walk upright than in dense tropical forest where they must travel from tree to tree.

The overall occupation pattern found in our study replicates the common bonobo distribution pattern; that they occur in patches (Inogwabini and Omari 2005, Kortlandt 1995, Kano 1984). These distributional patterns may be the result of a combination of ecological, historic and evolutionary factors such as: major flooding within the species range, epidemics (e.g. sleeping sickness), high hunting pressure, forest exploitation (Kortlandt 1995), variability in food availability (Malenky and Wrangham 1994, Malenky and Stiles 1991, Badrian and Malenky 1984, Kano and

Mulavwa 1984), topography, and forest and land use history (Inogwabini and Omari 2005).

The distribution of the bonobos in the Lake Tumba – Lake Maindombe hinterland is characterized by rivers and long distances between bonobo communities. Major rivers serve as natural barriers, and long physical distances between bonobo communities and large areas of open grassy savannahs may have impeded their interactions. The vegetation history of the region is not known yet, but colonizing species such as *Uapaca guineensis* occur in the middle of large savannah areas, indicating that gallery forests are naturally expanding and colonizing the savannahs. Evidence that the savannah-forest mosaic has been maintained by fires, used as a management tool since 1952 when the cattle raising industry came into the region, suggests that bonobo communities may have been kept separated by fires for at least the last 50 years. Logging activities have also fragmented the forest habitats in the region. Fragmentation by logging activities has been documented to have severe consequences on chimpanzees (White and Tutin 2000, White 1994) and may, therefore, accentuate effects induced by both rivers and long physical distances.

The combined impact of rivers and long physical distances between communities might indicate that bonobos in our discreet survey blocks may be distinct conservation units that exhibit a specific ecological or genetic character (Eriksson et al. 2004). Using blood samples from captive bonobos, Reinartz et al. (2000) and Reinartz (1997) found that bonobo geographical origins could be differentiated genetically. Therefore, the overall geographic range of the bonobos in the Lake Tumba – Lake Maindombe hinterland, long distances between groups and/or subgroups, and potential ecological and genetic variations warrant a species-targeted conservation strategy.

Low bonobo densities at the northern edges of the Lake Tumba – Lake Maindombe hinterland confirm earlier findings in the region (Horn 1980, 1976) and are somewhat similar to their density in some other research sites, e.g. the Lofeke-Lilungu-Ikomaloki region (Sabater Pi and Vea 1990). However, higher densities of bonobos found in this study were scientifically remarkable, because the presence of bonobos in this area had been, at best, debated or simply ignored.

The presence of bonobos in higher densities and larger group sizes in this study is puzzling and warrants an explanation. First, with > 10 tribes (including pygmies), higher human density (range: $6 - 24$ ind/km$^2$), and a rapidly increasing human population (3.8% per year, mean number of family members = 10 members/household; Colom et al. 2006), the region is occupied by people with diverse cultures. Accordingly, the distribution of the bonobos is not solely dependant upon the ecological factors described above. This becomes even more puzzling because the distribution and the abundance of bonobos in the southern portion of the landscape parallels human distribution, i.e. does not follow a classic straight line regression illustrating the relationship between wildlife occurrences versus human presence (Reinartz et al. 2006, Barnes et al. 1991).

Higher bonobo densities and larger group sizes in the southern portion of the Lake Tumba – Lake Maindombe hinterland may also be linked to cultural taboos of the Bateke, who view bonobos as their ancestors. As such, there has been an observed taboo for people living under the traditional authority of the great chief of the Bateke against hunting bonobos. That bonobos occurred in higher densities near villages also concords with the report from the Iyaelima area (Grossmann et al. 2008, Hart et al. 2008, Thompson et al. 2008).

Furthermore, there is a strong commitment by the cattle raising company named Société d'Organisation, de Participation et Management (ORGAMAN) to prevent armed and large-scale hunting in its estimated ca. 150,000 ha concession, thereby protecting the wildlife species living there, including bonobos. ORGAMAN has operated in the region since 1952. Cultural factors and indirect conservation activities by ORGAMAN have acted in tandem with ecological factors such as food availability, preferred habitats, etc. to ensure the presence of a resident population of bonobos.


## Threats to Bonobo Conservation in the Lake Tumba-Lake Maindombe Hinterland

Large group size and high density of bonobos in the southern portion of the Lake Tumba – Lake Maindombe hinterland should not detract from the fact that they, and many other wildlife species, are threatened by logging activities and hunting for subsistence and the bushmeat trade (Inogwabini 2005b).


### Logging Activities in the Area

The most important threat that the bonobos of the Lake Tumba – Lake Maindombe hinterland face is habitat loss induced by logging (Inogwabini 2005b). Most of the suitable habitat of the bonobos in the southern block of the area is almost entirely within logging concessions. Fourteen of the fifteen logging concessions of the landscape are in the southern section; even the 19 km² small scientific reserve of Mabli, created in 1948 under IUCN Category Ib, is now within a concession with an established permit that allows logging activities to commence any time the owner wills. Mabali was logged until 1975, but another cycle is expected soon. Our research teams encountered timber inventory teams that were working in the scientific reserve to enumerate extractable trees.

Logging activities will result in devastating effects on wildlife species because the outcome will be loss of habitat. Roads built to move timber from forests to the lake and rivers pose significant threats to bonobos and other wildlife species. Furthermore, camps to house logging workers and the sawing activities threaten wildlife species because settlement camps facilitate human demographic growth and movement in relatively remote areas, and lead to an increase in commercial hunting. Timber exploitation in the Lake Tumba – Lake Maindombe hinterland is focused on *Mellitia laurentii* (Wenge, a hard black wood species). Because it is logistically difficult to extract timber from other locations in DRC, many logging companies have recently rushed into the Lake Tumba – Lake Maindombe hinterland, which is relatively easy to access through an intricate river network. Most of these concessions were acquired after the 2002 moratorium was imposed on allocating new concessions. With the exception of two logging companies who are at the very early stages of the timber certification process, there are no management plans for the concessions.

**Bushmeat Trade**

The second most important threat facing bonobos of the Lake Tumba – Lake Maindombe hinterland is hunting; one of the human activities that represents the greatest threat to the maintenance of the biomes and wildlife species that reside therein (Bennett and Robinson 2000, Bowen-Jones and Pendry 1999). Human impact is essentially of two kinds: direct off-take of wildlife species for food or for cultural use. Off-take is for local subsistence or for commerce feeding into large markets in major towns such as Kinshasa, Mbandaka, Bikoro, and Inongo. In colonial times and shortly thereafter, there was trade in agricultural products (coffee, palm oil, rubber), which supported local people. However, after nationalization of the economy in 1972, the market system broke down and seriously affected the transport infrastructure, leading to a severe decline in agricultural activities across the entire region. After several years of economic and political chaos, hunting and fishing have become the only way for most people to gain monetary income.

Despite difficulties in transportation, the bushmeat trade has become an extraordinarily organized activity, promoted by social hierarchies based in major towns. The intensive demand for bushmeat to supply Kinshasa, Mbandaka, Inongo and other markets, spread over the region and depleted wildlife populations from many parts of the landscape. Data collected from local communities (Colom et al. 2006) using a combination of interviews and focus groups, indicated a mean annual monetary income of about $300/working adult generated by the bushmeat trade at the village level. This is compared to $60/working adult as the mean annual revenue for an administrative paid job in the area between Maindombe and the town of Mbandaka. In major towns such as Bikoro and Mbandaka, bushmeat is highly priced and generated incomes are consequently increased ten-fold (Colom et al. 2006), stimulating the bushmeat trade in the region.

Colom et al. (2006) also found an important trade in live animals involving diurnal animals, included bonobos, golden-bellied mangabeys, Angolan-pied colobus and birds such grey parrots, diverse species of eagles, and kingfishers. For example, all bonobos confiscated either in Mbandaka (4 individuals in 2005; WWF unpublished data), Bikoro (1 individual; WWF unpublished data), Kinshasa and Paris (1 individual, Jane Goodall Institute 2005) in the recent past had been captured for the pet trade. The pet trade has had an impact that is even more intractable than the bushmeat trade because this involves different participants, including the international market. As in other locations in the country, hunting pressure on wildlife has quadrupled with the persistent insecurity and war in the DRC (Aveling et al. 2003), because access to ammunition, automatic weapons, and other war paraphernalia has become prolific.

## Increasing Human Population

The large number of tribes across the DRC comprises a variety of cultures. Regardless of how different tribes value the bonobos, the alarming human population growth (3.8% per year; INS 1984) is likely to spark human-wildlife conflict. Human population growth may reach limits beyond what can be supported by natural resources alone. Indeed the decline in fish stocks and other wildlife species may parallel the increase of humans in the region (Inogwabini and Zanga 2006). This will affect bonobo conservation as the human demand for land and other natural resources will certainly affect land that otherwise would be reserved for bonobos. Therefore, sound natural resources management will have to team up with family planning programs in the region in order to address this vexing issue.

## Fire as Management Tool in Cattle Raising Concessions

Within the habitat types of the region, the savannah patches are an interesting ecotone displaying higher species diversity, including bonobos. The savannas have been exploited for cattle ranching since the late 1950s, with fire as a management tool. In order to have young shoots of savannah grasses and herbs as abundant as possible to feed the cow herds, the management of ORGAMAN has been burning three times a year to ensure that palatable herbs will always be available. This activity has maintained the current savannah-forest mosaic system.

Fire has been used in similar habitats in the region to maintain intact landscape mosaics. Satellite data from the Department of Geography, University of Maryland (UMD) and Observatoire Satellitale des Forêts d'Afrique Centrale (OSFAC) indicate that fire is most intense in the region from July through September, which corresponds to the long dry season. This has increased over the last five years due to the increase in numbers of cattle ranched. Preliminary results from an on-going fruit phenology study at the Malebo site (Territory of Bolobo) indicate that fruits are scarce during the long dry season, which corresponds with evidence of bonobos venturing into the savannahs to cross between forests (unpublished report to WWF 2006).

Fire may impede the bonobos foraging, as the situation infers. Thus, there is a need to develop a sound fire management system along with a gallery forest conservation program that takes into account different elements, including water quality and watershed management, human population needs in agricultural lands, logging activities, and cattle-raising necessities.

## *Opportunities for Conserving Bonobos in the Lake Tumba – Lake Maindombe Hinterland*

### Legal Framework

The changes in national conservation legislation are of prime importance in the protection of species in the region. Firstly, the new forest code has opened doors for privately owned and/or managed protected areas, with the local community land units becoming a legal category. Within this framework, there are several ways in which the conservation of the bonobos in the Lake Tumba – Lake Maindombe hinterland can be envisaged, starting from a conventional reserve of IUCN category VI down to community owned land units wherein bonobos are given full protection status.

### Cultural Considerations

The key for bonobo conservation in the Lake Tumba – Lake Maindombe hinterland is based on the model provided by the culture of the Bateke of the Bolobo, Bandundu. The region is one of a very few in the DRC where a traditional chief holds strong traditional authority, directly linked to the historical times of King Makoko, who signed the 1880 agreement putting his land under French protection. This action settled the land dispute between De Brazza and Henry Morton Stanley, representing the King Leopold II, who gained significant mineral-rich lands in the Democratic Republic of Congo (Pekenham 1992). Moreover, traditional authority is combined with politico-administrative constitutional power in the hands of a single individual, making the chief of Bateke the strongest authority in the region. People in the Chiefdom of the Bateke Plateau have mythical ties with their traditional hierarchy and comply with the chief's guidance, advice and orders.

Gallery forests in the Malebo region (26,520 km²) have been heavily logged in the last 25 years. Conflicts between industrial and/or anarchical artisanal timber exploitation have created a sense of disaster amongst the people for fear of what might occur if the forests were gone. In response, WWF started a community conservation program to protect the bonobos in the remaining gallery forests. Activities include the conservation of wildlife in the region, using bonobos as the umbrella species while sensitizing local communities to the importance of preserving their natural resources. This involves finding alternative sources of both protein and income through activities such as fish culture, improved agriculture techniques, revalorization of non-timber forest products as prime sources of cash revenue, and ecotourism for the future. Under the traditional authority, the community conservation program is a scenario wherein conservation activities are planned by local communities. WWF assists only with technical input, and the entire program is executed by the local communities. The project is building on the will of the local people, under traditional authority, to protect their wildlife and other natural resources from over-harvesting. It is also building on their commitment to improve their livelihood via sustainable techniques, if they become available. Since the arrival of the community conservation project in this region, the local chief publicly vowed to support sustainable resource management and to protect charismatic species such as bonobos and forest elephants.

## Tourism

The Malebo zone is located in the southernmost portion of the landscape at about 300 km north of Kinshasa, or a 50-minute flight on small aircraft. With the forest-savannah mosaic complex, the presence of bonobos in forests adjacent to villages, and several other wildlife species such a forest elephant and forest buffalo present, Malebo promises to be a valuable area for tourism as a site where people could see bonobos in the their natural habitat. There is a high likelihood that a tourism program in this region might attract the private sector to support both tourism and conservation activities.

**Acknowledgements** This publication was made possible through financial support provided by the US Agency for International Development through the Central African Regional Environmental Program - Congo Basin Forest Partnership. The authors' views expressed in this publication do not necessarily reflect the views of the United States Agency for International Development or the United States Government. We would like to thank the DRC government for permission to work in the Lake Tumba landscape. We thank the WWF National and Central African Regional coordination for unfailing support to the Lake Tumba project. The research was designed at the University of Kent under the supervision of Professor Nigel Leader-Williams. We thank the WWF local partners, particularly Mbou-Mon-Tour and its Executive President Mr. Bokika Jean-Christophe for providing support to our work and partnering with us for the community conservation work in the southern Lake Tumba landscape. The map on figure 13.1 is a partial reproduction of the paper published by Oryx. Constructive comments from two anonymous reviewers helped improve the chapter.

# References

Aveling C, Bofaya B, Hall JS, Hart JA, Hart TB, Inogwabini BI, Plumptre A, Wilkie D (2003) Democratic Republic of Congo – Environmental Analysis. Final Report – Prepared by the Wildlife Conservation Society. Submitted to USAID Washington DC (USA) and Kinshasa (DRC)

Badrian NL, Malenky RK (1984) Feeding ecology of *Pan paniscus* in the Lomako forest, Zaire. In: Susman RL (ed) The pygmy chimpanzee: evolutionary biology and behavior. Plenum Press, pp 325–346

Bailey RG (1986) The Zaïre River system. In: Davies and Walker (eds) The ecology of river systems. Junk Publishers, Dordrecht, Boston and Lancaster, pp 201–214

Banister KE (1986) Fish of the Zaïre system. In: Davies and Walker (eds) The ecology of river systems. Junk Publishers, Dordrecht, Boston and Lancaster, pp 215–224

Barnes RFW, Jensen KL (1987) How to count elephants in forests. African Elephant and Rhino Specialist group Technical Bulletin: 1– 6

Barnes RFW, Barnes KL, Alers MPT, Blom A (1991) Man determines the distribution of elephants in the rain forests of northern Gabon. African Journal of Ecology 29: 54–63

Bennett EL, Robinson JG (2000) Hunting of wildlife in tropical forests implications for biodiversity and forest peoples. The World Bank Environment Department, Biodiversity Series — Impact Studies, Paper No 76

Birdlife International (2005) Birdlife's online World Bird Database: the site for bird conservation Version 2 Cambridge, UK: Birdlife International. Available: *http://www.birdlife.org* (accessed 29/9/2006)

Bowen-Jones E, Pendry S (1999) The threat to primates and other mammals from bushmeat trade in Africa and how this could be diminished. Oryx 33: 233–246

Brown A (2005) Ecoregion 13: Lac Tumba. In: Thieme ML, Abell R, Stuassny MLJ, Skelton P, Lehner B, Teugels GG, Dinerestein E, Toham AK, Burgess N and Olson D. (eds) Freshwater ecoregions of Africa and Madagascar – A conservation assessment. Island Press, Washington USA, Covelo Italy and London UK

Buckland ST, Anderson DR, Burnham KP, Laake JL (1993) Distance sampling – Estimating abundance of biological populations. Chapman and Hall Publishing

Bultot F, Griffiths JP (1972) The equatorial wet zone. In: Griffiths (ed) Climates of Africa. Elsevier Publishing Company

Carroll R (1986) Status of the lowland gorilla and other wildlife in the Dzanga-Sangha region of Southwestern Central African Republic. Primate Conservation 7: 38–41

Colom A, Bakanza A, Mundeka J, Hamzal T, Ntumbandzondo B (2006) The socio-economic dimensions of the management of biological resources, in the Lac Télé – Lac Tumba Landscape, DRC Segment. A segment-wide baseline Socio-Economic study's Report submitted to WWF Lac Tumba (Kinshasa) and USAID Kinshasa

Cottam G, Curits JT, Hale BW (1953) Some sampling characteristics of a population of randomly disperse individuals. Ecology 34: 741 – 757

D'Août K, Vereeckc E, Schoonaert K, De Clercq D, Van Elsacker L, Aerts P (2004) Locomotion in bonobos (*Pan paniscus*): differences and similarities between bipedal and quadrupedal terrestrial walking, and a comparison with other locomotor modes. Journal of Anatomy 204: 353–361

Eriksson J, Hohmann G, Boesch C, Vigilant L (2004) Rivers influence the population genetic structure of bonobos (*Pan paniscus*). Molecular Ecology 13: 3425

Evrard C (1958) Recherches écologiques sur le peuplement forestier de sol hydromorphe de la Cuvette Centrale congolaise. ONRD/INEAC – Ministère Belge de l'éducation et de la culture, Bruxelles, Belgique

Fay JM (1997) The ecology, social organization, population, habitat and history of the western lowland gorilla (*Gorilla gorilla gorilla*) Savage and Wyman 1947 Ph.D. thesis, Washington University

Fenart R, Deblock R (1973) *Pan paniscus* et *Pan troglodytes*: craniométrie – étude comparative et ontogénique selon les méthodes classiques et vestibulaires. Tome 1, Musée Royal de l'Afrique Centrale, Tervuren, Belgique

Fruth B, Hohmann G (1994) Ecological and behavioural aspects of nest building in wild bonobo (*Pan paniscus*). Ethology 94: 113–126

Greig-Smith P (1964) Quantitative plant ecology. Second Edition. Butterworths, London – UK

Grossmann F, Hart JA, Vosper A, Ilambu O (2008) Range Occupation and Population Estimates of Bonobo in the Salonga National Park: Application to Large Scale Surveys of Bonobos in the Democratic Republic of Congo. In: Furuichi T, Thompson J (eds) The bonobos: behavior, ecology, and conservation. Springer, New York, pp 189–216

Hall JS, Inogwabini BI, Williamson EA, Omari I, Sikubwabo C, White LJT (1997) A survey of elephants (*Loxodonta africana*) in the Kahuzi-Biega National Park lowland sector and adjacent forest in eastern Zaire. African Journal of Ecology 35: 213–223

Hall JS, White LJT, Inogwabini BI, Omari I, Morland HS, Williamson EA, Walsh P, Sikubwabo C, Saltonstall K, Dumbo B, Kiswele K, Vedder, A, Freeman K (1998) A survey of Grauer's gorilla (*Gorilla gorilla graueri*) and chimpanzee (*Pan troglodytes schweinfurthi*) in Kahuzi-Biega National park and adjacent forests in eastern DR. Congo. International Journal of Primatology 19: 202–235

Hart JA, Grossmann F, Vosper A, Ilanga J (2008) Human hunting and its impact on bonobos in the Salonga National Park, D.R. Congo. In: Furuichi T, Thompson J (eds) The bonobos: behavior, ecology, and conservation. Springer, New York, pp 245–271

Hashimoto C, Tashiro Y, Kimura D, Enomoto T, Ingmanson EJ, Idani G, Furuichi T (1998) Habitat Use and Ranging of Wild Bonobos (*Pan paniscus*) at Wamba. International Journal of Primatology, Volume 19: 1045–1060

Horn AD (1976) A preliminary report on the ecology and behavior of the bonobo chimpanzee (*Pan paniscus* Schwarz 1929) and a reconsideration of the evolution of the chimpanzee. Ph.D. Thesis, Yale University

Horn AD (1980) Some observations on the ecology of the bonobo chimpanzee (*Pan paniscus* Schwarz 1929) near Lake Tumba, Zaire. Folia Primatologica 34: 145–169

Inogwabini BI (2005a) Preliminary conservation status of large mammals in the Lac Tumba-Lac Maindombe hinterland, with emphasis on identification of biologically important zones. Type-scripted report submitted to the United States Agency for International Development (USAID)/Central Africa Regional Program for Environment (CARPE), Kinshasa, Democratic Republic of Congo

Inogwabini BI (2005b) Mammals of the Lac Tumba: species and documented threat index Type-scripted report submitted to the United States Agency for International Development (USAID)/Central Africa Regional Program for Environment (CARPE), Kinshasa, Democratic Republic of Congo

Inogwabini BI, Omari I (2005) A Landscape-wide distribution of *Pan-paniscus* in the Salonga National Park, Democratic Republic of Congo. Endangered Species Update 22: 116–123

Inogwabini BI, Zanga L (2006) Les inventaires des poissons dans le Lac Tumba, la jonction Lulonga-Congo et la Ngiri au Paysage Lac Télé – Lac Tumba, Segment République Démocratique du Congo. Type-scripted report submitted to the Central African Regional Program for Environment (CARPE) of the United States Agency for International Development (USAID) and World Wide Funds for Nature (WWF)

Inogwabini BI, Hall JS, Vedder A, Curran B, Yamagiwa J, Basabose K (2000) Conservation status of large mammals in the mountain sector of Kahuzi-Biega National Park, Democratic Republic of Congo in 1996. African Journal of Ecology 38: 269–276

Inogwabini BI, Sandokan MB, Ndunda M (2006) A dramatic decline in rainfall regime in the Congo Basin: evidence from a thirty-four year data set from the Mabali Scientific Research Centre, Democratic Republic of Congo. The International Journal of Meteorology 31: 27–285

Inogwabini BI, Matungila B, Mbende L, Abokome M, Tshimanga T (2007) The great apes in the Lac Tumba landscape, Democratic Republic of Congo: newly described populations. Oryx 44(4): 1–7

Institut National des Statistiques (INS)(1984) Recensement scientifique de la population 1984: Zaïre et régions 1984 – 2000. Ministère du Plan et Amenagement du Territoire, Kinshasa

Jane Goodall Institute (2005) Trafic de bonobo. In *http://www.janegoodall.fr/htfr/ newsletter. htm#trafic* (Accessed 30 April 2007)

Kamdem AT, D'Amico J, Olson D, Blom A, Townbridge L, Burgess N, Thieme M, Abell R, Carroll RW, Gartlan S, Langrand, O, Mussavu MR, O'Hara D, Strand H (2006) (eds) A vision for the biodiversity conservation in Central Africa: biological priorities for conservation in the Guinean-Congolian forest and freshwater region. World Wide Funds for Nature, Washington DC, United States

Kano T (1984) Distribution of pygmy chimpanzees (*Pan paniscus*) in the central Zaire basin. Folia Primatologica 43: 36–52

Kano T, Mulavwa M (1984) Feeding ecology of the pygmy chimpanzees (*Pan paniscus*) of Wamba. In: Susman RL (ed) The pygmy chimpanzee: evolutionary biology and behaviour. Plenum Press, pp233–274

Kortlandt A (1995) A survey of the geographical range, habitats and conservation of the Pygmy chimpanzee (*Pan paniscus*): ecological perspective. Primate Conservation 16: 21–36

Krunkelsven E van, Inogwabini BI, Draulans D (2000) A survey of bonobos and other large mammals in the Salonga National Park, Democratic Republic of Congo. Oryx 34: 180–187

Malenky RK, Stiles EW (1991) Distribution of terrestrial herbaceous vegetation and its consumption by *Pan paniscus* in the Lomako forest, Zaire. American Journal of Primatology 23: 153–169

Malenky RK, Wrangham RW (1994) The relative importance of terrestrial herbs for bonobos and chimpanzees: comparative data from Lomako and Kibale. Bulletin of Chicago Academy of Science 15: 7

Marlier G (1958) Recherches hydrobiologiques au lac Tumba. Hydrobiologia 10: 352–385

McClanahan TR (1986) Quick population survey method using faecal droppings and a steady state assumption. African Journal of Ecology 24: 61–68

Mwanza N, Mulavwa M, Mola I, Yangozene K (2003) Confirmation of bonobo population around Lac Tumba. Pan Africa News 10(2) Web: *http://mahale.web.infoseek.co.jp/PAN/10_2/10(2)_07.html* (Accessed on November 29, 2005)

Pekenham A (1992) Scramble for Africa – White man's conquest of Dark Continent from 1876 to 1912. Perennial/ Time Warner Books

Reinartz GE (1997) Patterns of genetic biodiversity in the bonobo (*Pan paniscus*). Ph.D. Thesis, University of Wisconsin-Milwaukee, United States

Reinartz GE, Karron,JD, Phillips RB, Weber JL (2000) Patterns of microsatellite polymorphism in the range-restricted bonobo (*Pan paniscus*): considerations for interspecific comparison with chimpanzees (*P. troglodytes*). Molecular Ecology 9:315–328

Reinartz GE, Inogwabini BI, Ngamankosi M, Lisalama W (2006) Effects of forest type and human presence on bonobo (*Pan paniscus*) density in the Salonga National Park. International Journal of Primatology 27: 603–634

Sabater Pi J, Vea JJ (1990) Nest building and population estimates of the bonobo from the Lofeke-Lilungu-Ikomaloki region of Zaire. Primate Conservation 11: 43–48

Thompson JAM (1997) The history, taxonomy and ecology of the bonobo (*Pan paniscus* Schwarz 1929) with a first description of a wild population living in a forest/savanna mosaic habitat. Ph.D. Thesis, Oxford University, Oxford, UK

Thompson JAM( 2002) Bonobos of the lukuru wildlife research project. In Behavioural Diversity in Chimpanzees and Bonobos. Boesch C, Hohnmann G, and Marchant L (eds) Cambridge University Press. Pp 61–70

Thompson JM, Lubuta MN, Kabanda RB (2008) Traditional land-use practices for bonobo conservation. In: Furuichi T, Thompson J (eds) The bonobos: behavior, ecology, and conservation. Springer, New York, pp 227–244

Thompson-Handler N, Malenky R, Reinartz GE (1995) Action plan for the *Pan paniscus* - Report on free ranging populations and proposals for their preservation. Zoological Society of Milwaukee, Wisconsin, United States

Vancutsem C, Pekel JF, Kibambe LJP, Blaes X, de Wasseige C, Defourny P (2006) République démocratique du Congo – occupation du sol. Carte Géographique. Presses Universitaire de Louvain, Bruxelles, Belgique

de Waal FBM, Lanting F (1997) Bonobo – the forgotten ape. University of California Press, Berkeley, United States

Walsh PD, White LJT (1999) What it will take to monitor forest elephant populations. African Journal of Ecology 13: 1194–1202

White LJT (1994) The effects of commercial mechanised selective logging on a transect in lowland rainforest in the Lope Reserve, Gabon. Journal of Tropical Ecology, Volume 10: 313–322

White LJT, Edwards A (2000) Methods for assessing the status of animal populations. In: White L and Edwards A (eds) Conservation research in the african rain forests. New York: The Wildlife conservation Society pp 225–227

White LJT, Tutin CEG (2000) Why chimpanzees and gorillas respond differently to logging: a cautionary tale from Gabon. pp 449–462 in W. Weber, L.J.T. White, A. Vedder, and L. Naughton (eds) African rain forest ecology and conservation. Yale University Press, New Haven

# Changes in the Status of Bonobos, their Habitat, and the Situation of Humans at Wamba in the Luo Scientific Reserve, Democratic Republic of Congo

Gen'ichi Idani[1], Ndunda Mwanza[2], Hiroshi Ihobe[3], Chie Hashimoto[4],
Yasuko Tashiro[1], and Takeshi Furuichi[4]

## Introduction

Bonobos range over the area between the Congo-Lualaba River and the Kasai-Sankuru River in the Democratic Republic of the Congo (DRC; Kano 1984). Kano et al. (1996) stated that bonobos are more or less continuously distributed through out the northern part of their range, which is covered by tropical rain forest, and are distributed fragmentarily in the southern part, where the vegetation is a mosaic of forest and grassland. Kano (1992) estimated that the population of bonobos in the whole of the DRC to be ca. 50,000 in 1973. The number has decreased after several wars; however, the current population size is unknown.

Since 1973, when ecological, ethological and sociological studies of wild bonobos in Wamba began (Kano et al. 1996, Kano 1984, 1992), Japanese researchers conducted research seasonally through 1991, and developed much knowledge of the area (Furuichi et al. 1998). However, the studies were interrupted by repeated political instability. We had to discontinue our research due to rioting across the DRC in 1991. The research resumed in 1994, but we were forced to leave again due to the civil wars in 1996 and 1998–2002. Although the situation has remained unpredictable since the second war, we resumed research intermittently just after the end of the war in 2002.

We report on changes at Wamba, including the status of bonobos, their habitat, and the situation of humans, by comparing the condition after the civil wars with the early stage of the research. We also detail our efforts to assist the local community for promotion of the conservation of bonobos.

[1] *Great Ape Research Institute, Hayashibara Biochemical Laboratories, Inc., 952-2 Nu, Tamano, Okayama, 706-0316 Japan*

[2] *Research Center for Ecology and Forestry, Ministry of Scientific Research and Technology, D.R. Congo*

[3] *School of Human Sciences, Sugiyama Jogakuen University, Japan*

[4] *Primate Research Institute, Kyoto University, Japan*

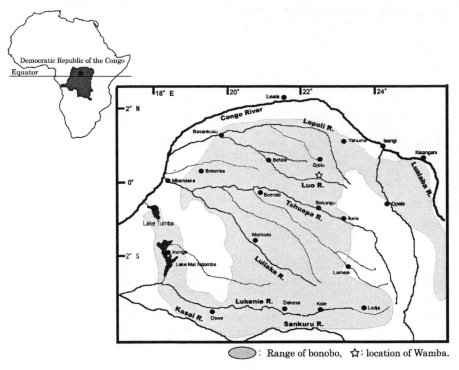

: Range of bonobo, ☆: location of Wamba.

**Fig. 14.1** Distribution of bonobos and location of Wamba.

## The History of Wamba

Wamba is located at 0°11′ 08″ N, 22°37′ 58″ E (Fig. 14.1), 80 km south of Djolu, the center of the region. To the south lie a swamp forest and the Luo River. To reach Wamba, we previously took a regular flight and then drove overland in our car. Currently, there is no regular flight, so we travel directly from Kinshasa to Djolu by charter flight and rent a truck. The roads of the Djolu area are in very bad condition due to the wars.

The village of Wamba comprises five hamlets along a north-south road (Fig. 14.2). Our base camp is in the hamlet of Yayenge, at the southern end of Wamba. The Bongando people, who live in the Wamba area, are Bantu farmers, though fishing, hunting, and gathering are also important daily activities (Kimura 1998). They are slash-and-burn agriculturists of manioc, cultivating within 1-2 km on either side of the main road. Wamba has about 1,000 residents (Kano et al. 1996).

As a study site, Wamba Forest has several idiosyncratic characteristics. The first is the high population density of bonobos. From 1974 to 1991, ca. 250 bonobos in six unit-groups (E1, E2, P, B, K and S) had at least parts of their home ranges within the Wamba Forest (the north sector of the Luo Scientific Reserve; Kano et al. 1996).

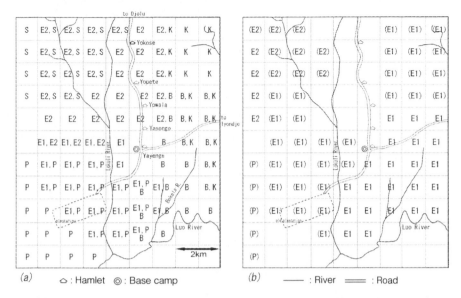

**Fig. 14.2** Change in home ranges on each group of the bonobo in the Wamba Forest. (a) Home ranges of six groups in 1995, (b) home ranges of three groups in 2005. Home range data in parentheses are based on information from co-researchers of Congolese and local research assistants.

Researchers observed them by following groups in the forest until 1976. The bonobos of the E group were habituated to feeding on sugar cane in 1977, and the P group was also provisioned beginning in 1979. E1 and E2 formed by the fission of E in 1984 (Idani 1990a, Furuichi 1987). E1, E2, and P were habituated by researchers who observed them directly, and most individuals could be identified. B and K had not been habituated, but a few members could be identified and were observed directly. In 1988, E1, E2, and P comprised 33, 54, and 39 individuals, respectively, whereas B and K were each estimated to have 80–100 members (Idani 1990a, Kano and Mulavwa 1984). On the basis of fragmentary data, S appeared to have >100 members.

The second characteristic of the study site is the co-existence of bonobos and humans. The local people have traditionally allowed the bonobos to co-exist with them because they had a strict taboo against eating them, believing that bonobos were their ancestors. Furthermore, the people of Wamba rely on the primary forest for subsidiary food sources such as wild animals and plants, materials for houses and various tools, and folk medicines. People also eat some of the major food plants of bonobos. Thus, both the bonobos and the primary forest are precious existences for the people of Wamba.

The third characteristic of this site is the extensive contact that we as researchers have maintained with the Wamba villagers. We have employed numerous temporary workers in addition to regular workers, and purchased local foods and materials for our huts, such as palm leaves for the roofing and sun-dried bricks. Moreover, we have been selling daily necessities at fair market value among the villagers, including clothes for women, salt, soap, machetes, and other goods, which are very difficult to obtain in the Wamba region. We hoped that villagers would benefit from our research activities. In other words, we hoped that they would consider bonobos to be useful to them.

## The Luo Scientific Reserve

Since bonobo research began in Wamba in 1974, we have performed continuing field studies (Furuichi et al. 1998). We suspect that no case of bonobo poaching occurred before 1983. During our absences between 1984 and 1987, hunters from outside of Wamba killed bonobos. Moved by these cases, we submitted a proposal to the Centre de Recherche en Sciences Naturelles (Research Center for Natural Science, CRSN), our counterpart in the country, suggesting that a reserve for bonobos should be established in the area of the Wamba Forest. We proposed the establishment of a reserve because all villagers would have had to leave the area if it was designated a National Park. The Luo Scientific Reserve was established in March 1990 (Fig. 14.3; Kano et al. 1996, Idani 1990b).

The Luo Scientific Reserve covers $481\,km^2$, with the northern sector encompassing the Wamba Forest ($147\,km^2$) and the southern sector containing the Ilongo Forest ($334\,km^2$). The Luo River, which is about 100m wide and >5m deep, separates the Wamba and Ilongo Forests. The river is too wide for the two bonobo populations to exchange members under normal climatic conditions. In Ilongo Forest, bonobos are as numerous as in the Wamba Forest, but human density is much lower (Hashimoto and Furuichi 2001). We have two research assistants working in the Ilongo Forest, and they record the ranging area, food remains, and other information about bonobos there. However, the bonobos are not habituated.

The vegetation in the Luo Reserve can be roughly divided into three types: secondary bush and forest, dry primary forest, and swamp forest (Kano and Mulavwa 1984). Although bonobos utilize all types of forest, the dry and swamp primary forests are most important for them because they provide major fruit food resources (Idani et al. 1994, Kano and Mulavwa 1984).

The goal of the Luo Reserve is to maintain the co-existence of bonobos and human inhabitants. Thus, the regulation of the reserve was designed to permit the traditional lifestyles of the villagers in Wamba. The following activities are prohibited to protect bonobos in the reserve: hunting of primates, using guns, wire snares or poison arrows, and clearing primary forest. Other activities such as collecting plants, traditional hunting, and cultivation of secondary forest are allowed in the reserve. The establishment of the reserve served to educate the local people and government officers about the illegal killing of bonobos. Nevertheless, the villagers disliked the establishment of the reserve because they disapprove any restriction of their activities. They also complained that the compensation money to regulation of

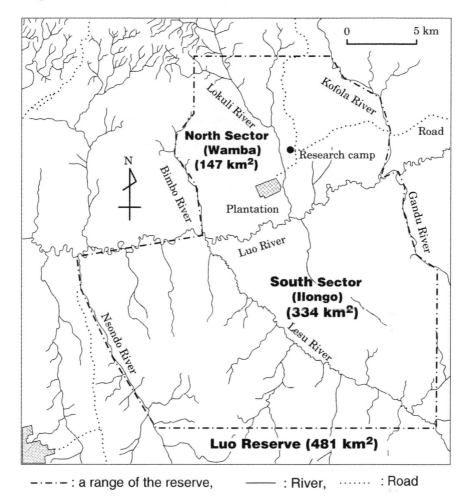

**Fig. 14.3** The Luo Scientific Reserve: the Wamba forest (147km²) on the north and the Ilongo forest (334km²) on the south.

the forest utilization paid to them was insufficient. Now the local people have begun to ignore the regulations of the reserve. Two official guards patrol the Luo Reserve; however, the presence of only two individuals with no compelling force does not provide effective protection over the wide range of the reserve.

## Decrease in the Bonobo Population and Missing Bonobo Groups

Two remarkable features have characterized the bonobo situation in the Wamba Forest during the past 15 years. First, each group decreased in size after the political disorder in 1991 (Fig. 14.4). For instance, E1 increased between 1976 and 1986

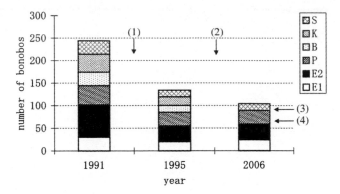

**Fig. 14.4** Democratic changes on each group of bonobos in the Wamba Forest. Arrow (1): decrease in group size during the political disorder, arrow (2): decrease in number of groups during two civil wars, arrow (3): presence and group size of S unconfirmed, arrow (4): group size of P group unconfirmed.

and thereafter remained at around 30 animals. Researchers provisioned them with sugar cane between 1977 and 1991 when the riots occurred. The supply of artificial food may have affected the survival and reproduction of bonobos when their food resources were scarce in the forest. By 1995, E1 had decreased to 20 animals. We have no datum on the demography of bonobos in Wamba during 1997–2002 due to the two civil wars, after which in 2004, 17 bonobos remained in E1 (Furuichi 2004).

Secondly, some bonobo groups went missing during the civil wars. In 1995, the home ranges of six groups overlapped around the hamlets of Wamba (Fig. 14.2). They had maintained similar home ranges since 1974, when bonobo research first began in Wamba. In 2005, we observed E1 at close range to the base camp. They mainly used swamp forest in the southern part of the Wamba Forest. During this period, E1 never used the west side of the main road, which they had used frequently before the war. Congolese co-researchers recorded the ranging area of E1 on the west side of the Lokuli River and a wide range on the east side of the main road in 2004. Combining these observations, we concluded that E1 had enlarged its home range markedly to the east and north of its previous home range.

Contrarily, we found no evidence of other groups in the Wamba Forest in 2005. We observed E2 at a site 3 hours on foot from our base camp, outside of the Wamba Forest. However, we found no direct or indirect evidence to indicate the survival of B and K. Research assistants have not been able to confirm their presence after the war till in 2005; thus, it is possible that they have disappeared. At the southwest edge of the Wamba Forest, we recorded some indirect evidence, including footprints and food remains, which may have been left by P, because E1 and other groups had never ranged there. E1 had ranged along the east side of the main road previously. A Congolese co-researcher observed a group in 2004 that appeared to have been P, but they were afraid of humans. P may have shifted their home range westward to avoid the war and human activities. The survival of S is also unconfirmed. We therefore conclude that the two civil wars greatly affected the population and home ranges of bonobos in the Wamba Forest.


## Problems of the Luo Reserve and Other Bonobo Habitats in D.R. Congo

The Bongando people in the Luo Reserve cultivate manioc by slash-and-burn agriculture. Although felling primary forest is prohibited in the Luo Reserve, slash-and-burn agriculture has been expanding rapidly in primary forests. We directly counted the number and size of fields cultivated by villagers in the hamlets of Yayenge and Yasongo, which are located in the southern part of Wamba (Fig. 14.5). Yayenge encompasses 108 ha comprising 91 fields, of which 7 fields (9 ha) are in primary forest, and 84 fields covering 98 ha were created by the clearing of secondary forest. Yasongo is 138 ha comprising 131 fields, of which 117 fields covering 129 ha are in secondary forest and 14 fields (9 ha) are in primary forest. As a result, 0.17% of the northern part of the Luo Reserve was cleared in only two hamlets.

We analyzed Landsat images to determine the distribution of vegetation types in the Wamba Forest in 1990 and 2003 (Fig. 14.6). Areas of dry forest, swamp forest, and secondary forest and/or fields are easily distinguished by color (Hashimoto et al. 1998, Kimura 1998). The image of 2003 shows many new fields, which were created during the intervening 13 years. The border of the primary and secondary forest in 1990 was converted into fields by 2003, and some new fields were located within the primary forest. Expansion of the fields is likely the result of poverty and hunger during and after the war.

Deforestation of primary forest is not only a problem in Wamba, but also across the whole of the bonobo's range. In addition to cultivation, many parts of tropical forest have been sold to timber companies (Schmidt-Soltau 2006, Thompson et al. 2003), with the deforestation to obtain timber is occurring in areas across the Congo Basin (Miles et al. 2005). We saw logging roads and cleared land from the chartered plane traveling from Kinshasa to Djolu. Suitable habitat for the bonobo is being rapidly destroyed.

Hunting bonobos is also a cause of population decrease (Dupain et al. 2000, Dupain and Van Elsacker 2001, Thompson 2001). Historically, eating bonobos was taboo for the people of Wamba. The Wamba Forest supported a high species richness and diversity of mammals, including elephants, buffaloes, hippopotami, bongos, sitatungas, leopards and many species of primates (Idani 1990b). If mammals and other meat resources are plentiful in the forest, the people of Wamba may not be willing to hunt bonobos for food. But now these animals are rare as a result of over-hunting.

During the study period in 2005, we often heard gunshots and also found numerous wire snares in the forest. A female bonobo that we observed in E1 had caught

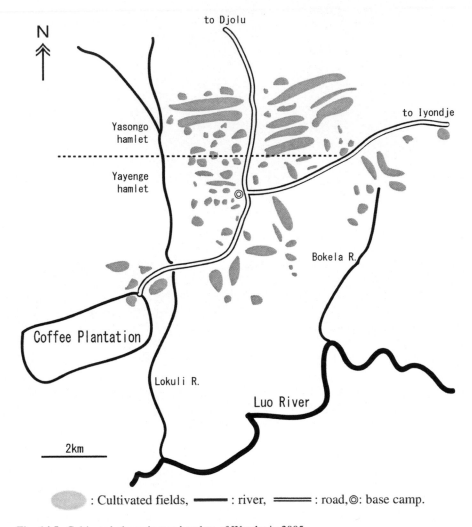

**Fig. 14.5** Cultivated places in two hamlets of Wamba in 2005.

her right hand in a wire snare, and a pangolin, also a protected species, was caught in a wire snare (Fig. 14.7). Moreover, in 1991, we received information that Wamba villagers hunted bonobos via poisoned arrows, and they ate bonobo meat in the village. Most people in the DRC now have diets of wild animals, including bonobos (Kano et al. 1996, Kano 1992). Even if the villagers do not eat bonobos, hunting by individuals from other areas may cause decreases in the bonobo population (Cowlishaw and Dunbar 2000).

Progressive economic deterioration in the DRC may have led local people to discard their previous lifestyles. The value of the DRC currency has continued to

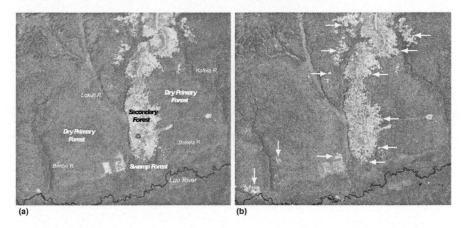

(a)                                                                              (b)

**Fig. 14.6** Landsat images to determine the distribution of vegetation types in the Wamba Forest. (a) An image in 1990, (b) an image in 2003: base camp, arrows show expanded fields.

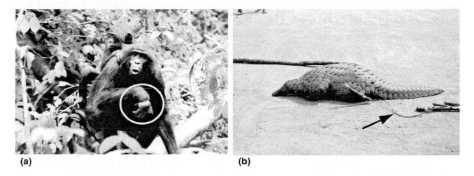

(a)                                                                              (b)

**Fig. 14.7** Victims by the wire snare. (a) A female bonobo who caught her right hand in a wire snare. A white circle shows a gaping wound with a wire snare. (b) A pangolin who was hunted by a wire snare. An arrow indicates a wire snare.

depreciate during the past 20 years. Recently, the number of available workers has decreased in the village and many men have moved to larger towns to obtain cash incomes instead of remaining in the villages; in particular, many youth emigrated from Wamba after the war. Economic deterioration has accelerated the movement of people, and the people who have returned from urban areas to Wamba have new habits. They may have accepted the idea that the bonobo flesh can be eaten, and it is possible that taboos against eating bonobos will disappear from Wamba in the near future.

## Support for the Local Community and Conservation of Bonobos

From the beginning of the research, we have supported public welfare to enhance the quality of life of the local people in Wamba and the surrounding areas. For example, we have maintained a primary school, contributed ballpoint pens, notebooks, blackboards and footballs for the pupils, and developed the educational program. We have provided materials and wages for workers to maintain a main road, bridges, and an airstrip in good condition. We built a dispensary for villagers. Moreover, we are building a medical facility and will provide appropriate medical equipment.

However, there is no end to the demands of the local people. In particular, they expect increased community development because they have received information regarding projects of large nongovernmental organizations (NGOs) moving into the area. They believe they will be able to receive large monetary donations from the NGOs. Most NGOs, however, do not know about local customs, cultures, nature, and people, and are rarely observed to be active in the Djolu area. Sometimes, the NGOs believe they are doing good things for the local people. They invest large sums of money for these projects, but much of the funds are not used to benefit local activities. As a result, the demands by villagers on us are escalating because they mistakenly believe that we are a large NGO and are granted large amounts of money. Unfortunately, we are just a handful of researchers, whereas the local community is a large population.

What more can be done? We understand the importance of conservation, and we would like to make every possible effort to preserve bonobos, in particular, and the local environment in general. Cooperation from the local community is indispensable for success in bonobo and environment conservation. At this time, however, local people are pressed with daily life for their own survival. Deforestation and poaching may not stop unless the quality of life of the local people is enhanced. We support their lives little by little at the least. We are organizing to build a clinic in Wamba, maintain the main road and bridges between Djolu and Wamba, and repair the airstrip to good condition at Djolu. We are working to improve the education program in the locality too, especially for the youngest generation. We believe that conservation activities are not temporal, using time and money and leaving nothing. Conservation should involve continuous activities over the next generation and into the future.

One of the things that we do is spread conservation information to the people of DRC and abroad. The natural environment is important for us and all living things, but human activity in the recent past has forced many species to the brink of disappearance. Evidence that has emerged over the past decades from a variety of disciplines has shown that the distance separating bonobos from humans is smaller than ever before envisaged. Long-term studies of wild bonobos have irreversibly reduced the man-made gap between bonobos and ourselves. As a result, most of the observed differences now appear quantitative rather than qualitative in nature. We must push forward conservation activities for the co-existence of bonobos and human beings.

**Acknowledgements** We are grateful to the Centre de Recherche en Écologie et Foresterie (CREF), Democratic Republic of the Congo, for permission to work in the Luo Scientific Reserve. We express our heartfelt thanks to Dr. Takayoshi Kano who led us to Wamba. We are grateful to all research assistants and workers of Wamba who helped us during the course of the research. We also thank Kansai Telecasting Corporation for supporting our conservation activity. This research was financed by the Japan Society for the Promotion of Science (JSPS) core-to-core program HOPE (#15001), the National Geographic Fund for Research and Exploration (#7511-03), the Toyota Foundation (#D04-B-285), and Japan Ministry of Environment Global Environment Research Fund (#F-061).

# References

Cowlishaw G, Dunbar R (2000) Primate conservation biology. University of Chicago Press, Chicago

Dupain J, Van Elsacker L (2001) The status of the bonobo (*Pan paniscus*) in the Democratic Republic of Congo. In: Galdikas B, Briggs N, Sheeran L, Shapiro G, Goodall J (eds) All apes great and small, Vol. 1: African Apes, pp 57–74

Dupain J, Van Krunkelsven E, Van Elsacker L, Verheyen RF (2000) Current status of the bonobo (*Pan paniscus*) in the proposed Lomako Reserve (Democratic Republic of Congo), Biological Conservation 94: 254–272

Furuichi T (1987) Sexual swelling, receptivity and grouping of wild pygmy chimpanzee females at Wamba, Zaire. Primates 20: 309–318

Furuichi T (2004) Current situation of bonobos endangered. Primate Research 20: 67–70

Furuichi T, Hashimoto C, Idani G, Ihobe H, Tashiro Y, Kano T (1998) Current situation of studies of bonobo (*Pan paniscus*) at Wamba, D. R. Congo. Primate Research 15: 115–127

Hashimoto C, Tashiro Y, Kimura D, Enomoto T, Ingmanson EJ, Idani G, Furuichi T (1998) Habitat use and ranging of wild bonobos (*Pan paniscus*) at Wamba. International Journal of Primatology 19: 1045–1060

Hashimoto C, Furuichi T (2001) Current situation of bonobos in the Luo Reserve, Equateur, Democratic Republic of Congo. Galdikas BMF., Briggs NE, Sheeran LK, Shapiro GL, Goodall J (eds) All Apes Great and Small, vol 1: African Apes, Plenum, pp 83–90

Idani G (1990a) Relations between unit-groups of bonobos at Wamba, Zaire; encounters and temporary fusions. Afri Study Monogr 11:153–186

Idani G (1990b) The Luo Reserve of pygmy chimpanzees in Wamba, Zaire. Journal of African Studies 37: 65–74

Idani G, Kuroda S, Kano T, Asato R (1994) Flora and vegetation of Wamba Forest, central Zaire with reference to bonobo (*Pan paniscus*) foods. Tropics 3: 309–332.

Kano T (1984) Distribution of pygmy chimpanzees (*Pan paniscus*) in the central Zaire basin. Folia Primatologica 43: 36–52

Kano T (1992) The last ape; behavior and ecology of pygmy chimpanzees. Stanford, University Press, California

Kano T, Mulavwa M (1984) Feeding ecology of the pygmy chimpanzees (*Pan paniscus*) of Wamba. In: Susman RL (ed) The Pygmy Chimpanzee: its evolutionary biology and Behavior. Plenum, New York, pp 233–274

Kano T, Bongoli L, Idani G, Hashimoto C (1996) The challenge of Wamba. In: Cavalier P (ed) Great Ape Project. Etica & Animali, Milan, pp 68–74

Kimura D (1998) Land use in shifting cultivation, the case of the Bongando (Ngandu) in central Zaire. Afr Stud Monogra, Supplementary Issue 25: 179–203

Miles L, Caldecott J, Nellemann C (2005) Challenges to great ape survival. In: Caldecott J, Miles L (eds) World Atlas of Great Apes and their Conservation. University of California Press, Berkeley, pp 217–241

Schmidt-Soltau K (2006) Projet d'urgence et de soutien au processus de réunification économique
    et sociale, Composante2: renforcement institutionnel. Repport final. Ministère du Plan Unité
    de coordination des projets et République Démocratique du Congo Justice-Paix-Travail, pp
    1–18
Thompson J (2001) The status of bonobos in their southernmost geographic range. In: Galdikas
    B, Briggs N, Sheeran L, Shapiro G, Goodall J (eds) All apes great and small, Vol. 1: African
    Apes, pp 75–81
Thompson J, Hohmann G, Furuichi T (2003) 2003 bonobo workshop: behaviour, ecology and
    conservation of wild bonobos. 1st Workshop Final Report, pp 1–44

# The Conservation Value of Lola ya Bonobo Sanctuary

Claudine André[1], Crispin Kamate[1], Pierrot Mbonzo[1], Dominique Morel[1], and Brian Hare[2]

## Introduction

Have you been to a football game lately? Think of the last time you were in an arena that seated fifty or even a hundred thousand people. That many people can make a lot of noise, but of course only represent a tiny piece of humanity today. If we could convince all the bonobos in the world to attend such a game, you could not come close to filling even the smallest professional football stadium. Our closest living relative is slipping off the precipice; their extinction in our own lifetime is a real possibility.

The best estimates of the current bonobo population in the wild are somewhere between 5,000–50,000 individuals; all live in the Democratic Republic of Congo (DRC), the only country in which they are found indigenously (Teleki and Baldwin 1979, Kano 1984, Van Krunkelsven 2001). While it might seem an administrative blessing to have bonobos concentrated in one single large country, this rare species still shares all the problems of population fragmentation, habitat loss, and victimization due to the bushmeat trade practiced by their African cousins. In addition, by being concentrated in one country, this species' survival is dependent upon the state of one single nation – for better or worse.

The ubiquitous threats to African apes seem particularly acute in the case of the bonobo as a result of DRC's ill fortune during the past decade. However, the DRC has begun recovering from a decade of wars and now has the chance to jump from an impoverished victim of an oft forgotten war between seven nations, to a regional power as it struggles to redevelop its shattered economy through what to many must seem like an infinite supply of natural resources (Clark 2002). What will the

---

[1] Lola Ya Bonobo, 'Petites Chutes de la Lukaya'", Kimwenza, Mont Ngafula, Kinshasa, Democratic Republic of Congo

[2] Duke University Department of Biological Anthropology and Anatomy, United States

T. Furuichi and J. Thompson (eds.), *The Bonobos: Behavior, Ecology, and Conservation*
© Springer 2008

increasing political stability and economic opportunity mean for the remaining wild bonobo populations? How is the future of the remaining bonobo populations linked to the fortunes of Congo? What methods are available and which ones should we utilize to assure their survival in the wild?

In this chapter, we outline how Lola ya Bonobo sanctuary plays a vital role both in offering lifelong care to bonobos who become orphans of the bushmeat trade, and in acting as an instrument for the conservation of the remaining wild bonobos. We present data on the arrival rate of bonobo orphans that suggest that the fate of wild bonobos is inextricably linked to DRC's path towards development. We therefore argue that the coming decade will be a crossroad for the wild bonobo, and that all methods available, however disparate, must be used to assure their survival. As a result, we conclude by considering the possibility of releasing sanctuary bonobos back into the wild as a possible future tool for the stabilization of wild bonobo populations.

## The Conservation Strategy for Apes

There is only one method for the protection of wild ape populations, and that is through the protection of ape habitat. There are three conventional steps to protecting this habitat: 1) work with a government to set aside as large a habitat area as possible where human activities that are detrimental to ape survival (e.g. hunting and logging) are banned, and such bans are consistently and effectively enforced; 2) work with the government and local population to implement programs for sustainable economic development and education in and around the protected area; and 3) demonstrate the direct economic value of the protected area to the government and local population – typically through tourism. Implementing these three steps has produced success stories where wild ape populations which were destined for extinction have been protected for decades through aggressive efforts to protect their habitat and health. The mountain gorillas arguably represent the most famous case of such success. This species, with its small population size, would likely be extinct today if its remaining habitat was not actively protected, attention had not been drawn to their plight, and they were not recognized as a valuable economic asset for attracting tourists.

Previous success gives hope for the future that we will continue to improve our ability to protect sustainable numbers of the remaining wild ape populations. However, the unfortunate reality is that protected area management in ape habitat countries has proven to be fraught with difficulties, and in most areas, including those with the highest levels of protection, wild ape populations are in decline (Jolly 2005, Butynski 2001). These difficulties are born from a complex of sources. Historically protected areas have not been gazetted based on population viability assessments but instead, and quite understandably, on a "bigger is better" philosophy. Meanwhile, protection comes in many different flavors. Laws are often either too weak to allow for appropriate enforcement or enforcement is too inconsistent to protect slow breeding ape populations that are especially vulnerable to acute hunting and logging pressures (White and Tutin 2001). Overall, many wild apes – including bonobos – live within unsustainable, genetically isolated populations that cannot depend on consistent protection from human threats (e.g., imminent threat from disease, hunting, logging etc). As a result, even some protected ape populations have all but disappeared, such as the gorillas of Kahuzi Biega, UNESCO World Heritage Site, DRC, the bushmeat trade has flourished, and thousands of orphaned infant apes have flooded markets across Africa over the past decades. If we are to save a place for wild bonobos (and other great apes), effective tools are needed to further strengthen the conventional protected area strategy and reverse the current trend.

## Sanctuaries for Conservation

African ape sanctuaries have evolved as one such supplemental tool by offering a second level of protection to wild ape populations when frontline conservation strategies failed to protect individuals from the bushmeat trade. As a member of the Pan African Sanctuary Alliance (PASA), Lola ya Bonobo sanctuary is one such sanctuary located just outside of Kinshasa, the capital of DRC. Lola ya Bonobo sanctuary has been in operation since 1996 and is the DRC's and the world's only sanctuary for orphaned bonobos.

Since 2002 the sanctuary has provided 30 hectares of primary tropical rainforest to the bonobos who live there. Previously the sanctuary, with smaller numbers of bonobos, was located in facilities at the American School in downtown Kinshasa. Currently, 53 bonobos range freely in three different social groups throughout the day at the sanctuary (see Table 15.1 and Fig. 15.1a). Typically, bonobos arrive as young infants and begin life at the sanctuary with close care from a substitute human mother, but are usually quickly ready to be integrated into a peer group, and shortly after into one of the large mixed-age social groups (See Fig. 15.1b). This means that the sanctuary bonobos can supplement their provisioned diet by navigating in order to forage on the dozens of edible plants available in the forest, can compete for mating opportunities among group mates, and can learn to avoid dangers such as stepping on poisonous snakes just as they would in the wild. As a result, the bonobos at Lola ya Bonobo sanctuary, living in their forested microcosm, are for the most part able to exhibit the full complement of naturally occurring behaviors observed in wild bonobos (in fact, they actually display some behaviors such as tool use that have not been observed in the wild!).

Because of the living conditions provided, the sanctuary can play a critical role by demonstrating the level of humane treatment that captive apes deserve, but at the same time, why do we also believe that sanctuaries like ours help protect wild apes? First, our sanctuary allows for the enforcement of national and international conservation laws aimed at preventing the trade in live bonobos. Second, the sanctuary acts as a mouth piece for conservation efforts in DRC by educating thousands of

**Table 15.1** The demograghy of the bonobos at Lola ya Bonobo by age and sex

| Age (# individuals) | Sex (# individuals) |
| --- | --- |
| 13-older (11) | Female (6) |
|  | Male (5) |
| 5–12 (23) | Female (8) |
|  | Male (15) |
| 0–5 (19) | Female (7) |
|  | Male (12) |
| Total (53) | Female (21) |
|  | Male (32) |

**Fig. 15.1** a) The bonobos at Lola ya Bonobo spend their days in 30 hectares (~75 acres) of primary tropical rainforest in which they display the majority of the species specific behaviors observed in wild bonobos. White arrows point to bonobos crossing a natural bridge; b) when orphans first arrive they are quickly integrated into a peer group and then as soon as possible into one of the large mixed age social groups.

Congolese visitors each year about the value of Congo's natural history, in particular the bonobo – their unique Congolese inheritance.

## Enforcement of Conservation Laws

There are two domestic laws in the DRC that protect bonobos and other wildlife: 1) *Ordonnance- loi n° 69- 041 du 22 août 1969 relative àla conservation de la nature* (Law number 69-041 of the 22<sup>nd</sup> of August 1969 with reference to the conservation of nature) which was passed in 1969 and states that the natural heritage of Congo must be protected.
2) *loi 82- 002 du 28 mai 1982 portant réglementation de la chasse* (Law 82-002 of the 28<sup>th</sup> of May 1982 with reference to hunting) passed in 1982 which states that the capture and trade of endangered species is prohibited within DRC. In addition, DRC is a signatory to the Convention on International Trade of Endangered Species (CITES) which prohibits the export of endangered species from DRC.

To our knowledge, there were no confiscations of illegally owned bonobos before Lola ya Bonobo sanctuary was founded. This means that before the sanctuary existed, it was impossible to use these conservation laws effectively to prevent the trade in live bonobos. Today we work hand in hand with the Ministry of Environment to confiscate any illegally owned bonobo in DRC – whether they are found in a bar, living caged with a chimpanzee, for sale on the side of the street, or living in someone's home as an ill-chosen pet.

As represented by the different modes by which orphans have arrived (Fig. 15.2), the confiscation process has evolved over the life of the sanctuary. Initial rescues resulted from the ad-hoc "persuasion" of prospective traders at the zoo or at the sanctuary. But we were not satisfied with this situation and soon realized the national authorities had to be engaged to develop a more systematic approach. Starting in 1997, the first official confiscations were conducted by the Inspectors of the Ministry of Environment. In response, live animal traders quickly adapted to these initial efforts and began avoiding Kinshasa, instead trying to sell orphans in Brazzaville – the neighboring capital of the Republic of Congo. However together with Project Protection Gorilla, another PASA sanctuary that is located just outside of Brazzaville, we have developed an ongoing collaboration with the Republic of Congo government to implement confiscations in Brazzaville as well. Therefore, there is now little hope for a trader to sell live bonobos in either of the Congolese capitals – where all their richest customers might be found – leaving little incentive to attempt to trade live bonobos in the region.

Meanwhile, more recently we have made progress in extending our reach into areas closer to bonobo habitat. Since 2004, together with the help of individual scientists and NGOs (African Wildlife Foundation and Bonobo Conservation Initiative), a number of infant bonobos have been identified and then confiscated with the help of local officials in urban markets in Mbandaka, Basankusu, and Lisala.

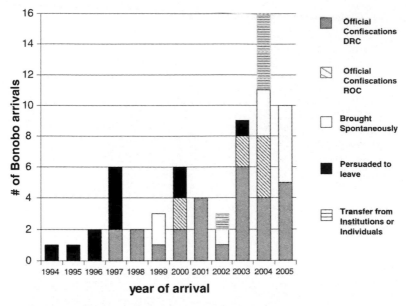

**Fig. 15.2** Modes by which orphan bonobos arrive at Lola ya Bonobo Sanctuary between 1994–2005. Official confiscation in the Democratic Republic of Congo (DRC) and the Republic of Congo (ROC), of bonobos brought spontaneously to the sanctuary, cases in which an individual had to be persuaded to turn over a bonobo and transfers from institutions and other individuals.

Our sanctuary's existence is also crucial for the enforcement of CITES laws against the international trade in bonobos. In 2006, an infant bonobo arrived in the carry-on luggage of a passenger at Paris' Charles De Gaule airport and was to be traded as a pet in Eurasia (the passenger was bound for Russia). An alert customs official discovered the bonobo and confiscated her after realizing she was being smuggled out of DRC; however, the customs office was hesitant to return the infant bonobo to DRC. It was only upon discovering that Lola ya Bonobo offered the infant bonobo the best home available, that they agreed to return the bonobo to its rightful home as CITES law requires. Therefore, we believe Lola ya Bonobo is valuable to wild bonobos in allowing enforcement of existing conservation laws that then act as a major deterrent of the illegal trade of live bonobos captured from the wild.

## Congolese Ownership of Bonobo Conservation

DRC has some of the largest remaining untouched tracts of tropical rainforest that can either be home to bonobos or their endangered cousins, chimpanzees and gorillas. No other African country boasts such an immense wealth of apes and ape habitat. Clearly, DRC's forests must remain at the forefront of the international community's conservation agenda. However, the will to conserve these unique resources must ultimately come from the Congolese themselves. In the end, only the Congolese can decide to conserve the bonobos and our other ape cousins living in Congo.

Before the existence of the Lola ya Bonobo sanctuary, there was no place in the DRC where a child or the average Congolese citizen could go and visit bonobos or discover the value of conserving their country's wild heritage. Today, the bonobos of Lola ya Bonobo act as ambassadors between their world and ours by giving thousands of ordinary Congolese the chance to come face to face with what they stand to lose – in 2005, that totaled over 14,000 people (see Fig. 15.3 for number of visitors to the sanctuary). The sanctuary's slogan is "conservation through education," and we have implemented a number of programs so that the sanctuary's bonobos have the chance to capture the hearts of every Congolese who encounters them.

Although the sanctuary is visited by people of all ages, our target audience is the children who visit. Many children visit the sanctuary with their families, but for those who would not otherwise have the opportunity, the sanctuary has reached out through its association of thirty-nine "Kindness Clubs" (each at a different school in Kinshasa, see Table 5.2), and by hosting visits by school groups. The Kindness Clubs exist to promote kindness to animals by motivating members to take practical

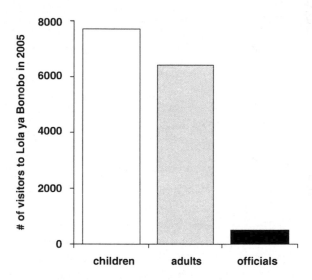

**Fig. 15.3** Illustrates the number and type of visitors that came to see the bonobos at Lola ya Bonobo Sanctuary in 2005. All children were Congolese students (this does not include children under 12 who visit with parents, as we never charge admittance for this age group and have no record for this group) and the majority of government officials were Congolese as well (including the Vice-President of DRC in charge of reconstruction and development). We most often host official visits for members of Ministry of the Environment, their diplomatic guests, and delegations from foreign embassies. We also organize a trip once a month for Congolese civil servants to visit either from ICCN, the Ministry of Environment, the Directorate of Resources, the Office of the Secretary General, and the CITES DRC office. In all cases, Lola ya Bonobo offers to pay for transportation, food and drink. In 2005 we also hosted dignitaries from all great ape range countries who attended the GRASP Inter-governmental meeting in September 2005.

**Table 15.2** The name and location of the schools where Lola ya Bonobo runs a Kindness club in which children learn about conservation and welfare. Kinshasa has 24 communes, with the 26 kindness clubs being located in 18 different schools located in 7 communes of the east, west, north, south and centre of Kinshasa city

| N° | Schools | Communes |
|---|---|---|
| 1 | EDAP/UPN (Secondary school) | Ngaliema (west) |
| 2 | Lycée BOSANGANI (Secondary school) | Gombe (north) |
| 3 | C.S. Mgr MOKE (Primary school) | Kalamu (centre) |
| 4 | E.P. Lycée BOSANGANI (Primary school) | Gombe (north) |
| 5 | Institut BOBOKOLI (Secondary school) | Ngaliema (west) |
| 6 | Institut BOBOKOLI (Secondary school) | Ngaliema (west) |
| 7 | Institut. du Mont- AMBA UNIKIN (Secondary school) | Lemba (south) |
| 8 | E.P. Lycée TOBONGISA (Primary school) | Ngaliema (west) |
| 9 | Institut du Mont- AMBA UNIKIN (Primary school) | Lemba (south) |
| 10 | E.P. St Cyprien (Primary school) | Ngaliema (west) |
| 11 | E.P. Martyrs de l'Ouganda (Primary school) | Ngaliema (west) |
| 12 | Lycée St Joseph (Secondary school) | Kimbanseke (east) |
| 13 | Lycée St Joseph (Secondary school) | Kimbanseke (east) |
| 14 | E.P. St Cyprien (Secondary school) | Ngaliema (west) |
| 15 | UNIKIN (University) | Lemba (south) |
| 16 | E.P. Lycée BOSANGANI (Primary school) | Gombe (north) |
| 17 | Institut des Beaux Arts (Secondary school) | Gombe (north) |
| 18 | Collège St Frederic (Secondary school) | Kimbanseke (east) |
| 19 | Collège St Frederic (Secondary school) | Kimbanseke (east) |
| 20 | Collège St Frederic (Secondary school) | Kimbanseke (east) |
| 21 | Lycée de Kimwenza (Secondary school) | Mont- Ngafula (south) |
| 22 | Collège Pierre Bouvet (Primary school) | Selembao (south-west) |
| 23 | Collège Pierre Bouvet (Secondary school) | Selembao (south-west) |
| 24 | Lycée de Kimwenza (Secondary school) | Mont- Ngafula (south) |
| 25 | C.E. Les gazelles (Primary school) | Kalamu (centre) |
| 26 | C.E. Les gazelles (Secondary school) | Kalamu (centre) |

actions to improve animal welfare and conservation. We do this through regular visits to schools by our education staff and by sponsoring trips to the sanctuary. Funding for our school program also allows large groups of school children from the poorest areas of Kinshasa to visit the sanctuary by providing them with transportation and lunch during the day (~ 50% of school groups that visit are from poorer districts that require financial aid to pay for their visit).

Arriving at the sanctuary, the children are greeted by one of our education staff members. The children are brought to our education center where they learn the basics of bonobo life, the risks to bonobos associated with the bushmeat trade, and the role they can play in protecting bonobos and Congo's wildlife. To help children understand how similar bonobos can be to them, we show them a short video in which the famous bonobo Kanzi works together with Sue Savage-Rumbaugh in solving all sorts of complicated problems; in addition, we inform them about the illegal bushmeat trade. Impressed by Kanzi, the children then leave on a guided tour around the sanctuary's

2.5 km trail system so that they encounter the bonobos playing in the ponds or chasing each other through the canopy of the trees, just as they would in the wild. Children, as well as adults, commonly make remarks about how they never realized humans and bonobos could be so similar.

Over the years, we have tried to improve our ability to convey our messages regarding the conservation of bonobos and their habitat by conducting pre- and post-visit surveys (see Appendix I for example). With our surveys we have learned that children retain our conservation messages best if they are presented with them in class a few days before they visit the sanctuary (it seems with the excitement of being at the sanctuary itself, it is more difficult for children to retain the messages; see Figs. 15.4 and 15.5 for the results of pre- and post-visit surveys that suggest our programs have been successful at communicating these messages). Therefore, an education officer from the sanctuary visits each school group taking a portable LCD projector and laptop so that he can make a presentation in preparation for the children's sanctuary visit shortly after. Between the pre-visit seminar and the experience of visiting the sanctuary's bonobos, the children of Kinshasa are learning the value of conserving their country's unique, 100% Congolese ape. Overall, we believe Lola ya Bonobo sanctuary also has value for wild bonobos by giving

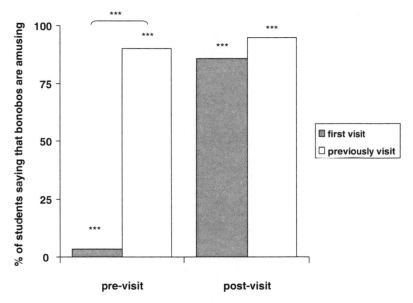

**Fig. 15.4** The results of pre- and post-visit survey questions for 200 children who visited the sanctuary for the first time and 200 children who have visited the sanctuary previously. The children were posed the question of whether they thought bonobos were a) frightening, b) amusing, c) dangerous, or d) beautiful. The figure represents the percentage of children responding that the bonobos were amusing. After their first visit, children changed significantly from choosing to describe bonobos as amusing at below chance levels to significantly describing them as amusing at above chance levels – this preference then persisted when they returned on a second visit (Chi-square ***p<0.001).

**Fig. 15.5** Pre- and post-visit survey results for a) 200 children who visited the sanctuary for the first time and b) 200 children who had visited the sanctuary previously. The children were asked to respond true or false to questions: 1) bonobos do not make good pets, 2) bonobos are not an endangered species, 3) hunting and snares are dangerous for bonobos, 4) planting trees is something you can do to help bonobos, and 5) bushmeat trade threatens bonobos with extinction. Children responses improved significantly after they visited the sanctuary. Interestingly, before visiting the sanctuary, children responded significantly above chance that bonobos made good pets, while after their visit they responded that bonobos did not make good pets (Chi-square ***$p<0.001$ **$p<0.01$ *$p<0.05$).

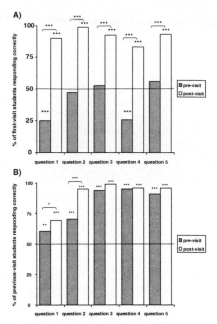

Congolese citizens, and in particular children, the opportunity to meet our bonobo ambassadors who have the best chance to instill the will for conservation in the Congolese.

In addition to the education that takes place at the sanctuary, even the actual confiscation process of live bonobo orphans serves as an invaluable education opportunity for the civil servants responsible for the enforcement of environmental laws. For example, recent confiscations in Mbandaka, Basankusu, and Lisala provided an opportunity for the education of law enforcement officials (and other people) closer to the source of the bushmeat trade. This type of education will prove to be crucial, as many live animal traders arriving in Kinshasa with live bonobos and bonobo meat have official documents from the veterinary services of the Ministry of Agriculture authorizing them to bring "gorilla meat" to sell in Kinshasa. Nothing more clearly illustrates the need for education regarding endangered primate species and the laws protecting them among civil servants working in provincial towns closest to the actual habitat of these endangered animals. Our continued efforts in this direction will also afford wild bonobos an additional level of protection.

## Evaluating the Impact of Lola ya Bonobo

As with almost any conservation project, it is difficult to put numbers together to measure the exact level of protection that sanctuaries like Lola ya Bonobo provide to wild apes by enabling enforcement of laws against the live animal trade and through educating Congolese about the value of bonobos. However, just as we have monitored how best to present our conservation message to children through pre- and post-visit surveys, we are also interested in testing whether the sanctuary is doing its job in reducing the trade in live bonobos and increasing awareness about and respect for bonobos among the Congolese. Fig. 15.6 presents the number of individuals that were confiscated each year while Fig. 15.2 presents the way in which they were confiscated.

What immediately becomes apparent, is that the arrival rate of orphaned bonobos has increased threefold since the sanctuary opened. In addition, you see two peaks of confiscation during the main conflict periods of 1997 and 2000 in which six bonobos arrived at the sanctuary each year. So perhaps not surprisingly, there is a sharp increase as the two wars raged within the bonobos' territory. However, what is most disturbing is that this rate has actually increased with the cessation of hostilities. Since 2002, when the Lusaka peace accord was signed, there has been another increase in orphan infants needing sanctuary. We have received 8–10 orphans in each of the last three years. Possible explanations for this increase in arrivals after the peace accord include, 1) acute changes, such as the continued presence of soldiers in the bonobo habitat, increased communication, and trade since navigation on the Congo re-opened in 2002; 2) the resumption of forest exploitation since small unregulated companies with little knowledge or concern for environmental laws arrived; and 3) more lasting changes related to population displacement,

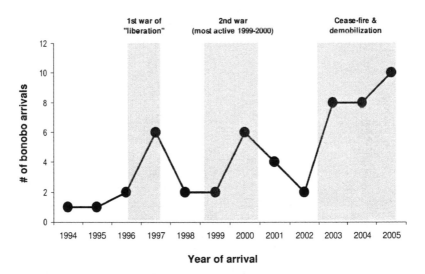

**Fig. 15.6** The number of orphan bonobos that arrived at Lola ya Bonobo relative to the socio-political atmosphere in DRC. During the two war periods, all the bonobo orphans reportedly arrived from the Kasai region – specifically from Salonga. More recently, the arrival pattern is more mixed, with orphans originating from Mbandaka/Lomako as well. We suspect that when the Congo river was largely closed to transport, the orphans traveled from the South of Salonga and Bandundu area via the Kasai River to Kwamouth and then to Kinshasa. Now orphans can also be transported via the Congo river from the Lomako area.

weakening in taboos against hunting bonobos, and increases in available arms – all occurring within the remaining bonobo ranges.

Although the last three years have seen the largest number of arrivals to the sanctuary, perhaps there is reason for hope. Our data on the mode by which infants have arrived at the sanctuary (Fig. 15.2) also suggest the possibility that the increase in arrivals is at least partly due to increased awareness of Lola ya Bonobo and our effort to prevent the trade of live bonobos, as evidenced by a number of "spontane-ous" rescues. Spontaneous rescues occurred when those illegally holding bonobos as pets or for sale became aware through neighbors that they were breaking the law and these individuals came voluntarily to the sanctuary to turn over the bonobo in their possession. Some of these neighbors were children who had visited the sanc-tuary with their schools and acted as "ambassadors" for the sanctuary by actively encouraging the turn over of the bonobo to the sanctuary. Although this type of confiscation remains exceptional, our hope is that our continued efforts will make such cases the rule. Regardless, it is clear that there is a tremendous increase in awareness about the plight of the bonobo among those living in Kinshasa. Hopefully, as word of the sanctuary's efforts continue to spread and more Congolese participate in our education program, it will become increasingly difficult to attempt to trade live bonobos.

Overall, monitoring of bonobo arrivals at the sanctuary provides valuable data regarding the status of wild bonobos relative to the political climate of the country, which complements information obtained by colleagues working in situ (Reinartz and Bila-Isia 2000). In addition, it gives some reason for hope that our education efforts are beginning to have a direct effect on the live animal trade.

## *Releasing the Future for Wild Bonobos*

Given the current state of the wild bonobo population and the likelihood that things will become worse before they potentially become better, should we develop new ways to use sanctuaries as weapons for bonobo conservation? We believe one potential way that sanctuaries and, in particular, Lola ya Bonobo as the world's lone bonobo sanctuary can provide another tool for wild bonobos is in developing a method for the release of sanctuary bonobos back into the wild.

As Congo continues to expand its use of natural resources, bonobo habitat will be increasingly threatened, and populations will become more genetically isolated from one another (Reinartz and Bila-Isia 2000). Some bonobo habitat areas may also become depopulated due to human activities related to hunting or even disease transmission (Dupain et al. 2000, Walsh et al. 2003). As with a number of other animal species (Kleinman 1986, Griffith et al. 1989), it has been suggested that the release of sanctuary apes into areas bordering a genetically isolated but unique ape population, or even into a depopulated region linked through corridors to another populated region, could help stabilize critically threatened wild ape populations (Tutin et al. 2001, Goossens et al. 2002).

While in practice such release programs largely remain an artform due to the difficulty in pre- and pos-release monitoring (Breitenmoser et al. 2001), there are cases in which the utility of release has been demonstrated in saving inbred wild populations from extinction (e.g., Madsen et al. 1999). Therefore, because of the impending threats to wild bonobo populations and the need for new tools for their conservation, we begin to outline our own proposed bonobo release project. We hope that through the use of new technologies, such as new genotypic techniques (Goossens et al. 2002), and with careful planning and monitoring, we can help create an effective methodology for the release of sanctuary bonobos. If we can take the art out of release, we believe the release of sanctuary bonobos could provide a powerful new weapon in the fight to conserve wild bonobo populations.

## Previous Releases of Captive Primates

Release of captive animals into the wild has recently become a method used more commonly to stabilize wild populations, with the number of release programs increasing by 300% during 1993–1997 (Sedon and Soorae 1999). Populations of captive Oryx, ferrets, and red wolves have all recently been successfully reintroduced into the wild in an attempt to increase genetic variation, while preventing wild populations from crashing (Stanley-Price 1989, Moore and Smith 1991, Clark 1994).

A number of captive primate releases have been conducted as well in the recent past, producing mixed results at significant financial costs (Stoinski et al. 2004). Perhaps more than one hundred orangutans have been released into the forests of Sumatra and Borneo since the 1960s. Unfortunately, few records have been kept, so it is difficult to know the impact of this program. It is likely that survival rates were low (Yeager 1997). An initial attempt to release captive chimpanzees in Senegal was abandoned after attacks by wild conspecifics raised concerns for the safety of the individuals to be released (Brewer 1978). Dozens of golden lion tamarins were released into the tropical forests of Brazil following IUCN guidelines for release. With only 200 wild individuals remaining, but a burgeoning captive population available in zoos, the release program was viewed as the best option available to assure the survival of this species in its natural habitat (Kleinman et al. 1986). Captive animals were rigorously screened for disease and given a few months to live in "artificial jungles" at zoos in order to acquire skills (climbing trees, foraging etc.) that were thought to be essential for their survival. Pre- and post-release data of the wild population showed that the release program was successful in that many captive animals have reproduced successfully (Stoinski et al. 2004). However, the cost of the program has been significant with a budget of $120,000 per year (Beck et al. 1991), while the released tamarins and their offspring both show behavioral deficiencies that result in higher mortality than wild tamarins, and thus, may affect their long-term survival (Stoinski et al. 2004).

More recently, 20 captive chimpanzees were reintroduced into the wild following IUCN guidelines by HELP Congo, another PASA sanctuary working in the

Republic of Congo. A site was carefully evaluated and chosen where the released population would have limited contact with wild populations, individuals were screened for disease that might threaten wild populations as well as the individuals, and released individuals with and without radio collars were followed by observers (Ancrenaz et al. 1995, Tutin et al. 2001). After three years post reintroduction, at least 70% of released individuals survived (this survival rate could be as high as 90% since some mortality was not confirmed and disappearance could be due to outmigration), and more recently, females have successfully given birth (Beck 2007). Presumably in the future, this population will begin to contribute to the genetic diversity of the surrounding chimpanzee communities (Goossens et al. 2002), but meanwhile the release itself has generated significant media and governmental attention that has allowed for an increase in protection of the areas surrounding the release site. Importantly, the success of the project was not due to a large budget, but is attributed to the hard work of professionals and non-professionals who volunteered to help with the release (Tutin et al. 2001).

## The First Bonobo Release

Bonobos remain the only great ape species for which a method for release of captive individuals into the wild has not been developed. We believe that it is an important step to take in the near future to assure that release can be used as a tool in bonobo conservation. While there are significant risks involved in such a strategy, including the long-term survival of the released bonobos, the disease risks they may potentially pose to wild populations, and the long-term costs of such projects, we feel that such risks can be managed effectively and thus may be worth taking, if the current situation in bonobo habitat areas continues to decline. In the long run, the strategic release of sanctuary bonobos may provide an important technique, as it has for other critically endangered mammals (Kleinman 1989), to stabilize remnant wild populations with outside genetic material or to repopulate forests that were emptied of wild bonobos due to previous human activities. In addition, the individual interest stories that such a program will create should generate increased attention to the plight of wild bonobos, while offering the possibility to develop unique tourist experiences for future bonobo enthusiasts.

Thankfully, as reviewed above, we have many people who have gone before us in releasing captive animals, so we can heed some important lessons learned from previous release programs. Therefore, in preparation for a proposed bonobo release, we have begun identifying long-term partners within both the conservation and business community who are interested in providing support for the project with financing and expertise.

We are currently designing plans for a soft release in which a stable social group of 15–20 bonobos from the sanctuary will be released within a large forest block that prevents the possibility of contact with wild populations through natural barriers (i.e. an island) or even fencing. We are following IUCN guidelines closely to help evaluate potential release sites and we have identified a number of scientists who will help us follow and improve upon the IUCN guidelines regarding pre- and post-release health checks, genotyping strategies, and behavioral monitoring. We will then prevent contact with wild populations largely to limit any possibility of disease transmission between populations due to premature immigration. But, if the released bonobos are able to sustain themselves, artificially constructed barriers could eventually be removed to allow for immigration. In addition, such barriers supplemented by tracking collars on a number of key individuals will also give us a chance to intervene if certain individuals have difficulty adapting to life in the wild.

Although we will design such a safety net to protect the released bonobos, we have reason to predict that they will adapt rapidly to life in the wild. Stoinski et al. (2004) concluded from their systematic comparison of wild and released tamarins, that released captive tamarins have behavioral deficiencies because they were not allowed enough time to adapt to a simulated wild environment; instead of years, they only had a few months to gain survival skills before being released. However, as in the case of the HELP Congo chimpanzees who lived on large forested islands before their release, the bonobos at Lola ya Bonobo have been living in large stable social groups within a sizable forest enclosure for years, and thus have much experience foraging for dozens of plants they will also find available in the wild. Therefore, given the success of the chimpanzees from the HELP Congo release project in quickly adapting to life in the wild and the similar pre-release experience our bonobos have to those chimpanzees, we are optimistic that our sanctuary bonobos will also adapt quickly. However, in a soft release phase, depending on the site, we can potentially provision with food and even intervene in extreme circumstances (e.g. disease outbreak).

In addition, our bonobo release program will also have an enormous advantage over any chimpanzee release program simply because of the differences in these species social systems – bonobos do not display lethal forms of aggression seen in chimpanzees, and have behaviors used to effectively reduce social tension that are not found in chimpanzees (Hare et al. 2007, Kano 1992, Wrangham 1999). Therefore by controlling immigration, and due to the bonobos less aggressive nature and the pre-release environment, our release program should result in even higher survival rates than those seen in chimpanzees (Tutin et al. 2001), assuming human activities are effectively controlled.

We are currently assessing whether we can provide consistent and effective protection for the release area through the presence of tourist activities and education programs that can not only generate sustainable revenue for the project and the surrounding communities, but allow for countless education opportunities for Congolese living in and around bonobo habitat areas. In addition, we are also investigating whether our local and international partners will help us in maintaining active patrols around our release site.

As we write this, we are still in the initial phases of our planning, so our review only includes some of the strategies we will use in designing our release, with our ultimate goal being the development of a systematic method for the release of bonobos that can supplement conventional protected area strategies in use today.

With an effective release method available, we will have one more important tool
to make sure that wild populations remain viable.

## Summary

Without an appropriate facility to receive great apes confiscated from the pet
trade, application of existing trade and detention laws, including CITES, is not
possible. Lola Ya Bonobo sanctuary in Kinshasa, DR Congo, was created by
Friends of Bonobos in Congo (ABC) in response to this need and focuses its
efforts on protecting bonobos (*Pan paniscus*), which are our species' closest living relative, but also among the most endangered primate species. In order to
stymie the trade in live bonobos, ABC is pro-active in working with the Ministry
of Environment to ensure the legal confiscation of all infant bonobos reported for
sale in the streets of Kinshasa.

In addition, the sanctuary's slogan is "conservation through education." By hosting
thousands of Congolese each year at the sanctuary, we are able to convey the value
of conserving bonobos and the tropical forest on which they depend. While our
records of confiscations over the past ten years show that there has been a threefold
increase in confiscations that is tightly linked to political instability in the DRC,
there is evidence that the sanctuary has raised awareness of the bonobos' plight
among Congolese – perhaps the most important way we can protect bonobos for
the future. In this chapter we outlined the continued value of Lola ya Bonobo for
the protection of wild bonobos, while discussing some of the pros and cons of
potentially releasing sanctuary bonobos back into the wild as another tool for managing
wild bonobo populations.

By working with civil servants to enforce Congolese and international laws banning
the trade in bonobos, and by presenting over ten thousand Congolese each year
with the opportunity to personally visit and learn about bonobos in the country's
capital, we have argued that Lola ya Bonobo offers an added level of protection to
wild bonobos. Therefore, we believe sanctuaries like Lola ya Bonobo will continue
to play an important conservation role by supplementing conventional protected
area strategies.

Further, we have presented arrival data on bonobo orphans at the end of the
bushmeat chain that suggest that the two periods of most intense conflict over the
past decade co-occurred with an influx of orphans. And perhaps most disturbing is
that another significant increase in arrivals has occurred since the Lusaka peace
accord was signed in 2002. We believe this data has bearing on the situation within
the bonobo habitat, and thus suggests that while our efforts along with those of
many others are making sure that the bushmeat trade does not go unchecked (in
particular making poaching for the sake of trading live bonobos intractable), the
aftermath of the peace process may present us with the conservation community's
greatest challenge to date. Thus, it seems that bonobos are at a crossroad – with
their future tied to how well we respond to the way Congo's development affects

their continued survival. Luckily, together as a conservation community, we have far more tools available to us now that peace has returned.

We have briefly reviewed the possibility of developing the release of sanctuary bonobos back into the wild as one such tool. While we acknowledge there are significant risks involved in such a strategy, we feel that such risks can be managed so that the potential benefit will far outweigh the costs. If we can succeed in developing a systematic method for releasing bonobos, sanctuary bonobos will provide a plan B for conservation, if in the short run, things continue to deteriorate in habitat areas. Sanctuary bonobos can be used to stabilize crashing populations or repopulate habitat areas where bonobos no longer exist – even if this scenario only plays out long into the future.

Using strategies new and old, Lola ya Bonobo stands together with the conservation community and looks forward to working with all parties involved for the conservation of bonobos and DRC's natural resources. It is with our combined efforts – no matter how disparate the methods might seem – that we will assure that our closest relatives will remain wild in Congo. In addition, our teamwork will also help develop a powerful tool kit that the next generation of conservationists can continue to utilize and develop further to assure that they too are successful in protecting bonobos on the next watch.

**Acknowledgements**  We would like to thank the editors for inviting us to contribute to this volume and for their work in putting it together. We appreciate the helpful comments by an anonymous reviewer. We appreciate Vanessa Woods allowing us to use the photos in Fig. 15.1. We would also like to thank the generous funding provided by a number of welfare groups, businesses, individuals, and others for their continued financial support of Lola ya Bonobo sanctuary. Particularly relevant to this chapter, is the financial support of Disney's Wild Animal Kingdom that provided for the evaluation of our education program. Please see our website at *www.lolayabonobo.org* for a full list of our sponsors. The research of the last author (B.H.) is supported by a Sofja Kovalevskaja award received from the Alexander von Humboldt Foundation and the German Federal Ministry for Education and Research.

# References

Ancrenaz M, Paredes M, Leroy E (1995) Veterinary report on the health screening of H.E.L.P.'s chimpanzees Conkouati Congo. H.E.L.P., Pointe-Noire, Republic of Congo

Beck B (2007) The conservation value of African Sanctuaries. Understanding Chimpanzees: Chimpanzee Minds Conference, Frankpark Zoo, Chicago

Beck B, Kleinman D, Dietz J, Castro I, Carvalho C, Martins A, Rettberg-Beck B (1991) Losses and reproduction in reintroduced golden lion tamarins *Leontopithecus rosalia*. Dodo Journal of the Jersey Wildlife Preservation Trust 27:50–51

Breitenmoser U, Breitenmoser-Wusten C, Carbyn L, Funk S (2001) Assessment of carnivore reintroductions. In: Gittleman J, Funk S,Wayne R, Macdonald D (eds) Carnivore conservation. Cambridge Univ Press, Cambridge, pp 241–281

Brewer S (1978) The chimps of Mt. Assirik. Knopf, New York

Butynski T (2001) Africa's great apes: current taxonomy distribution numbers conservation status and threats. In: Beck B, Stoinski T, Hutchins M, Maples T, Norton B, Rowan A, Stevens E,

Arluke A (eds) Great Apes and Humans: the ethics of coexistence. Smithsonian Institution Press, Washington D.C., pp 3–56

Clark J (2002) The African Stakes of the Congo War. Palgrave Macmillan, New York

Clark T (1994) Restoration of the endangered black-footed ferret (*Mustela nigrepes*): a 20-year overview. In: Bowles ML, Whelan C (eds) Restoration and Recovery of Endangered Species. Cambridge Univ Press, Cambridge, pp 272–297

Dupain J, Van Krunkelsven E, Van Elsacker L, Verheyen R (2000) Current status of the bonobo (*Pan paniscus*) in the proposed Lomako Reserve (Democratic Republic of Congo). Biological Conservation 94:265–272

Griffith B, Scott J, Carpenter J, Reed C (1989) Translocation as a species conservation tool: status and strategy. Science 245:477–480

Goossens B, Funk S, Vidal C, Latour S, Jamart A, Ancrenaz M, Wickings E, Tutin C, Bruford M (2002) Measuring genetic diversity in translocation programs: principles and application to a chimpanzee release project. Animal Conservation 5:225–236

Hare B., Melis A., Woods V, Hastings S, Wrangham R (2007) Tolerance allows bonobos to out-perform chimpanzees on a cooperative task. Current Biology 17: 1–5

Jolly A (2005) The last great apes? Science 309:1457

Kano T (1984) Distribution of pygmy chimpanzees *Pan paniscus* in the Central Zaire Basin. Folia Primatologica 43:36–52

Kano T (1992) The Last Ape: Pygmy chimpanzee Behavior and Ecology. Stanford Univ Press, Stanford

Kleinman D (1989) Reintroduction of captive mammals for conservation: guidelines for reintro-ducing endangered species into the wild. BioScience 39:152–161

Kleinman D, Beck B, Eitz J, Ballou J, Coimbra-Fiho A (1986) Conservation program for the golden lion tamarin: captive research and management ecological studies educational strate-gies and reintroduction. In: Gipps J (ed) Beyond captive breeding: reintroducing endangered mammals to the wild. Clarendon Press, Oxford, pp 263–278

Madsen T, Shine R, Osson M, Wittzell H (1999) Conservation biology-restoration of an inbred adder population. Nature 402:34–35

Moore D, Smith S (1991) The red wolf as a model for carnivore reintroduction. In: Gipps J (ed) Beyond captive breeding: Reintroducing endangered mammals to the wild. Clarendon Press, Oxford, pp 263–278

Reinartz G, Bila-Isia I (2000) Bonobo survival and a wartime conservation mandate. Conference Proceedings, The Apes: challenges for the 21st Century, Brookfield Zoo, pp 52–56

Sedon P, Soorae P (1999) Guidelines for subspecific substitutions in wildlife restoration projects. Conservation Biology 13:177–184

Stanley-Price M (1989) Animal reintroduction: the Arabian oryx in Oman. Cambridge University Press, Cambridge

Stoinski T, Beck B, Bloomsmith M, Maple T (2004) A behavioral comparison of captive-born reintroduced golden lion tamarins and their wild born offspring. Behaviour 140:137–160

Teleki G, Baldwin L (1979) Known and estimated distribution of extant chimpanzee populations *Pan troglodytes* and *Pan paniscus* in Equatorial Africa. IUCN, Gland

Thomas-Handler N, Malenky R, Reinartz G (1995) Action plan for *Pan paniscus*: report on free ranging populations and proposals for their preservation. The Zoological Society of Milwaukee County, Milwaukee

Tutin C, Ancrenez M, Paredes J, Vacher-Valles M, Vidal C, Goossens B, Bruford M, Jamart A (2001) Conservation Biology framework for the release of wild-born orphaned chimpanzees into Conkouati Reserve Congo. Conservation Biology 15:1247–1257

Van Krunkelsven E (2001) Density estimation of bonobos (*Pan paniscus*) in Salonga National Park Congo. Biological Conservation 99:387–391

Walsh P, Abernethy K, Bermego M, Beyers R, de Wachter P, Akou M, Huijbregts B, Mambounga D, Tham A, Kibourn A, Lahm S, Latour S, Maisels F, Mbina C, Mihindou Y, Obiang S, Effa E, Starkey M, Telfer P, Thibault M, Tutin C, White L, Wilkie D (2003) Catastrophic ape decline in western equatorial Africa. Nature 422:551

White LJT, Tutin CEG (2001) Why chimpanzees and gorillas respond different to logging: a cautionary tale from Gabon. In: Weber W, White LJT, Vedder A, Naughton-Treves L (eds) African rain forest ecology and conservation: an Interdisciplinary perspective. Yale University Press, New Haven, pp 449–462

Wrangham RW (1999) Evolution of coalitionary killing. Yearbook of Physical Anthropology 42:1–30

Yeager C (1997) Orangutan rehabilitation in Tanjung Putting National Park, Indonesia. Conservation Biology 11:802–805

# Appendix I

## Questionnaire Pre- and Post-Test for Congolese Children from Figure 15.4

1. Bonobos do not make good pets. TRUE, FALSE
2. Bonobos are not an endangered species. TRUE, FALSE
3. Bonobos are not important for the forest. TRUE, FALSE
4. Do you think that bonobos are protected? TRUE/YES, FALSE/NO
5. Check the word which in your opinion best matches bonobos:
   a. frigthening
   b. amuzing
   c. beautiful
   d. dangerous
6. Illegal hunting and snares are dangerous for bonobos. TRUE, FALSE
7. Planting trees is something you can do to help bonobos. TRUE, FALSE
8. Bushmeat trade exposes bonobos to the risk of extinction. TRUE, FALSE

## Questionnaire Pre- and Post-Test for Congolese Children from Figure 15.5

Identification of the participant:
Name:                              Class:
Age:                               School:

Questions:
A. TRUE/FALSE QUESTIONS: "Do you think that…."

1. Bonobos make good pets.
2. Bonobos are currently in danger in their natural habitat.
3. Bonobos live in many countries of Africa.
4. The number of bonobos in the forest has been stable over the past decade.
5. Bonobos are more closely related to gorillas than to humans.
6. Bushmeat trade does not necessarily expose bonobos to the risk of extinction.

7. The survival of the Bonobo in the forest depends on our actions today.
8. It is not our responsibility to protect the Bonobo.

B. CHECK ALL THE CORRECT RESPONSES

9. What can you do to help the Bonobo?
    a. Learn more about the bonobo to become a well-informed advocate.
    b. Explain to my relatives, friends, and neighbors that the bonobo is unique to
       the DRC.
    c. Reassure those who don't know too much about bonobos, so that they leave
       them alone in their forest,
    d. Only eat bushmeat that has been well cooked to prevent diseases.
    e. Discourage my relatives, friends, and neighbors from eating monkeys, apes,
       and other kinds of bushmeat.
    f. Cut only the biggest trees in the forest, but leave the shrubs for the bonobos
       to hide in.

# Index